Fundamentals of Adhesion and Interfaces

OF RELATED INTEREST

ADHESION MEASUREMENT OF FILMS AND COATINGS
Editor: K.L. Mittal

PLASMA SURFACE MODIFICATION OF POLYMERS:
RELEVANCE TO ADHESION
Editors: M. Strobel, C.S. Lyons and K.L. Mittal

CONTACT ANGLE, WETTABILITY AND ADHESION
Festschrift in Honor of Professor Robert J. Good
Editor: K.L. Mittal

SILANES AND OTHER COUPLING AGENTS
Festschrift in Honor of Dr. Edwin P. Plueddemann
Editor: K.L. Mittal

ACID-BASE INTERACTIONS:
RELEVANCE TO ADHESION SCIENCE AND TECHNOLOGY
Festschrift in Honor of Dr. Frederick M. Fowkes
Editors: K.L. Mittal and H.R. Anderson, Jr.

Journal of ADHESION SCIENCE and TECHNOLOGY
Editors: K.L. Mittal and W.J. van Ooij

FUNDAMENTALS OF ADHESION AND INTERFACES

Editors:
D.S. Rimai, L.P. DeMejo and K.L. Mittal

///VSP///
Utrecht, The Netherlands, 1995

VSP BV
P.O. Box 346
3700 AH Zeist
The Netherlands

First published in 1995

ISBN 90-6764-197-9

CIP-DATA KONINKLIJKE BIBLIOTHEEK, DEN HAAG

Fundamentals

Fundamentals of adhesion and interfaces / D.S. Rimai, L.P.
DeMejo, K.L. Mittal. (eds) - Utrecht : VSP
Papers presented at the symposium'Fundamentals of
Adhesion and Interfaces', held in Chicago at the August
1993 meeting of the American Chemical Society (ACS),
under the auspices of the Division of Polymer Chemistry
- Originally publ. in: Journal of Adhesion Science and
Technology, v. 8, no. 11(1994) and vol. 9, nr. 8
(1995)
ISBN 90-6764-197-9 bound
NUGI 812
Subject headings: adhesion / interfacial sciences.

Printed in The Netherlands by Koninklijke Wöhrmann BV, Zutphen.

Contents

Fundamentals of Adhesion and Interfaces, pp. vii–viii
D. S. Rimai, L. P. DeMejo and K. L. Mittal (Eds)

Preface

Surfaces are everywhere and the importance of their interactions transcends individual technologies. Some of the technological areas where surfaces play a vital role are friction, wear, triboelectrification, surface contamination control in microelectronics, bioadhesion, particle adhesion, pharmacology, agriculture, electrophotography, electronics packaging, semiconducter fabrication, and, of course, adhesive bonding. Accordingly, the fields of adhesion and interfacial sciences have become quite multidisciplinary in recent years.

In addition to technological advances pushing the development of adhesion and interfacial sciences, recent experimental and computational tools have helped to pull these areas forward as well. For example, the advent of the atomic force apparatus and cantilever force microscopy has allowed the interaction mechanisms between materials to be probed directly. Novel spectroscopic techniques and improved equipment has led to a better understanding of bond formation and breaking upon contact and separation. In addition, the recent advances in computer technology have facilitated both the acquisition of data and the enhancement of theoretical understanding of adhesion and interfacial phenomena.

Because of the rapid advances and importance of the fields of adhesion and interfacial sciences, a three-day symposium entitled 'Fundamentals of Adhesion and Interfaces' was held in Chicago at the August 1993 meeting of the American Chemical Society (ACS), under the auspices of the Division of Polymer Chemistry. This symposium was divided into 6 sessions in which invited speakers gave presentations on recent developments. In addition there was a poster session in which contributed papers were presented.

The papers in this volume are based on selected presentations given at the aforementioned symposium. These papers were originally reviewed and published in the *Journal of Adhesion Science and Technology* in two special issues: Volume 8, number 11 (1994) and Volume 9, number 8 (1995). In order to facilitate wide dissemination of the material, we have decided to also issue these papers as a hardbound volume.

This symposium would not have been possible without the support, encouragement, and contributions of many organizations and individuals. First, we would like to thank all the individuals who participated in this symposium, and especially those who contributed to this volume. The financial support of the Division of Polymer Chemistry, and the Petroleum Research Fund of the American Chemical Society is gratefully acknowledged. We are also thankful to Drs. J. Pochan and C. Frank for their guidance and moral support in organizing this symposium.

Of all the people who contributed to this symposium and conference proceedings, we would especially like to acknowledge two individuals. First, Mr. Ray Bowen organized and ran the poster session, paying attention to detail that allowed that session to be the success that it was. Second, Mrs. Sherrie Coleman cheerfully performed many of the behind-the-scenes tasks which are necessary for a symposium to be successful and for high quality proceedings to be published.

D. S. Rimai
L. P. DeMejo
Office Imaging
Eastman Kodak Company
Rochester, NY

K. L. Mittal
J. Adhesion Science and Technology
Hopewell Junction, NY

Fundamentals of Adhesion and Interfaces, pp. 1–23
D. S. Rimai, L. P. DeMejo and K. L. Mittal (Eds)

Mechanics of particle adhesion

D. S. RIMAI,[1,*] L. P. DEMEJO[1] and R. C. BOWEN[2]

[1] *Office Imaging Research and Technology Development, Eastman Kodak Company, Rochester, NY 14653-6402, USA*
[2] *Analytic Technologies Division, Eastman Kodak Company, Rochester, NY 14650, USA*

Revised version received 15 November 1993

Abstract—The adhesion of particles to surfaces is accompanied by deformations of the materials arising from adhesion force-induced stresses. These deformations, which can be the result of an elastic, a non-linear elastic, a viscoelastic, or a plastic response of the materials to the stresses, can significantly affect the forces needed to remove the particles from the substrate. The mechanics of adhesion-induced deformations between micrometer-size particles and various substrates are discussed in this paper. Examples of elastic and plastic deformations resulting from the adhesion forces are presented. The experimental results are analyzed in terms of various adhesion theories, which, under certain circumstances, permit the calculation of the thermodynamic work of adhesion for contacting solids. The ranges of validity of these theories and their predictions are discussed. Finally, adhesion-induced deformations which are not currently explicable in terms of these theories are presented.

Keywords: Adhesion; particle; deformation; work of adhesion.

1. INTRODUCTION

The adhesion and removal of fine particles to substrates is a topic of great interest, both scientifically and technologically. For example, the adhesion of medicinal particles to specific sites is important to the pharmaceutical industry [1]. The removal of particles from silicon wafers is extremely important in the semiconductor industry [2]. In xerography, the ability to successfully transfer toner particles from the photoconductor to the receiver requires an understanding of the mechanics of particle adhesion [3, 4]. The adhesion of latex particles to surfaces is extremely important in the paint industry. Particle adhesion is also important to such diverse technologies as graphic arts and agriculture. Finally, the thermodynamic work of adhesion, which is not in general possible to determine for contacting solids, can be calculated from the contact radius between a particle and a substrate.

*To whom correspondence should be addressed.

The adhesion of particles to substrates has also been a target of intense scientific interest for over 60 years. Specifically, adhesion-induced deformations were first postulated independently by Bradley [5, 6] and Derjaguin [7]. Derjaguin calculated the contact radius, a, between a particle and a substrate by assuming that the particles would acts as Hertzian indenters with the applied load, P^0, arising from van der Waals interactions. He found that for a rigid particle and a compliant substrate under these assumptions,

$$a^3 = \frac{3}{4} P^0 \left[\frac{1 - \nu^2}{E} \right] R, \tag{1}$$

where ν and E are the Poisson's ratio and Young's modulus of the substrate, respectively; R is the radius of the particle; and P^0 is given by the relationship

$$P^0 = \frac{\hbar\bar{\omega}}{8\pi z_0^2} R, \tag{2}$$

where z_0 is the separation distance between the particle and the substrate (typically approximately 4 Å for van der Waals-bonded crystals), and $\hbar\bar{\omega}$ is the Hamaker coefficient.

Krupp [8] was the first to postulate that the stresses arising from the surface forces could be sufficiently large so as to exceed the elastic limit of one of the materials. He attempted to generalize the original model by Derjaguin to include plastic flow by assuming that the contact radius could be divided into two distinct regions: an inner region of radius a_1, which would be subjected to higher stresses and therefore would deform plastically; and an annular concentric ring extending from a_1 to the extent of the total contact radius a_0, which would be subjected to lower stresses and would, accordingly, deform elastically.

He then calculated the radii assuming a phenomenologically determined time-dependent hardness, $H(t)$, finding that

$$\pi a_0^2 = \frac{P^0}{H(t)} + \frac{1}{3} \left[\frac{\pi^{3/2}(1 - \nu^2) R H(t)}{2E} \right]^2, \tag{3}$$

and

$$\pi a_1^2 = \frac{P^0}{H(t)} - \frac{1}{3} \left[\frac{\pi^{3/2}(1 - \nu^2) R H(t)}{2E} \right]^2. \tag{4}$$

To allow the contact radius to reach a limiting size, Krupp further assumed that $H(t)$ would approach an asymptotic value of the order of $10^{-3}E$ within approximately 30 min.

Johnson, Kendall, and Roberts [9] (hereafter referred to as JKR) recognized that both compressive and tensile interactions contributed to the size of the contact radius. Assuming that the materials respond elastically and that all interactions occur within the contact zone, they found that for a spherical particle in contact with a planar substrate

$$a^3 = \frac{R}{K} \left\{ P + 3w_A \pi R + \left[6w_A \pi R P + (3w_A \pi R)^2 \right]^{1/2} \right\}, \tag{5}$$

where w_A is the thermodynamic work of adhesion and is related to the surface energies, γ_1 and γ_2, of the two materials and their interfacial energy, γ_{12}, by

$$w_A = \gamma_1 + \gamma_2 - \gamma_{12}. \tag{6}$$

P represents any externally applied load in equation (5) and

$$K = \frac{4}{3\pi(k_1 + k_2)}, \tag{7}$$

where

$$k_i = \frac{1 - \nu_i^2}{\pi E_i}, \tag{8}$$

and E_i and ν_i are the Young's modulus and Poisson's ratio of the ith material, respectively. Implicit in the JKR model is the assumption that the stresses become infinite at the edge of the contact area.

In the absence of any externally applied load, equation (5) reduces to

$$a^3 = \frac{6w_A \pi R^2}{K}. \tag{9}$$

Alternatively, the JKR theory predicts that upon application of a negative (tensile) load, separation occurs at a finite contact radius $a_s \sim 0.64a_0$, where a_0 is the contact radius established with no applied load. Moreover, the load needed to effect particle/substrate separation, P_s, is given by

$$P_s = -\frac{3}{2} w_A \pi R. \tag{10}$$

In other words, the JKR model predicts that the force needed to remove a particle from a surface is independent of the modulus of the materials. This may seem intuitive if one thinks in terms of 'elastic rebound' effects [8] wherein the work done in initially deforming the materials during the particle entrainment process is recovered upon the removal of the particle. However, this prediction is frequently incorrect. As discussed by Johnson *et al.* [9], this may be due to difficulties in determining when thermodynamic equilibrium is reached. However, as previously proposed [10] and discussed both within this paper and elsewhere [11, 12], hysteresis in the stress-strain curves associated with placing and removing a particle from a surface due to the relatively short range of the surface forces can also account for the effects of the Young's moduli of the materials on particle adhesion.

An alternative approach for calculating the contact radius between a particle and a substrate which also allows for tensile interactions was proposed by Derjaguin, Muller, and Toporov [13] (hereafter referred to as the DMT model). Assuming that the shape of the contact region would be Hertzian, the DMT model predicts that half of the interactions occurs outside of the contact zone. This is in contrast with the JKR model, which assumes that all interactions occur within the radius of contact.

Tabor compared the assumptions and predictions of the JKR and DMT models [14] and showed that the contact radius and pull-off forces, as predicted by the DMT model, were approximately 1/2 and 4/3 of those predicted by the JKR model. Following a somewhat heated debate in the literature [15, 16], Muller *et al.* [17, 18] showed that the discrepancy between the two models arises from the differing assumptions regarding the occurrence of a non-zero potential outside the contact radius. Moreover, they showed that the JKR and DMT models each have a region of validity, which can be defined in terms of a dimensionless parameter μ, such that

$$\mu = \frac{32}{3\pi}\left[\frac{2Rw_{\mathrm{A}}}{\pi E^* z_0^3}\right]^{1/3}, \tag{11}$$

where

$$\frac{1}{E^*} = \frac{(1-\nu_2^2)}{E_1} + \frac{(1-\nu_2^2)}{E_2}. \tag{12}$$

If $\mu > 1$, corresponding to large particles, high surface energy materials, and low elastic moduli, the adhesion-induced deformations are described by the JKR theory. Conversely, if $\mu < 1$, the DMT model is valid. Further discussion of the JKR and DMT models has been presented by Tsai *et al.* [19]. The JKR-DMT transition has recently been the subject of investigation by Maugis [20].

Finally, Maugis and Pollock (hereafter referred to as MP) generalized the JKR model to include plastic deformations [21]. They found that in the case of plastic deformations, the contact radius a is related to the particle radius R and any externally applied load P by

$$P + 2\pi w_{\mathrm{A}} R = \pi a^2 H, \tag{13}$$

where the hardness H is related to the yield strength Y of the material by

$$H = 3Y. \tag{14}$$

Therefore, in the absence of any externally applied load, elastic and plastic adhesion-induced deformations can be distinguished by the power law dependences of the contact radius on the particle radius. Once the nature of the deformation is determined, it is possible to calculate w_{A} for two contacting solids.

Adhesion force-induced deformations have also been studied experimentally. For example, nanoidenters [22–29] and crossed mica [30] and polymer-coated mica sheets [31] have been used to simulate particles on surfaces. Researchers such as Dahneke [32], Chaudhri and Yoffe [33], and Johnson *et al.* [9] have studied adhesion-induced deformations of particles. More recently, the results of systematic determinations of the contact radii resulting from surface forces between micrometer-size particles and various substrates have been reported [10, 34–44]. In certain cases, it was found that the contact radius varied with the particle radius and Young's modulus according to the predictions of the JKR model. However, even here, evidence suggestive of hysteretic effects between the deposition and removal of particles arising from the short-range nature of the surface forces was found. In other instances, the contact radius was found to vary with the particle radius, as predicted by the MP theory. For yet other systems, behavior suggestive of a JKR-DMT transition was observed [41]. Finally, unexplained phenomena, which are not predicted by any of the more commonly used adhesion theories, have also been observed. These effects include anomalous power law dependences of the contact radius on the particle radius, time-dependent creep effects, and evidence suggestive of total engulfment of the particle by the substrate.

This paper reviews theories and recent experimental findings aimed at characterizing adhesion-induced deformations of materials. Evidence for elastic and plastic flow is presented. In addition, some observations suggestive of a JKR-DMT transition are reported. The ranges of validity of the various theories are discussed. For those examples where a theory adequately predicted the response, the thermodynamic work of adhesion was calculated. Finally, unexplained phenomena including creep effects, anomalous power law dependences of the contact radius on the particle radius, and new evidence suggestive of total engulfment of the particles by the substrate are presented.

2. EXPERIMENTAL

Spherical or spheroidal particles of various materials including cross-linked polystyrene, glass, polyvinylidene fluoride (PVF_2), tin, and gold were deposited onto different substrates and the particle-substrate contact radii were measured using scanning electron microscopy (SEM). In general, the particles were deposited on the substrates by gently sprinkling them from a height of less than 1 cm. However, particles having average radii greater than 10 μm were deposited by slowly lowering the substrate towards the particles until the particles until the particles were attracted to the substrate. This procedure ensured that the kinetic energy of the particles was small [45–47] compared with the energy needed to produce the observed deformations.

Preparation of the particles and surfaces are discussed in detail elsewhere [34–44]. In general, however, the polystyrene particles and polymeric substrates were produced in-house. Care was taken to avoid unnecessary contamination of the surfaces. The glass, tin, and gold particles were purchased and used as received. While this precludes obtaining accurate values for the works of adhesion for clean materials, the results obtained reflect the 'real world' and the physics discussed is applicable to clean surfaces also.

Generally, samples were imaged in an SEM using secondary electron emission. After the particles were deposited on the substrates, the samples were mounted onto cross-sectional stubs to allow viewing at an 88° tilt angle. The samples were stored for approximately 2 weeks to allow contact-induced stresses to relax. The time dependence of the deformations was also determined for selected samples to ensure that the relaxation times were adequate [40, 48, 49]. Prior to viewing, the samples were coated with approximately 10 nm of a 60 : 40 gold/palladium alloy by sputtering with argon to avoid beam-induced charging and minimize Joule heating. As discussed elsewhere [38], these artifacts typically take several minutes to appear and are readily distinguishable. Short viewing times were used to further reduce these artifacts. More recently, the results obtained using standard SEM techniques have been confirmed using low-energy SEM and field emission SEM [48]. This permitted operation near the unity point (where one electron from the beam causes the emission of one secondary electron), thereby eliminating the need for a conductive coating, reducing Joule heating, and improving resolution. Contact radii were measured using particles and substrates that were optically smooth in order to minimize any effects on adhesion due to surface roughness.

3. RESULTS

Typical SEM micrographs of the contact region between particles and surfaces are shown in Figs 1 and 2. Figures 1A and 1B show cross-linked polystyrene particles, having radii of approximately 1.7 and 5.1 μm, on a polished silicon substrate. Most of the micrographs were taken at a relatively low magnification. However, near the contact zone, the magnification was increased. This technique allowed precise determinations of both the contact radius and the particle radius to be made in a single scan. Magnification scales are shown adjacent to each view. The contact area is due to the deformation of the polystyrene, which, having a Young's modulus of approximately 2.6×10^9 Pa and a yield strength of approximately 10.8×10^6 Pa, is much more compliant than the rigid silicon. Corrections to the apparent contact radius to account for the effects of the conductive coating are discussed elsewhere [37, 39].

Figure 2 shows examples of different-size glass particles having average radii of approximately 11.2 (Figs 2A and 2C) and 19.2 μm (Figs 2B and 2D) on different-hardness polyurethane substrates having Young's moduli at room temperature and 1 Hz of 3.83 MPa (Figs 2A and 2B) and 41.1 MPa (Figs 2C and 2D), respectively. In this instance, the finite contact area is due principally to the compliance of the substrate.

Several features are apparent in these sets of micrographs. The contact area is seen to increase with increasing particles radius, but at a rate that is less than linear. In contrast, the contact radius decreases with increasing elastic modulus, as can be seen by comparing Figs 2A and 2C, or 2B and 2D. In addition, small menisci, suggestive of tensile interactions, can be observed, particularly in Fig. 2A. For a particle of a given size and material on an elastomeric substrate, the size of the meniscus can vary considerable [10]. The reasons for these variations are presently unknown. However, it is possible that these variations may be due to surface contamination of inhomogeneities.

Whether an adhesion-induced deformation is elastic or plastic has considerable bearing on the amount of energy recovered upon removal of the particle from the surface. As can be seen by comparing equations (9) and (13), these types of deformations can be readily distinguished by comparing the power law dependence of the contact radius on the particle radius. Examples of both cases have been reported [10, 38, 43].

Adhesion-induced plastic deformation is illustrated by the case of polystyrene spheres having radii between approximately 1.5 and 10 μm in contact with a silicon substrate.

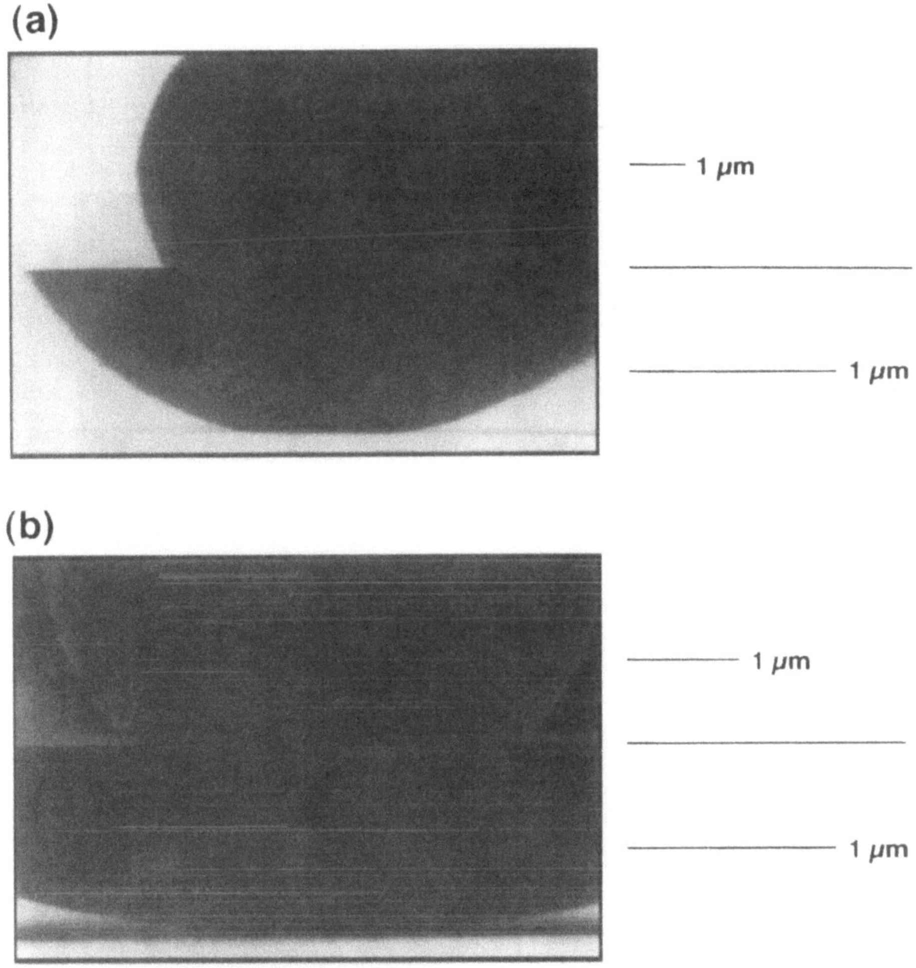

Figure 1. SEM micrographs of cross-linked polystyrene spheres having radii of 1.7 μm (1A) and 5.1 μm (1B) on a silicon substrate. The micrographs were taken at a tilt angle of 88°. Most of each particle was scanned at a relatively low magnification. However, near the contact region, the magnification was increased to improve resolution. Scale bars are shown adjacent to the particles.

The silicon substrate is rigid compared to the polystyrene so that any observed deformations are due to the flattening of the spheres. The Young's modulus and the yield strength of the bulk polystyrene, determined using a Sintech/20 tensile tester at a strain rate of 1.7×10^{-3} s^{-1} at room temperature, were found to be 2.55×10^9 and 10.8×10^6 Pa, respectively. The power law dependence of the contact radius on the particle radius was found, from a least-squares fit analysis of the log of the data, to be 0.42 ± 0.13. This is consistent with the predictions of the MP model, which assumes a plastic response, but not consistent with those of the JKR theory.

Figure 3 shows a plot of the contact radius of the polystyrene spheres as a function of the particle radius to the 0.5 power. It is seen that, within the error of the mea-

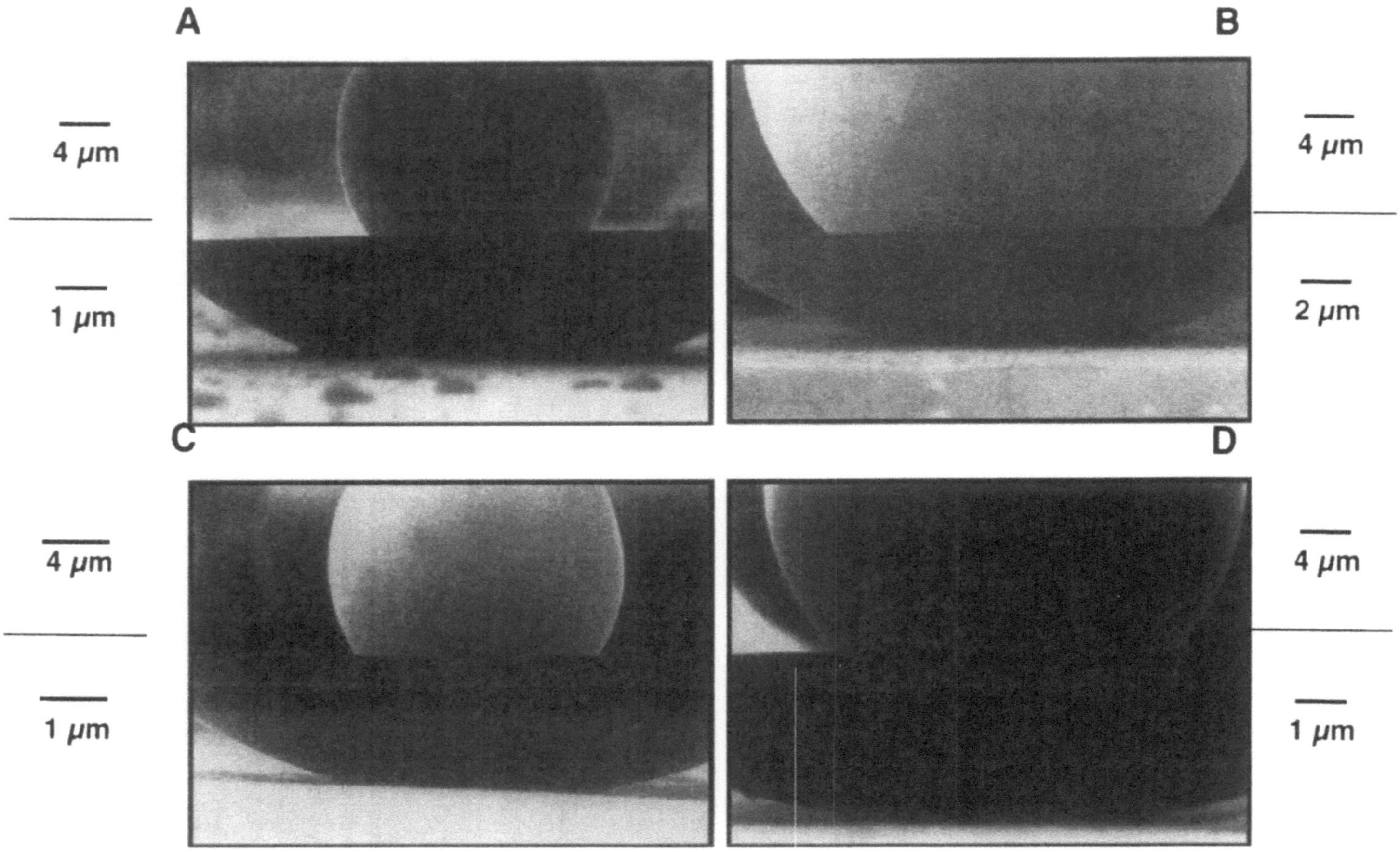

Figure 2. SEM micrographs of glass particles having radii of approximately 11.2 μm (A, C) and 19.2 μm (B, D) on polyurethane substrates having Young's moduli at room temperature and 1 Hz of 3.83 MPa (A, B) and 41.7 MPa (C, D). The scaling behavior of the contact radius with the particle radius and the substrate Young's modulus, as well as the height of the meniscus (most readily visible in A), are consistent with the predictions of the JKR model.

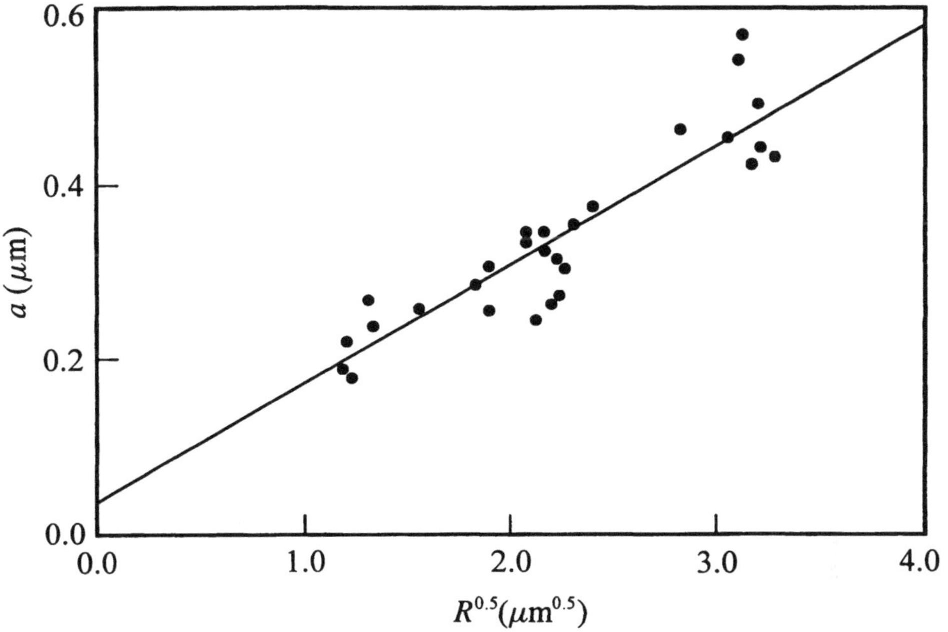

Figure 3. The contact radius as a function of the particle radius for polystyrene particles on silicon. These results are consistent with the predictions of the Maugis–Pollock theory, which assumes a plastic response.

surements, the extrapolated least-squares fit line through the data intersects the origin. This suggests that any externally applied loads, such as those arising from gravitation, triboelectrification, or beam-induced charging, are small when the particle is in contact with the substrate. The work of adhesion, calculated from the slope of this line and equation (13), is 0.32 J/m^2. In calculating w_A it was assumed that the yield strength of the particle was the same as that of the bulk polymer. That assumption is questionable and variations in the yield strength, Y, will affect the value calculated for w_A.

It is worthwhile to calculate the average pressure exerted by the particle on the substrate and to determine whether the stresses are sufficiently large so as to exceed the elastic limit of the polystyrene. Following the arguments of Tabor [14], the initial attractive force, F_0 between the substrate in contact with the undeformed particle is related to the work of adhesion and the radius of the particle by

$$F_0 = 2\pi R w_A. \tag{15}$$

The enhancement of that force due to the finite contact area can be estimated using the analysis of Bowling [50] (and assuming van der Waals interactions). Accordingly, the force of attraction of the deformed particles, F_d, is related to F_0 by

$$F_d = \frac{F_0 a^2}{z_0 R}, \tag{16}$$

where z_0 is the separation distance for a van der Waals-bonded solid and is typically about 4 Å. The average pressure, p, is just F_d divided by the contact area, or

$$p = \frac{2w_A}{z_0}. \tag{17}$$

Upon substituting the value of $z_0 = 4$ Å and $w_A = 0.32$ J/m^2 determined in this experiment into equation (17), it is found that p is approximately 10^9 Pa. This is close to the modulus of the polystyrene and well in excess of the yield strength of the material. Similarly, using the map shown in Fig. 2 of [21], it is also seen that fully plastic deformations are expected. Therefore, it is likely that the particles deformed plastically and the observed power law dependence of the contact radius on the particle radius and the adhesion-induced stresses are consistent.

Elastic response to the adhesion-generated stresses is illustrated by the case of glass particles having radii between approximately 5 and 60 μm on a polyurethane substrate having a Young's modulus of approximately 5×10^6 Pa. In this instance, the glass, having a Young's modulus of approximately 7×10^{10} Pa, is essentially rigid and the finite contact region is due to the compliance of the substrate.

As done previously, the power law dependence of the contact radius on the particle radius was found by linearly regressing log (a) onto log (R). In this case, it was found that a varied as $R^{0.68}$, which is consistent with an elastic response. Moreover, as is seen in Fig. 4, the least-squares fit line through the data of the contact radius versus the particle radius to the 2/3 power intersects the origin, within the limits of error of the experiment. This, again, argues that the effects of any externally applied forces are small. The calculated value for the work of adhesion is 0.17 J/m^2.

As previously discussed, the substrates are tilted on edge after having contacted the particles and are viewed in that manner. Moreover, prior to viewing, the system would be allowed to relax for approximately 2 weeks to ensure that it was in equilibrium. Therefore, a particle on the substrate would experience a peeling force due to gravity. In most cases, this force is sufficiently small to be negligible. However, if the particles are sufficiently large, gravity becomes a significant externally applied load. This is the case for glass particles having radii slightly in excess of 100 μm. Here, most particles separated from the polyurethane upon raising the substrate. Some particles, however, remained in contact with the substrate. A typical micrograph of the contact zone between such a particle and the polyurethane substrate is shown in Fig. 5. However, the contact radius of those particles with the substrate (12.5 ± 2.9 μm) was about the same as that for particles having a radius of approximately 60 μm (13.2 ± 0.55 μm). This effect allows the testing of a prediction of the JKR model, namely that of separation at a finite contact radius.

The force needed to detach a particle from a substrate, as predicted by the JKR model, can be calculated by substituting the appropriate values of w_A (0.17 J/m^2) and R (100 μm) into equation (10), assuming the occurrence of only elastic deformations. Accordingly, P_s was found to be approximately 8×10^{-5} N. The gravitational force peeling the particle from the substrate can be approximated by a lever with the

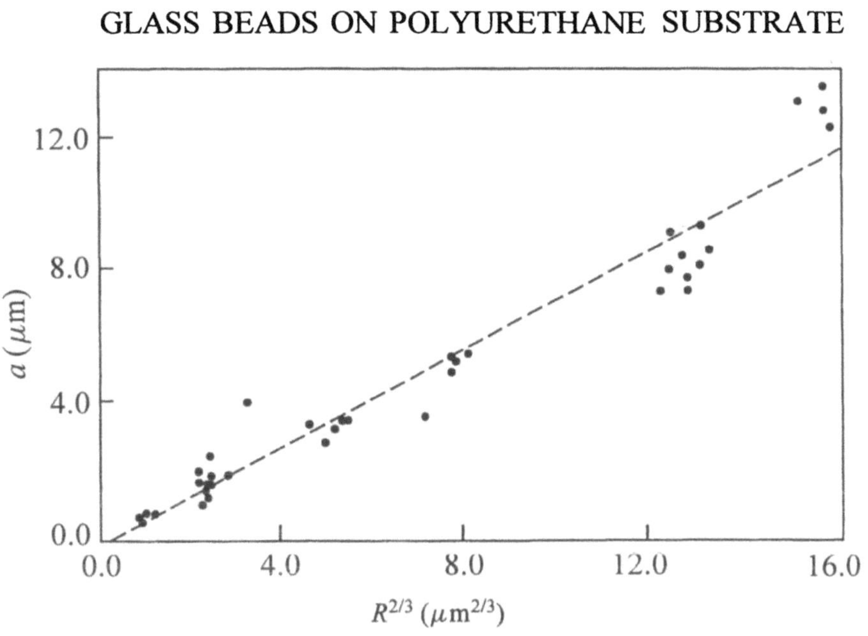

Figure 4. The contact radius as a function of the particle radius for glass beads on a polyurethane substrate. The power law dependence is consistent with the predictions of the JKR model.

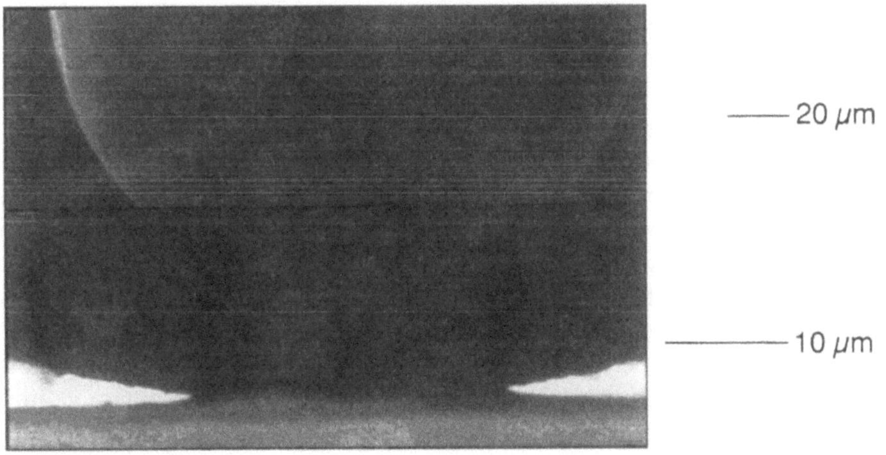

Figure 5. The contact zone between a glass particle being subjected to a negative load and a polyurethane substrate just prior to separation. The contact radius is approximately 0.7 of that expected if no load had been applied. The separation from a finite contact is consistent with predictions of the JKR model. However, the distension of the substrate can give rise to hysteretic effects which can account for the variations in separation forces needed with varying Young's modulus.

fulcrum at the edge of the particle–substrate contact. The weight of these particles is approximately 10^{-7} N. The effect of the levering action is to multiply this force by a factor of approximately 100, or $P = -10^{-5}$ N. This is less than, but within the same magnitude as, the theoretically predicted separation force. It should be noted that the observed particles have not yet been subjected to a force sufficiently large to remove them from the substrate. Moreover, the contact radius of these particles is approximately 68% of that expected for these particles in the absence of gravitational forces. This is close to, but slightly greater than the critical contact radius upon which separation occurs, as predicted by the JKR model. These results suggest that, within its range of validity, the JKR theory seems to accurately describe particle adhesion.

However, whereas the JKR theory predicts that the separation force is independent of the Young's modulus of the materials, it is frequently found that more compliant materials tend to be more adherent. The reason for this discrepancy can be understood by considering hysteretic effects. An example of such an effect is apparent in Fig. 5. As can be seen, the substrate appears to be distended around the particle. It would appear that as a negative force is being applied to the particle, work is also being done on the substrate. An equal extension of the substrate as the particle approaches would not be observed (although a smaller distension may be present at close range) because of the short-range nature of the surface forces. In other words, owing to the relatively short range of surface forces, the particle experiences little attraction to the substrate prior to contact. However, as the particle comes in close proximity to the substrate, the force of attraction increases rapidly and the particles snaps into contact with the substrate. If a negative force is then exerted on the particle in an attempt to detach it from the substrate, the stresses cause the substrate to deform past the point where the particle initially contacted it. Such effects have been observed experimentally by attaching particles to the cantilever of an atomic force apparatus and measuring the force as a function of the displacement upon attaching and removing a particle from a substrate [11, 51, 52]. Similar behavior has also been predicted theoretically using molecular dynamic modeling [53].

As can be seen from the previous discussion, particle adhesion can be frequently described in terms of well-established theoretical models. In cases where the stresses due to adhesion are sufficiently great so as to exceed the elastic limit of the material, the behavior of the system appears to be predicted by the model proposed by Maugis and Pollock [21]. In cases where the materials behave elastically, the JKR [9] or DMT [13] model may be appropriate. Which of these two models is the appropriate one to use can be determined from equation (11) or by microscopic determination as to whether the separation of the particle and the substrate occur at zero or finite contact radius, as was done in this study. Determination of whether the deformations are elastic, plastic, or other, and subsequently determining w_A, can be readily accomplished by determining how the contact radius scales with various parameters such as the particle radius.

Thus far, examples of particle–substrate adhesion, which can be readily described by theories that assume either and elastic or a plastic response of the materials, have been presented. However, systematic examination of the contact regions between different types of particles and substrates shows many instances where proposed theories do not adequately predict the behavior of the system. For example, hysteretic behavior and its effects on particle adhesion even in the case of a purely elastic, JKR, response has already been discussed. Other examples will be presented forthwith.

Consider, for example, polystyrene and glass particles, having radii between approximately 0.5 and 6.0 μm, on elastomeric (polyurethane) substrates. For such systems, linear regression of $\log(a)$ on $\log(R)$ showed that a varied as $R^{3/4}$. Typical plots of the contact radius as a function of the particle radius to the 3/4 power for polystyrene and glass particles in contact with elastomeric substrates (Young's modulus approximately 3–5 MPa) are shown in Figs 6A and 6B, respectively. Although this behavior is anomalous, it is similar to that found in the related problem [54] of frictional of polymeric materials [55]. In that case, the contact area, and hence the friction force, was found to vary with the applied load to the 3/4 power rather than the 2/3 power predicted by elastic theory.

Considering the period of time allowed for stresses to relax, it is unlikely that the anomalous dependence of the contact radius on the particle radius is due simply to viscoelastic effects. Perhaps this behavior arises from non-linear elasticity or from partial plasticity [21] due to the large surface force-induced stresses on the materials. Further theoretical and experimental studies are needed to clarify this problem.

Another topic of importance is the time dependence of the force needed to remove a particle from a substrate. That the removal force frequently increases with time has been well established [56]. This may be due to an increase in the particle–substrate contact area with time, as would be expected if plastic flow or viscoelastic creep occurred.

In a recent paper, DeMejo *et al.* [44] reported that anomalous behavior, including unexpectedly large menisci, interparticle bridging by the substrate polymer, and severe encapsulation, was observed for a variety of particles on an alternating block polyester–polydimethylsiloxane (PDMS) copolymer. Subsequently, Vrtis *et al.* [57] reported that upon removal of tin and gold particles by amalgamation with mercury from that substrate material, evidence of craters and accompanying menisci, such as those shown in Fig. 7, were observed. The size of these craters did not appear to change significantly for periods up to 2 weeks after the particles had been removed. These results suggest that the forces of adhesion may be inducing creep of the substrate polymer in contact with the particles.

In order to determine whether creep was occurring, particles of gold and polystyrene were deposited onto the polyester–PDMS substrate and the contact region between the particles and the substrate was observed as a function of time as soon as possible (approximately 15 min after the particles were deposited) and at various intervals up to 4 weeks later. The Young's moduli [58] of the polystyrene (3×10^9 Pa) and gold (8×10^{10} Pa) were substantially higher than that of the substrate (9×10^6 Pa). Therefore, any observed deformations were assumed to be due to the compliance of the substrate. To ensure that the gold/palladium coating on the samples did not affect the results, portions of the same samples were coated immediately prior to viewing at each time interval.

Figure 8 shows typical examples of polystyrene particles on the polyester–PDMS substrate at intervals of 15 min (A), 4 days (B), 2 weeks (C), and 1 month (D) after deposition. As can be seen, the contact radius between the particles and the substrate

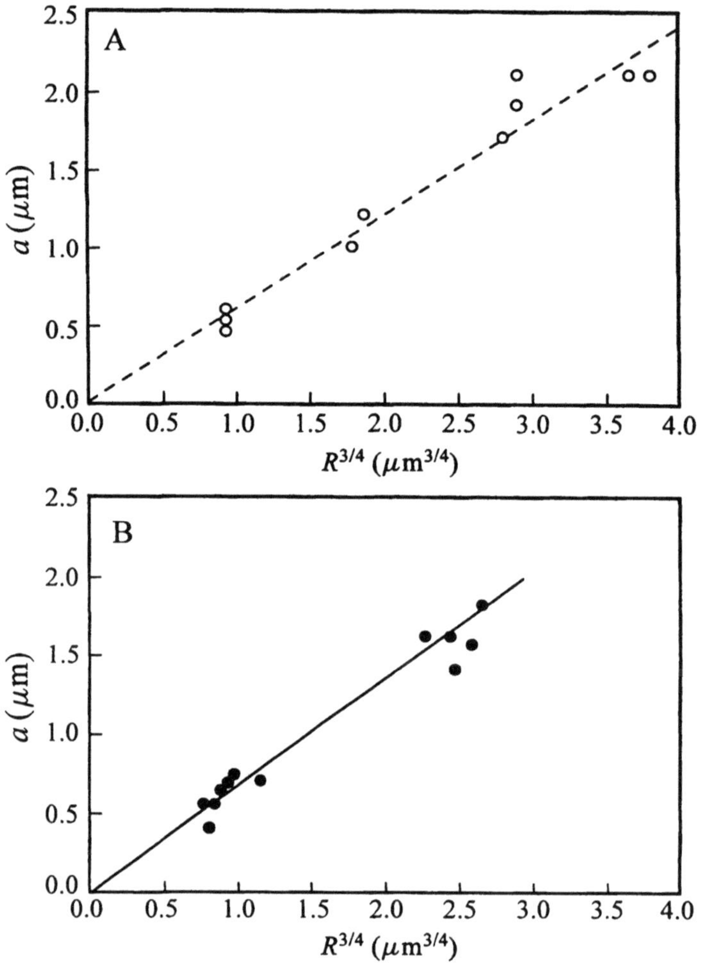

Figure 6. The contact radius of polystyrene (A) and glass (B) particles on an elastomeric substrate as a function of the particle radius for particles having radii between approximately 0.5 and 6.0 μm. The 3/4 power law dependence is not currently understood.

increases approximately 50% during the first 4 days. Subsequently, there is little change in that parameter. Therefore, there does appear to be some time dependence of the contact area for this system. It should be noted that the time dependences of the contact radii appear to be strongly dependent on the system examined. For example, while this substrate appears to have a relaxation time of the order of days, another elastomeric substrate had a relaxation time of the order of less than a few minutes [40]. Similarly, Rimai *et al.* have observed relaxation times for cross-linked polystyrene particles having radii between approximately 1 and 10 μm in contact with a silicon substrate to be less than a few minutes [48]. However, Busnaina, using submicrometer polystyrene latex spheres in contact with a silicon wafer, has observed relaxation times of the order of days [56]. Because the various theories describing particle adhesion assume that thermodynamic equilibrium has been established, it is extremely important

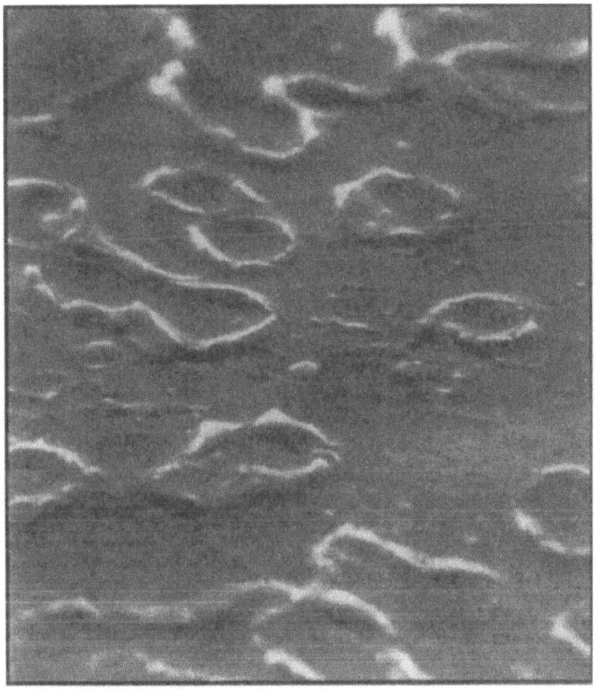

Figure 7. SEM micrograph of a polyester–PDMS copolymer substrate after gold particles were removed by amalgamation with mercury. Craters and menisci where the particles have been removed are visible.

to ensure that any relaxation effects have occurred prior to measuring the contact radii. To do otherwise could invalidate the determination of which, if any, theory describes the system under examination and can result in erroneous estimations of w_A.

Another interesting feature revealed in Fig. 8 is the time dependence of the apparent interparticle bridging by the substrate. As can be seen from Fig. 8A, little, if any, bridging is present a short time after the particles have been deposited onto the substrate. However, after 4 days (Fig. 8B) distinct interparticle bridging was present. The bridging continued to increase for approximately 2 weeks after the particles were deposited. Little enhancement to the bridging was seen with longer contact times. Similar behavior has been observed with 1.0 μm gold particles [59]. This behavior is highly reminiscent of creep and, together with the observation of the craters after the particles were removed, is further evidence that the stresses generated by the surface forces are sufficient to cause certain materials to deform plastically.

As discussed previously in this paper, the fundamental assumptions of the JKR and DMT models are different. In particular, the JKR model assumes that all interactions

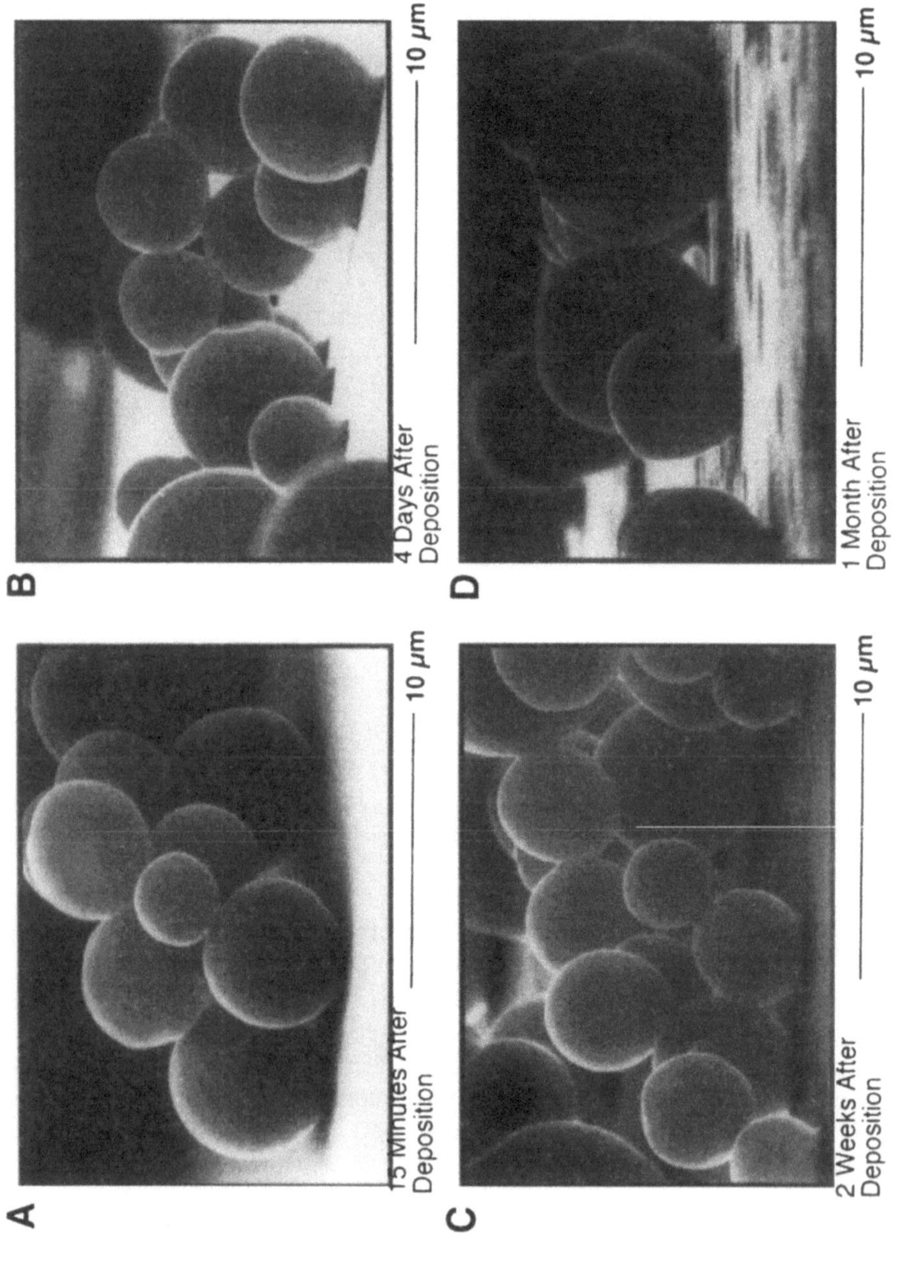

Figure 8. Polystyrene particles on an alternating block polyester–PDMS copolymer 15 min (A), 4 days (B), 2 weeks (C), and 1 month (D) after deposition. Creeping of the substrate up the particles is apparent.

occur within the contact zone. This results in a singularity in the potential at the boundary. On the other hand, the DMT model assumes that the shape of the contact zone is Hertzian. Consequently, half of the interaction between a particle and a substrate occurs within the contact radius and half occurs outside. As shown by Tabor [14], these theories predict different maximum pull-off forces. In addition, the JKR model predicts that the particle and substrate will separate with a finite contact radius, whereas the DMT model predicts that separation will occur when the contact radius vanishes.

As discussed previously, the transition between the JKR and DMT theories was shown to depend [17, 18] on a parameter, μ, which is related to the particle radius, the Young's moduli of the materials, and the work of adhesion by equation (11). If $\mu > 1$, the JKR model is the appropriate choice. If, on the other hand, $\mu < 1$, the DMT model is used. The transition region where $\mu = 1$ is a subject of current research [60, 61].

In practice, using μ to define which theory is the appropriate choice can be tricky. In particular, μ depends on w_A. However, until the deformations are sufficiently characterized, w_A can be difficult to determine. If just one measurement of the contact radius for a given 'particle' radius is made, such is frequently the case when crossed cylinder or nanoindenters [22–31] are used to simulate particle adhesion, it may difficult to distinguish whether JKR, DMT, or MP behavior is exhibited. Indeed, evidence has been presented in this paper where none of these theories appears at times to properly describe the response of the system.

Because of the different fundamental assumptions of the JKR and DMT theories, it is possible to differentiate JKR and DMT responses to the adhesion-generated stresses by examining the shape of the edge of the contact zone [31]. Even here, however, caution must be exercised. For example, Figs 9A and 9B show the contact region between a glass particle and a polyurethane substrate. The height of the meniscus is obviously much greater than those shown in Fig. 2, despite the similarity of the materials. The expected height of the meniscus can be calculated, assuming the validity of the JKR model, using the method proposed by Tabor [14]. Accordingly, the height of the meniscus around the contact zone is given by the sum of the height of the Hertzian gap, h_c, plus the JKR tensile contribution, h_t, according to the following equations:

$$h_c = \frac{8\sqrt{2}}{3\pi} \frac{a_0^2}{R} \left[\frac{r}{a} - 1\right]^{3/2}, \tag{18}$$

$$h_t = \frac{4}{3} \frac{a_0^2}{R} \left(1 - 2\pi \sin^{-1} \frac{a}{r}\right), \tag{19}$$

and

$$h = h_c = h_t \approx \left[\frac{R\left(\frac{w_A}{2}\right)^2}{E^2}\right]^{1/3}, \tag{20}$$

where r is an arbitrary distance from the center of the contact zone, R is the particle radius, a is the contact radius, a_0 is the contact radius under zero applied load, E is the Young's modulus of the more compliant material, and w_A is the work of adhesion. Equation (20) was derived from equations (18) and (19), with w_A related to a_0 by equation (9).

For the glass particles and the substrate under consideration, $w_A = 0.17$ J/m^2, $E = 5$ MPa, and $R = 3$ μm, equation (20) predicts that h should be of the order of 0.1 μm. This value is in good agreement with the measured height of most of the glass particles such as that shown in Fig. 2A, and, in general, appears to scale appropriately with the size of the particle. However, it is approximately a factor of 5 too small for the meniscus shown in Fig. 9. The cause of this variation is not presently known. However, possibilities include the tensile interactions exceeding the tensile

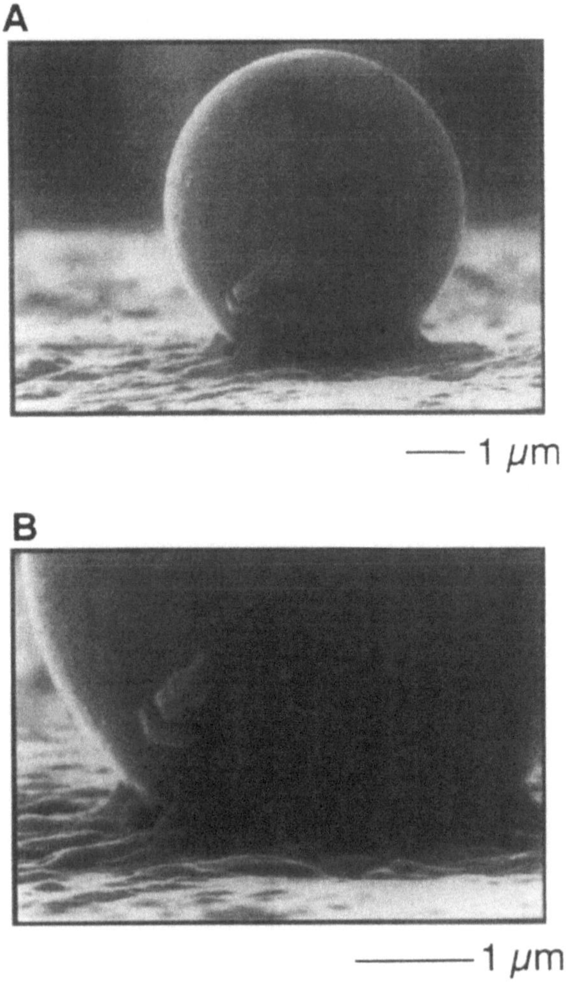

Figure 9. SEM micrographs of the contact region between a glass particle and a polyurethane substrate at low (A) and high (B) magnifications. Scale bars are shown beneath each micrograph. The height of the meniscus is much larger than that predicted by the JKR model.

strength of the polyurethane, variations in surface constituencies, contamination, and surface roughness [51, 62].

Even more dramatic is the case where large-scale deformation has occurred, which appears to be the case for the polyester–PDMS substrate. In this instance, the measured sizes of the menisci, for a variety of particles in contact with this substrate, are approximately an order of magnitude greater than that predicted by equation (20).

Finally, it is interesting to consider the case where the contact radius equals the size of the particle radius. Under such circumstances, would the particle continue to embed into the substrate and, conceivably, even become engulfed by the substrate? Gent predicted [63] that for contacting materials obeying the JKR model, the critical radius of the particle, R_c, at which engulfment would occur is given by

$$R_c \approx 7(\gamma_{12}/E). \tag{21}$$

Alternatively, for a material describable by the MP model,

$$R_c \approx 2\gamma_{12}/3Y. \tag{22}$$

As can be seen from equations (21) and (22), it is energetically more favorable for smaller particles to become engulfed than it is for larger particles. In addition, engulfment is more likely to occur with more 'compliant' materials irrespective of whether the compliance is due to the elastic response of the material or is from yielding.

Examples of adhesion force-induced deformations where the contact radius equalled the particle radius have been reported for particles of polyvinylidene fluoride (PVF_2), having radii of approximately 0.15 μm, on various polyester–PDMS copolymers [37, 44]. More recently, glass particles having nominal radii of 0.8, 4.0, and 11.0 μm were deposited onto substrates consisting of coatings of polystyrene, whose elastic moduli had been varied by adding various concentrations of plasticizer up to 20% by weight, on Kodak ESTAR Clear Base. The coatings ranged from 10 to 50 μm thick.

Figures 10A and 10B show nominal 4.0 μm and 11.0 μm radius glass particles, respectively, on a 50 μm thick coating of polystyrene/20% plasticizer. The larger particles appear to be approximately 40% embedded and the contact radius is almost equal to the particle radius. While the data do not permit an accurate determination of R_c at this time, they do suggest that R_c is less than, but not small compared to, 11 μm. Indeed, when he radius of the particles was decreased to approximately 4.0 μm, the particles appeared to become approximately 90% embedded, suggesting that total embedment may also occur. Such an effect suggests that materials having certain desired surface properties can be made by using surface forces to embed specific materials into the matrix material. Further research in this area is needed to more fully understand this effect.

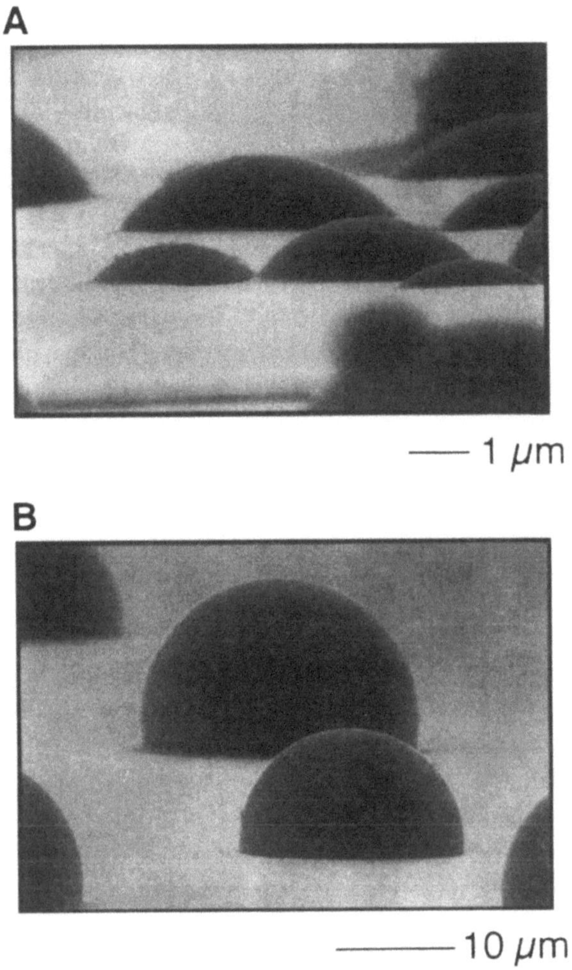

Figure 10. SEM micrographs of nominal 4.0 μm (A) and 11.0 μm (B) radius glass particles on a plasticized polystyrene substrate. Extreme embedding of the particles into the substrate is visible, especially for the smaller particles.

4. CONCLUSIONS

The adhesion of particles to a surface is a complex problem which depends on a number of parameters of the particle–substrate system. These include the size of the particles, the work of adhesion, and the mechanical properties of the materials such as their Young's moduli and yield strengths. The way in which parameters relate to particle adhesion has been the subject of much theoretical work.

The stresses generated by the forces of adhesion between a particle and a substrate result in the materials deforming. Depending on the nature of the materials and the magnitude of the stresses, these deformations can exceed the elastic limit of the materials, thereby resulting in a plastic deformation. In order to determine parameters for contacting solids such as the thermodynamic work of adhesion, the nature of the deformation must first be determined. This is most readily done by measuring the dependence of the contact radius on the particle radius.

Even after determining the nature of the deformation, further characterization may be necessary. For example, although the JKR and DMT models both assume an elastic response and allow for tensile interactions, they predict different contact radii and different removal forces. In addition, the JKR theory predicts that separation of the particle from the substrate occurs at a finite contact radius, whereas the contact radius must vanish before separation occurs according to the DMT theory. Therefore, differentiating between these two theories is readily done by examining the contact region between the particle and the substrate near separation.

Finally, many effects of adhesion are not explainable using current theories. These include hysteretic effects, the effect of Young's modulus on the particle removal force, rheological creep behavior, anomalous power law dependences of the contact radius on the particle radius, and particle engulfment. These effects can be readily differentiated from better understood behavior by examining parameter variations on particle–substrate contact. Moreover, to further our understanding of this field, further studies of this nature are needed.

Acknowledgements

We would like to thank M. Zaretsky, T. Tombs, R. Moore, D. Quesnel, R. Reifenberger, A. Busnaina, W. Light, W. Vreeland, and J. Fitzgerald for their assistance.

REFERENCES

1. K. K. Lam and J. M. Newton, *Powder Technol.* **73**, 267 (1992).
2. V. B. Menon, L. D. Michaels, R. P. Donovan and D. S. Ensor, in: *Particles on Surfaces 2: Detection, Adhesion, and Removal*, K. L. Mittal (Ed.), pp. 297–306. Plenum Press, New York (1989).
3. N. S. Goel and P. R. Spencer, in: *Adhesion Science and Technology, Part B*, L.-H. Lee (Ed.), p. 763. Plenum Press, New York (1975).
4. C. J. Mastrangelo, *Photo. Sci. Eng.* **26**, 194 (1982).
5. R. S. Bradley, *Philos. Mag.* **13**, 853 (1932).
6. R. S. Bradley, *Trans. Faraday Soc.* **32**, 1088 (1936).
7. B. V. Derjaguin, *Kolloid Z.* **69**, 155 (1934).
8. H. Krupp, *Adv. Colloid Interface Sci.* **1**, 111 (1967).
9. K. L. Johnson, K. Kendall and A. D. Roberts, *Proc. R. Soc. London, Ser. A* **324**, 301 (1971); K. Kendall and J. C. Padget, *Int. J. Adhesion Adhesives* **2**, 149 (1982).
10. L. P. DeMejo, D. S. Rimai and R. C. Bowen, *J. Adhesion Sci. Technol.* **5**, 959 (1991).
11. D. M. Schaefer, M. Carpenter, R. Reifenberger, L. P. DeMejo and D. S. Rimai, *J. Adhesion Sci. Technol.* **8**, 197–210 (1994).
12. D. J. Quesnel, D. S. Rimai and L. P. DeMejo, *J Adhesion Sci. Technol.* (submitted).
13. B. V. Derjaguin, V. M. Muller and Yu. P. Toporov, *J. Colloid Interface Sci.* **53**, 314 (1975).
14. D. Tabor, *J. Colloid Interface Sci.* **58**, 2 (1977).
15. B. V. Derjaguin, V. M. Muller and Yu. P. Toporov, *J. Colloid Interface Sci.* **67**, 378 (1978).
16. D. Tabor, *J. Colloid Interface Sci.* **67**, 380 (1978).
17. V. M. Muller, V. S. Yushchenko and B. V. Derjaguin, *J. Colloid Interface Sci.* **77**, 91 (1980).
18. V. M. Muller, V. S. Yushchenko and B. V. Derjaguin, *Colloids Surfaces* **7**, 251 (1983).
19. C.-J. Tsai, D. Y. H. Pui and B. Y. H. Liu, *Aerosol Sci. Technol.* **15**, 239 (1991).

20. D. Maugis, *J. Colloid Interface Sci.* **150**, 243 (1992).
21. D. Maugis and H. M. Pollock, *Acta Metall.* **32**, 1323 (1984).
22. D. Maugis and M. Barquins, *J. Phys. D* **11**, 1989 (1978).
23. D. Maugis and M. Barquins, *J. Phys. D* **16**, 1843 (1983).
24. H. M. Pollock, P. Shufflebottom and J. Skinner, *J. Phys. D* **10**, 127 (1977).
25. H. M. Pollock, *J. Phys. D* **11**, 39 (1978).
26. M. D. Pashley and D. Tabor, *Vacuum* **31**, 619 (1981).
27. R. Mouginot and D. Maugis, *J. Mater. Sci.* **20**, 4354 (1985).
28. R. Mouginot, *J. Mater. Sci.* **22**, 989 (1987).
29. R. Mouginot, *J. Am. Ceram. Soc.* **71**, 658 (1988).
30. R. G. Horn, J. N. Israelachvili and F. Pribac, *J. Colloid Interface Sci.* **115**, 480 (1987).
31. W. W. Merrill, A. V. Pocius, B. V. Thakker and M. Tirrell, *Langmuir* **7**, 1975 (1991).
32. B. Dahneke, *J. Colloid Interface Sci.* **40**, 1 (1972); see also G. L. Brown, *J. Polym. Sci.* **22**, 423 (1956).
33. M. M. Chaudhri and E. H. Yoffe, *Philos. Mag. A* **44**, 667 (1981).
34. L. P. DeMejo, D. S. Rimai and R. C. Bowen, *J. Adhesion Sci. Technol.* **2**, 331 (1988).
35. D. S. Rimai, L. P. DeMejo and R. C. Bowen, *J. Appl. Phys.* **65**, 755 (1989).
36. L. P. DeMejo, D. S. Rimai and R. C. Bowen, in: *Particles on Surfaces 2: Detection, Adhesion, and Removal*, K. L. Mittal (Ed.), pp. 49–58. Plenum Press, New York (1989).
37. R. C. Bowen, D. S. Rimai and L. P. DeMejo, *J. Adhesion Sci. Technol.* **3**, 623 (1989).
38. D. S. Rimai, L. P. DeMejo and R. C. Bowen, *J. Appl. Phys.* **66**, 3574 (1989).
39. D. S. Rimai, L. P. DeMejo and R. C. Bowen, *J. Appl. Phys.* **68**, 6234 (1990).
40. R. C. Bowen, L. P. DeMejo, D. S. Rimai and W. B. Vreeland, *J. Appl. Phys.* **70**, 3360 (1991).
41. D. S. Rimai, L. P. DeMejo, W. Vreeland, S. R. Gaboury and M. W. Urban, *J. Appl. Phys.* **71**, 2253 (1992).
42. D. S. Rimai, L. P. DeMejo, W. B. Vreeland, R. C. Bowen, S. R. Gaboury and M. W. Urban, *J. Appl. Phys.* **73**, 668 (1993).
43. D. S. Rimai, R. S. Moore, R. C. Bowen, V. K. Smith and P. E. Woodgate, *J. Mater. Res.* **8**, 662 (1993).
44. L. P. DeMejo, D. S. Rimai, J. Chen and R. C. Bowen, *J. Adhesion* **39**, 61 (1992).
45. L. N. Rogers and J. Reed, *J. Phys. D* **17**, 677 (1984).
46. S. Wall, W. John and S. L. Goren, in: *Particles on Surfaces 2: Detection, Adhesion, and Removal*, K. L. Mittal (Ed.), pp. 19–34. Plenum Press, New York (1989).
47. K. L. Johnson and H. M. Pollock, Unpublished results.
48. D. S. Rimai, R. S. Moore, R. C. Bowen, V. K. Smith and P. E. Woodgate, Unpublished results (1992); see also L. P. DeMejo, D. S. Rimai and R. C. Bowen, Unpublished results (1992).
49. L. P. DeMejo, D. S. Rimai, J. Chen and R. C. Bowen, *J. Adhesion* **39**, 61 (1992).
50. R. A. Bowling, in: *Particles on Surfaces 1: Detection, Adhesion, and Removal*, K. L. Mittal (Ed.), pp. 129–142. Plenum Press, New York (1988).
51. D. M. Schaefer, M. Carpenter, B. Gady, R. Reifenberger, L. P. DeMejo and D. S. Rimai, *J. Adhesion Sci. Technol.* (in press).
52. D. M. Schaefer, M. Carpenter, R. Reifenberger, L. P. DeMejo and D. S. Rimai, *Polym. Prepr.* **34**, 322 (1993).
53. D. J. Quesnel, D. S. Rimai and L. P. DeMejo, *Polym Prepr.* **34**, 320 (1993).
54. J. N. Israelachvili, Y.-L. Chen and H. Yoshizawa, *Polym. Prepr.* **34**, 282 (1993).
55. F. P. Bowden and D. Tabor, *Friction*, pp. 83–85. Anchor Press, Garden City, NY (1973).
56. A. Busnaina, *Polym. Prepr.* **34**, 314 (1993).
57. J. K. Vrtis, C. D. Athanasiou, R. J. Farris, L. P. DeMejo and D. S. Rimai, *J. Adhesion Sci. Technol.* **8**, 929 (1994).
58. D. W. Van Krevelen, *Properties of Polymers*. Elsevier, Amsterdam (1976).
59. L. P. DeMejo, D. S. Rimai, J. Chen and R. Bowen, *J. Adhesion* (in press).

60. D. Maugis and B. Gauthier-Manuel, *Polym. Prepr.* **34**, 318 (1993).
61. D. Maugis, *J. Colloid Interface Sci.* **150**, 243 (1992).
62. K. N. G. Fuller and D. Tabor, *Proc. R. Soc. London, Ser. A* **345**, 327 (1975).
63. A. Gent, Private communications (1992).

Fundamentals of Adhesion and Interfaces, pp. 25–34
D. S. Rimai, L. P. DeMejo and K. L. Mittal (Eds)

The role of adhesion in the impact of elastic spheres

K. L. JOHNSON[1] and H. M. POLLOCK[2,*]

[1] *University Engineering Laboratory, Cambridge CB2 1PZ, England*
[2] *School of Physics and Materials, Lancaster University, Lancaster LA1 4YB, England*

Revised version received 25 April 1994

Abstract—Many experiments have shown that when small particles impact a fixed surface, the coefficient of restitution can be influenced by adhesion forces. At low enough velocities the particles may be 'captured', so that no rebound occurs, and several theories exist in the literature which attempt to combine the effects of adhesion with inelastic deformation of the solids. In this paper we make use of the Johnson–Kendall–Roberts–Sperling (J-K-R-S) theory of adhesive elastic contact to examine interaction between adhesion and inelastic deformation in the impact of predominantly elastic solids.

Keywords: Adhesion; particles; impact; restitution; surface energy; viscoelasticity.

1. INTRODUCTION

The rate at which particles accumulate at a surface, for example during the transport of powders or the filtration of gases, is strongly affected by the sticking probability, which in general depends upon the impact velocity as well as the mechanical properties of the materials involved. Many experiments have shown that when small particles impact a fixed surface, the coefficient of restitution can be influenced by adhesion forces. At low enough velocities the particles may be 'captured', so that no rebound occurs, and several theories exist in the literature which attempt to combine the effects of adhesion with inelastic deformation of the solids [1–5, 20]. In this paper we make use of the Johnson–Kendall–Roberts–Sperling (J-K-R-S) theory of adhesive elastic contact [6, 7] to examine interaction between adhesion and inelastic deformation in the impact of predominantly elastic solids. We start by considering the normal impact of a solid sphere of radius R with a half-space, both of which are perfectly elastic, and will show that even in this case some energy dissipation would be expected; dynamic effects are involved even in the case of zero impact velocity.

*To whom correspondence should be addressed.

2. THE J-K-R-S THEORY

At any stage during the impact the bodies are mutually compressed by a displacement δ, giving rise to a contact area of radius a. This deformation is caused by the inertia of the sphere, $m(\mathrm{d}^2\delta/\mathrm{d}t^2)$, which is just balanced by the total force at the interface, P. (P represents the net effect of compressive stress at the centre of the contact region and peripheral tensile stress due to adhesion forces. In the case of static contact, an *external* load of value P would produce the given equivalent value of δ.) Displacements and forces are normalised as follows:

$$\bar{\delta} = \delta\left(\frac{\pi^2 w^2 R}{3K^2}\right)^{-1/3}, \qquad \bar{a} = a\left(\frac{3\pi w R^2}{K}\right)^{-1/3}, \tag{1}$$

$$\overline{P} = \frac{P}{3\pi w R} \tag{2}$$

and for consistency, energies are normalised by

$$\left(\frac{9\pi^5 w^5 R^4}{K^2}\right)^{1/3}, \tag{3}$$

where w is the Dupré energy of adhesion, equal to $\gamma_1 + \gamma_2 - \gamma_{12}$ (γ_1 and γ_2 are the surface energies and γ_{12} is the interfacial energy), and

$$\frac{1}{K} = \frac{3}{4}\left(\frac{1-\nu_1^2}{E_1} + \frac{1-\nu_2^2}{E_2}\right),$$

where E is Young's modulus and ν is Poisson's ratio.

The energy of the system at any time includes the stored elastic strain energy U_E and the surface energy $U_\mathrm{S} = -\pi w a^2$, and their sum,

$$U_\mathrm{I} \equiv U_\mathrm{E} + U_\mathrm{S}, \tag{4}$$

may be regarded as the internal energy.* Under equilibrium conditions, changes in the internal energy may be regarded as areas under an equilibrium curve of load against displacement [8]:

$$\Delta U_\mathrm{I} = \Delta(U_\mathrm{E} + U_\mathrm{S}) = \int(\mathrm{d}U_\mathrm{E} - G\,\mathrm{d}A) = \int_{\mathrm{equil}} P\,\mathrm{d}\delta, \tag{5}$$

where G is strain energy release rate, A is contact area and in this situation, G is equated to w. It is also convenient to make use of a variable parameter, termed the

*For changes at fixed grips (imposed displacement rather than imposed load), U_I is the thermodynamic potential [8].

'apparent Hertz load' $P_1 = (K/R)a^3$ [6], since each of the other variables is a single-valued function of P_1. In accordance with the J-K-R-S theory, for the main part of the process P_1 is given in normalised form by [6, 8] $\overline{P_1} = 1 + \overline{P} + \sqrt{1 + 2\overline{P}}$, but for part of the negative range of load, involving contact areas smaller than the value obtained at the point of separation at fixed load (i.e. $\overline{P_1} < 1/2$), it must be defined by $\overline{P_1} = 1 + \overline{P} - \sqrt{1 + 2\overline{P}}$. In both cases we have

$$\overline{P} = \overline{P_1} - \sqrt{2\overline{P_1}}, \tag{6}$$

$$\bar{\delta} = \frac{\overline{P_1} + 2\overline{P}}{\overline{P_1}^{1/3}} = \frac{3\overline{P_1} - \sqrt{8\overline{P_1}}}{\overline{P_1}^{1/3}}, \tag{7}$$

$$\bar{a} = \overline{P_1}^{1/3}, \tag{8}$$

$$\overline{U_S} = -\overline{P_1}^{2/3}, \tag{9}$$

$$\overline{U_E} = \frac{\frac{6}{5}\overline{P_1}^2 + 2\overline{P_1} - \left(2\overline{P_1}\right)^{3/2}}{\overline{P_1}^{1/3}}, \tag{10}$$

and

$$\overline{U_I} = \frac{\frac{6}{5}\overline{P_1}^2 + \overline{P_1} - \left(2\overline{P_1}\right)^{3/2}}{\overline{P_1}^{1/3}}. \tag{11}$$

From expressions (7) and (11) with $\overline{P_1}$ being used as the plotting parameter, the variation of $\overline{U_I}$ with $\bar{\delta}$ is shown in Fig. 1. Figure 2 (curve I) shows $\overline{U_I}$ as a function of $\overline{P_1}$.

3. ELASTIC IMPACT

By referring to Fig. 1 we can follow the events during a perfectly elastic impact. In view of the short range of adhesion forces, they will have no appreciable effect until the surfaces are on the point of contact (O in Fig. 1). This is an unstable point. As soon as the surfaces touch, they will snap into contact at the point A on the equilibrium curve. Since this action takes place at elastic wave speed, the change in $\bar{\delta}$ from zero in this time is negligible. Thölén [9] has presented evidence from zero-velocity contact experiments that the associated release of energy may be converted to energy of vibration. We assume that the energy $U_I(\mathrm{A})$ is entirely dissipated in wave motion, so that the sphere gains no significant additional kinetic energy through being accelerated by surface forces. Subsequent compression, until the sphere comes to rest at some point of maximum compression B, is reversible, as is the following expansion back along the same curve BCAE...

Clearly the criterion for separation of the bodies after impact is whether or not the internal energy, on recoil, reaches point D on the curve. If it does not, the system will oscillate along the curve, but if the incident kinetic energy T_i is large enough, U_I will

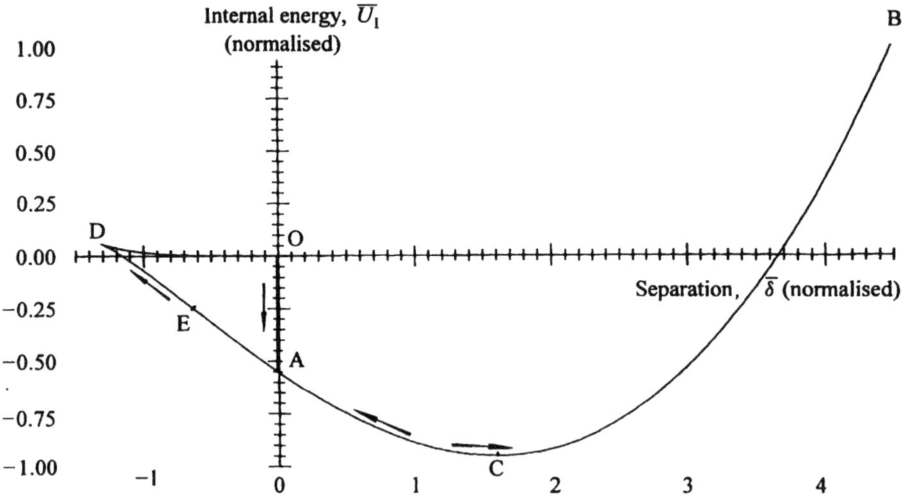

Figure 1. Variation in normalised internal energy $U_E + U_S$ as a function of normalised separation during a perfectly elastic impact, as predicted by the J-K-R-S model of elastic sphere/flat contact with adhesion. A, initial non-equilibrium contact; C, point of static equilibrium at zero load; E, static equilibrium at maximum negative imposed load; D, static equilibrium at maximum negative imposed displacement. Between D and O, stable equilibrium is not possible. The gradient of the curve is given by the function $(2 - 7\sqrt{2P_1} + 6P_1)/(6 - \sqrt{2/P_1})$: note the cusp at D, where $P_1 = 1/18$. The compression A—B and the separation B—D are reversible. Energy is dissipated in snap-on O—A and in snap-off D—O.

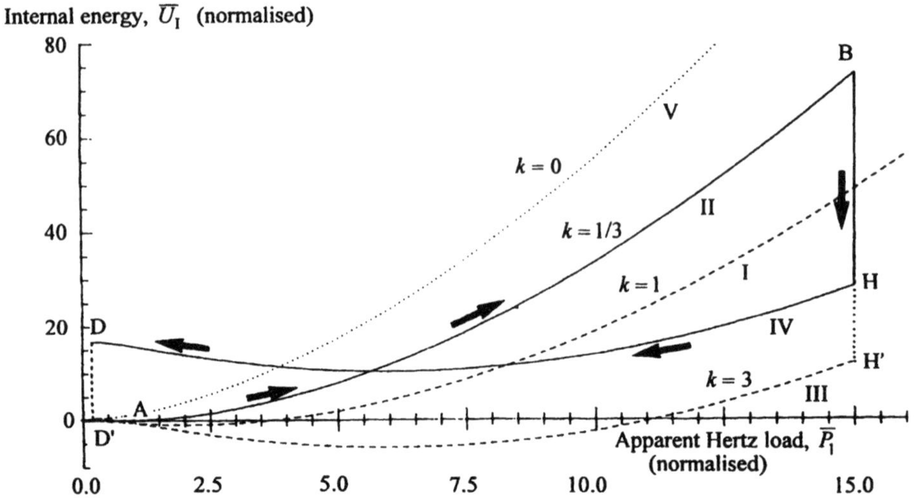

Figure 2. Interaction between adhesion and inelastic deformation in an impact cycle. In compression A—B the effective adhesion energy $w_f = k''w$, $(k'' < 1.0)$: in separation H—D, $w_f = k'w$, $(k' > 1.0)$. During the unloading phase B—H the contact area and adhesion energy remain constant.

reach D, where the negative value of $\bar{\delta}$ indicates that a neck of material is being stretched out by adhesion forces. D, the 'fixed grips' limit, is another unstable point, and here $U_I(D)$ is also dissipated in vibration. If T_i and T_r are the incident and rebound kinetic energies of the sphere, we have $T_i = U_I(B) - U_I(A)$ and $T_r = U_I(B) - U_I(D)$, so that $T_i - T_r = U_I(D) - U_I(A)$. Now $\overline{U_I}(D) = 0.058$ and $\overline{U_I}(A) = -0.555$, so that

$$T_i - T_r = 0.613\left(\frac{9\pi^5 w^5 R^4}{K^2}\right)^{1/3}. \tag{12}$$

Capture will therefore occur if

$$T_i = \frac{1}{2} m V_i^2 < 0.613\left(\frac{9\pi^5 w^5 R^4}{K^2}\right)^{1/3}, \tag{13}$$

where m $(= 4\pi\rho R^3/3)$ is the mass of the sphere and V_i is the incident velocity. It follows from equation (13) that the critical capture velocity varies as $R^{-5/6}$. If V_i is less than this value, the particle will oscillate about the minimum energy position C.

One may predict how the kinetic energy of the rebounding particle will vary with the incident kinetic energy. From equation (12) we see that the measured coefficient of restitution $\sqrt{T_r/T_i}$ will not be constant, but will vary according to the relation

$$\frac{T_r}{T_i} = 1 - \frac{0.61}{\overline{T_i}}. \tag{14}$$

Comparing equations (13) and (14) with the experimental data of Wall *et al.* [5] on the impact of ammonium fluorescein particles on targets of Tedlar (a deformable fluorocarbon polymer) and of molybdenum, we see several discrepancies:

First, with increasing incident velocity, the coefficient of restitution $\sqrt{T_r/T_i}$ tended towards a value of about 0.8, rather than unity. Wall *et al.* in Fig. 12 of their paper [5] described this behaviour by means of an equation similar to equation (14) but containing a second adjustable parameter* e:

$$\frac{K_r}{K_i} = e^2 + \frac{\Delta E}{K_i}. \tag{15}$$

This behaviour is the expected consequence of *bulk* inelasticity of the particle [9, 10], which may be taken into account by a theory such as that of Rogers and Reed [4]. This aspect of the problem will not be considered further in this paper.

Secondly, the critical capture velocity was observed to vary with radius R by an index < -1.0, rather than $-5/6$. We shall return to this point later.

Finally, fitting equation (14) to the data yielded values of $w = 0.58$ Jm^{-2} for ammonium fluorescein/molybdenum and 1.35 Jm^{-2} for ammonium fluorescein/Tedlar. These values are about an order of magnitude too large compared with measurements on comparable materials by other methods, and an attempt to fit the model to Dahneke's data for polystyrene latex onto polished quartz [1] leads to a similar conclusion. This suggests some significant inadequacy in the theory.

*Not the same as our coefficient of restitution $\sqrt{T_r/T_i}$.

4. INTERACTION OF ADHESION WITH INELASTIC DEFORMATION

As mentioned above, introducing a *bulk* hysteresis loss factor into the model improves the fit with experiment at higher incident velocities, but does not change the apparent surface energy. The clue to the discrepancy lies in the infinite tensile traction at the edge of the contact area predicted by the J-K-R-S theory. As noted by Fichman and Pnueli [11], it should cause some highly localised inelastic deformation even at the lightest load. This effect has been investigated for viscoelastic materials by Greenwood and Johnson [12, 13] and by Barquins and Maugis [14, 8], and for elastic–plastic materials by Tvergaard and Hutchinson [15]. These studies (supported by experiments [16, 17]) reveal a significant difference between growth of the contact area when the surfaces are coming together and contraction of the area when they are peeling apart since during peeling only, the crack undergoes a drag force that is proportional to w. The variations in P, a and δ appear to follow the J-K-R-S relationships quoted above, but with the strain energy release rate G equated to a modified adhesion energy of fracture w_f, instead of to w, such that

$$w_f = k \cdot w, \tag{16}$$

where the loss factor k is equal $k' > 1.0$ during peeling and $k'' < 1.0$ during compression. With viscoelastic materials, k is a function of the loss modulus of the material (and hence, of temperature) and also of the rate of deformation [8, 13]. A high value of k' during peeling is associated with a low value of k'' during compression (closing), such that the effect of surface forces during closing becomes imperceptible [12, 17]. In a study of interfacial fracture of elastic–plastic solids Tvergaard and Hutchinson [15] found that the total work of fracture w_f also followed equation (16), where w, in this case, is the work done in the fracture process zone. During peeling, the factor k takes account of the work dissipated in anelastic deformation in the surrounding material. It is a function of the ratio of two quantities, the theoretical strength σ_0 (itself proportional to w), and the yield stress of the material σ_y, which is why $k' > 1.0$. The important point to note from these results is that the total work to separate the surfaces is the *product*, rather than the sum, of the true work of adhesion w and a factor k which takes account of the energy dissipated in inelastic deformation of the material adjacent to the point of separation. The consequences of this result for impacting particles may be followed from Fig. 2.

It will be assumed that the processes of attachment and separation still follow the J-K-R-S relationships with an effective adhesion energy w_f which is less than w when the surfaces are coming together, and greater than w when they peel apart. Accordingly we need to amend the definitions of the normalised variables. The master curves as given by equations (7–11) should be re-written so as to include in their normalisation factors the adhesion energy that applies to the particular state of the system in question, i.e. w_f where appropriate instead of w in equations (1–3). (We then use the notation $(\overline{P_1})_{\text{master}}$, $(\overline{U})_{\text{master}}$, etc.) In addition, we will need to describe states with different values of w_f by separate curves on one diagram rather than a single master curve. To this end, we re-express $\overline{P_1}$ and $\overline{U}$ in terms of a single value of energy w. To convert from $(\overline{P_1})_{\text{master}}$ to $\overline{P_1}$, etc. we have

$$\overline{P_1} = \frac{P_1}{3\pi w R} = \frac{1}{3\pi w R} \cdot \left(\overline{P_1}\right)_{\text{master}} \cdot 3\pi w_f R = k\left(\overline{P_1}\right)_{\text{master}}, \tag{17}$$

and likewise,

$$\overline{U} = k^{5/3}\left(\overline{U}\right)_{\text{master}}. \tag{18}$$

We will also need $\overline{U_S}$ in terms of $\overline{P_1}$: from equations (18) and (9) it follows that

$$\overline{U_S} = k^{5/3} \cdot \left(- \left(\overline{P_1}\right)^{2/3}_{\text{master}} \right),$$

which from equation (17) gives

$$\overline{U_S} = -k^{5/3}\left(\frac{\overline{P_1}}{k}\right)^{2/3} = -k\overline{P_1}^{2/3}. \tag{19}$$

During compression, $k'' < 1.0$ (1/3, say) and the scaling relations (17) and (18) may be used to plot curve II in Fig. 2 from curve I. B again is the point of maximum compression, when the particles come to rest. We then drop from B to H, with a and hence P_1 remaining unchanged, until the release of elastic energy is sufficient to initiate peeling (point H), with $k' > 1.0$ (3, say) during separation. Equilibrium peeling then follows curve IV (note that points on this curve do not satisfy J-K-R-S formulae such as equation (11)). To plot this curve, we first construct curve III (from H′ to D′) using the scaling equations (17) and (18) with $k = k'$. Comparing states represented by curves IV and III, it is clear that for a given value of a (and hence P_1), the geometry of the contact, and hence U_E, will be the same. The two curves differ only in the values of U_S, and thus curve IV is the same as curve III displaced upwards by an amount $\overline{U_S}(\text{H}) - \overline{U_S}(\text{H}')$. But $U_S(\text{H}) = U_S(\text{B})$ since a and P_1 are the same for B and H, so that the upward displacement of curve IV in normalised form (using equation (19)) is

$$\overline{U_S}(\text{B}) - \overline{U_S}(\text{H}') = (k' - k'')\overline{P}^{2/3}_{1(\text{B,H})}. \tag{20}$$

The difference between incident and rebound kinetic energies then becomes

$$\overline{T_i} - \overline{T_r} = \overline{U_I}(\text{D}) - \overline{U_I}(\text{A}) \tag{21}$$

$$= \overline{U_I}(\text{D}') + (k' - k'')\overline{P}^{2/3}_{1(\text{B,H})} - \overline{U_I}(\text{A}) \tag{22}$$

$$= 0.058k' + 0.555k'' + (k' - k'')\overline{P}^{2/3}_{1(\text{B,H})}. \tag{23}$$

This energy difference, unlike that given by equation (12), thus varies with the maximum compression imposed. Equation (23) reduces of course to equation (12) for perfect elasticity when $k' = k'' = 1.0$.

If $k'' \ll 1.0$, compression takes place according to the Hertz relationship (curve V), so we can express $\overline{P}_{1(B)}$ in terms of T_i, since

$$\overline{T_i} = \overline{U_E}(B) = \frac{6}{5}\overline{P}^{5/3}_{1(B)}. \tag{24}$$

Neglecting the small snap-off term $0.058k'$, equation (23) then becomes

$$\overline{T_i} - \overline{T_r} = k'\overline{P}^{2/3}_{1(B)} = k' \cdot \left(\frac{5}{6}\overline{T_i}\right)^{2/5}, \tag{25}$$

so that

$$\frac{T_r}{T_i} = 1 - 0.93k'\overline{T_i}^{-3/5}, \tag{26}$$

i.e.

$$\frac{V_r^2}{V_i^2} = 1 - 2.91\frac{k'w}{R}\rho^{-3/5}K^{-2/5}V_i^{-6/5}, \tag{27}$$

where ρ is the density of the sphere. Noting that $(k'w)$ is the work of separation w_f, it is easily shown that this result is consistent with equation (12) of Roberts and Thomas [16], which correlated well with observations of steel balls bouncing on rubber.* Comparing equation (26) (or (27)) with the results of Wall *et al.* [5] we note the following points:

1. It follows from equation (27) that, as with the purely elastic model (equation (13)), the critical velocity for capture varies as $R^{-5/6}$, i.e. more slowly than was found by experiment.
2. To fit equation (26) to individual data points, the required values of $k'w$ are reasonable. For example, for the ammonium fluorescein/Tedlar data ([5], Fig. 12), the point ($T_r/T_i = 0.3$, $1/T_i = 10^{12}$ J^{-1}) is consistent with the material parameters $k'w = 0.92$ Jm^{-2}, $R = 3.5$ μm, and $K = 1.1 \times 10^9$ Nm^{-2}, on the basis of equation (26).
3. The coefficient of restitution $\sqrt{T_r/T_i}$ is predicted to vary more slowly with incident velocity (equation (26)) than according to the purely elastic model (equation (14)). As regards this variation, the data available do not allow us to distinguish confidently between the two predictions: in either case, if we use the appropriately fitted values of w or w_f, experiment and theory give reasonable agreement as regards the slope of plots of (coefficient of restitution)^2 against T_i^{-1} (equation (14)) or $T_i^{-3/5}$ (equation (26)).

*Note added in proof: equation (24) is equivalent to equation (12) of Tsai *et al.* [20], who, however, wrongly include a factor of $2^{2/3}$ in their equation (13).

5. CONCLUSIONS

1. It follows from the J-K-R-S theory that, in the presence of adhesion forces, even perfectly elastic impacting spheres undergo some energy loss due to an irreversible 'snap-on' when they first make contact and an irreversible 'snap-off' when they separate.
2. Perfect elasticity, however, is not easily realised or even approximated. The singularity in traction at the edge of the contact area the J-K-R-S theory ensures some inelastic deformation in that region, which leads to 'apparent adhesion energies of fracture' that are different when the surfaces are peeling apart than when they are coming together. The way in which this effect leads to a reduction in rebound energy is explained.
3. Predictions of the coefficient of restitution on this basis agree well with the results of experiments with steel balls bouncing on rubber, but less well with smaller spheres and less compliant materials, although the required values of elastic modulus, adhesion energy and viscoelastic loss factor are of the right order of magnitude.
4. To a close approximation, this viscoelastic loss model predicts that 1 minus the square of the coefficient of restitution varies as the incident kinetic energy to the power $(-3/5)$, and that the critical capture velocity varies with particle radius to the power $(-5/6)$. For very small impacting particles, the J-K-R-S theory becomes inappropriate and the situation is better described by the Derjaguin–Muller–Toporov (D-M-T) model [18]. This suggests a possible reason for the discrepancy in the variation of capture velocity with radius between our theory and experiments on small particles. An analysis of the transition between the J-K-R-S and D-M-T approximations has recently been presented by Maugis [19]. This will be considered in relation to the impact problem in a later paper.

REFERENCES

1. B. Dahneke, *J. Colloid Interface Sci.* **51**, 58–65 (1975).
2. R. V. Latham, A. S. Brah, K. Fok and M. O. Woods, *J. Phys. D: Appl. Phys.* **10**, 139–150 (1977). R. V. Latham and A. S. Brah, *J. Phys. D: Appl. Phys.* **10**, 151–167 (1977).
3. J. Aboudi, *Comput. Methods Appl. Mech. and Eng.* **20**, 135–150 (1979).
4. L. N. Rogers and J. Reed, *J. Phys. D: Appl. Phys.* **17**, 677–689 (1984).
5. S. Wall, W. John, H.-C. Wang and S. Goren, *Aerosol Science and Technology* **12**, 926–946 (1990).
6. K. L. Johnson, K. Kendall and A. D. Roberts, *Proc. R. Soc.* **A 324**, 301 (1971).
7. G. Sperling, Dissertation. Karlsruhe Technical High School (1964).
8. D. Maugis and M. Barquins, in: *Adhesion and Adsorption of Polymers*, L.-H. Lee (Ed.), pp. 203–277. Plenum Press, New York (1980).
9. A. R. Thölén, *Phys. Stat. Solid.* **a60**, 153 (1980).
10. D. Maugis and H. M. Pollock, *Acta Metall.* **32**, 1323–1334 (1984).
11. M. Fichman and D. Pnueli, *ASME J. Appl. Mech.* **52**, 105–108 (1985).
12. K. L. Johnson, in: *Theoretical and Applied Mechanics*, W. T. Koiter (Ed.), pp. 133–138. North-Holland, Amsterdam (1977).
13. J. A. Greenwood and K. L. Johnson, *Phil. Mag.* **A43**, 697–711 (1981).
14. M. Barquins and D. Maugis, *J. Phys. D: Appl. Phys.* **11**, 1989–2023 (1978).
15. V. Tvergaard and J. W. Hutchinson, *J. Mech. Phys. Solids* **40**, 1377–1397 (1992).

16. A. D. Roberts and A. G. Thomas, *Wear* **33**, 45–64 (1975).
17. D. Maugis and M. Barquins, *C R Acad. Sci. Paris* **B287**, 49–52 (1978).
18. B. V. Derjaguin, V. M. Muller and Y. P. Toporov, *J. Colloid Interface Sci.* **53**, 314–326 (1975).
19. D. Maugis, *J. Colloid Interface Sci.* **150**, 243–269 (1992).
20. C.-J. Tsai, D. Pui and B. Y. H. Liu, *Aerosol Sci. Technol.* **12**, 497–507 (1990).

Fundamentals of Adhesion and Interfaces, pp. 35–48
D. S. Rimai, L. P. DeMejo and K. L. Mittal (Eds)

Surface roughness and its influence on particle adhesion using atomic force techniques

D. M. SCHAEFER,[1] M. CARPENTER,[1] B. GADY,[1] R. REIFENBERGER,[1] L. P. DEMEJO[2] and D. S. RIMAI[2,*]

[1] *Department of Physics, Purdue University, W. Lafayette, IN 47907, USA*
[2] *Office Imaging Research and Technology Development, Eastman Kodak Co., Rochester, NY 14653-6402, USA*

Revised version received 5 September 1994

Abstract—The surface force interactions between individual 8 μm diameter spheres and atomically flat substrates have been investigated using atomic force techniques. The lift-off force of glass, polystyrene, and tin particles from atomically smooth mica and highly oriented pyrolitic graphite substrates was determined as a function of the applied loading force in an inert nitrogen environment. While the relative magnitudes of the measured lift-off force were found to scale as expected between the various systems studied, the absolute values were a factor of ~50 smaller than expected from the Johnson, Kendall, and Roberts theory. The surface topography of representative spheres was characterized with atomic force microscopy, allowing a quantitative assessment of the role that surface roughness plays in the adhesion of micrometer-size particles to substrates. Taking into account the radius of curvature of the asperities measured from the atomic force scans, agreement between the measured and theoretical estimates for the lift-off forces was improved, with the corrected experimental forces about a factor of 3 smaller than theoretical expectations.

Keywords: Particle adhesion; atomic force microscope (AFM); surface force interaction.

1. INTRODUCTION

Recently, the measurement of the removal force of an individual, micrometer-size polystyrene sphere from a silicon substrate has been reported using techniques borrowed from atomic force microscopy [1]. The techniques developed in this prior work are particularly useful for investigating the force between a sphere and a flat surface in the limit of zero applied load, a regime dominated by the surface interaction forces between the two objects. By affixing a suitable micrometer-size sphere to an atomic force cantilever, the interaction between the sphere and a suitable substrate can be investigated as a function of a number of variables which include the sphere

*To whom correspondence should be addressed.

diameter, sphere–substrate separation, and the ambient environment. The results of these initial experiments on a polystyrene sphere above a silicon substrate allowed a measure of the lift-off force of individual polystyrene spheres from silicon. Using similar techniques, Mizes *et al.* [2] have also measured the force on spheres suspended above substrates as a function of the applied electric field. The advantage of atomic force techniques is that new studies of adhesion between micrometer-size particles and substrates are now possible. For instance, because the adhesion force may vary as a function of time for certain materials, a measurement of adhesion as a function of the contact time would provide new information that is inaccessible using more conventional experimental techniques [3–7].

The initial study of polystyrene on silicon [1] has cleared the way for systematic studies of adhesion and removal of small particles to and from surfaces as a function of a variety of parameters including the particle size, substrate material, and surrounding gas environment. In what follows, we report measurements made on well-characterized systems and compare these measurements with the Johnson, Kendall, Roberts (JKR) theory of the adhesion of spheres to flat substrates. In this way, the reliability of the atomic force techniques can be better assessed.

A shortcoming of the initial study of a polystyrene sphere on a silicon substrate was that no information about the surface roughness was available to correlate with the measured lift-off forces. Surface roughness is often invoked to explain discrepancies between measured and expected adhesion properties, but a quantitative measurement of this effect is difficult to achieve, especially as the size of the adhering object approaches micrometer dimensions.

Because surface roughness is known to play an important role in determining the adhesion of particles to substrates, further studies to elucidate this aspect of adhesion would be worthwhile. In order to address this tissue, we have intentionally chosen to measure the lift-off forces of nominal 8 μm diameter spheres of glass, polystyrene, and tin from atomically flat substrates of mica and highly oriented pyrolitic graphite (HOPG). The importance of surface roughness in the adhesion of these particles to substrates has been assessed by using atomically flat substrates and by characterizing the surface roughness of the particles with atomic force microscopy.

2. EXPERIMENTAL

2.1. Particle and substrate preparation

Spherical, crosslinked, polystyrene particles having a narrow size distribution were prepared as described elsewhere [1]. The glass spheres used in this study were obtained from the Duke Scientific Co., whereas the tin spheres were purchased from Aldrich Chemicals. No attempt was made to clean the spheres before attachment to the atomic force cantilevers.

The mica and HOPG substrates used in this study were prepared by cleaving a fresh surface immediately before inserting into the atomic force apparatus described below.

2.2. Particle attachment

The experiments were performed on individual, size-selected particles of 8 μm nominal diameter attached to cantilevers (available from Park Scientific Instruments, Sunnyvale, CA 94089) using a microscopic drop of Norland Optical Cement No. 68 to the tip of the cantilever with the aid of an optical microscope system. Experiments showed that this cement remained viscous for a sufficiently long time to accurately position the sphere on the cement with the aid of a micromanipulator. The particle was pressed into the cement to allow close contact of the particle with the cantilever. The cement was subsequently cured by exposure to ultraviolet (UV) light for approximately 20–25 min.

2.3. AFM apparatus

The AFM used in these studies was a custom-built instrument. The clean substrate was magnetically mounted onto a segmented PZT-5A piezoelectric tube. A second piezoelectric tube was used to hold the cantilever and provided the option of oscillatory motion of the tip for noncontact mode of operation and cantilever calibration. Detection of the cantilever displacement was done using a laser deflection method [8–10]. A focused laser beam was reflected off the back of the cantilever into a United Detector Model SPOT2D split photodiode position sensing detector (PSD). The output of the PSD was digitized by a 68 030 CPU-based computer system similar to that described previously [11]. The computer was used to handle all data acquisition and system control. This AFM presently can operate in constant force, constant height, and noncontact (attractive) modes.

The $X-Y$ calibration for large displacements was performed by scanning a diffraction grating with a periodicity of 278 nm. Smaller scale calibration was accomplished by measuring the known lattice constant of highly oriented pyrolitic graphite (HOPG). The loading forces were inferred by calibrating the cantilevers as described below.

To correct for hysteresis in the piezo tube that inevitably occurs during the acquisition of the force versus displacement data, an additional small mirror was attached to the sample holder. Using a second laser system, the mirror motion could be determined. This was then used to linearize the substrate motion during analysis after the data had been acquired.

Finally, in order to facilitate studies on the influence of the surrounding gaseous environment, the entire AFM apparatus was mounted inside a small stainless steel chamber that could be evacuated to forepump pressures (~500 mTorr) and back-filled with dry nitrogen gas. The pumping system was designed to minimize back-streaming of oil from the vacuum line. Using this chamber, the mounted sphere and substrate could be outgassed under vacuum conditions before back-filling the chamber with dry nitrogen gas and performing the adhesion measurements.

2.4. Cantilever calibration

To understand the data obtained in a quantitative way, a conversion between the position-sensitive detector output and the actual motion of the cantilever is required.

The cantilever displacement calibration was performed by first measuring the deflection of each bare cantilever as a function of the motion of a hard sapphire substrate that was brought into contact with it. For a hard substrate, the cantilever deflection should equal the calibrated motion of the substrate. After correcting for hysteresis as described above, a reliable calibration between the photodetector response (i.e. cantilever deflection) and substrate motion was obtained. The reproducibility of the calibration was checked by repeating this procedure on a number of different cantilevers. The variation in the calibration constant between the cantilevers was small, indicating that no large, uncontrollable errors were introduced during the mounting and alignment of the system. It follows that by knowing the cantilever spring constant k, the displacement of the cantilever can be converted to the force exerted between the cantilever and the substrate.

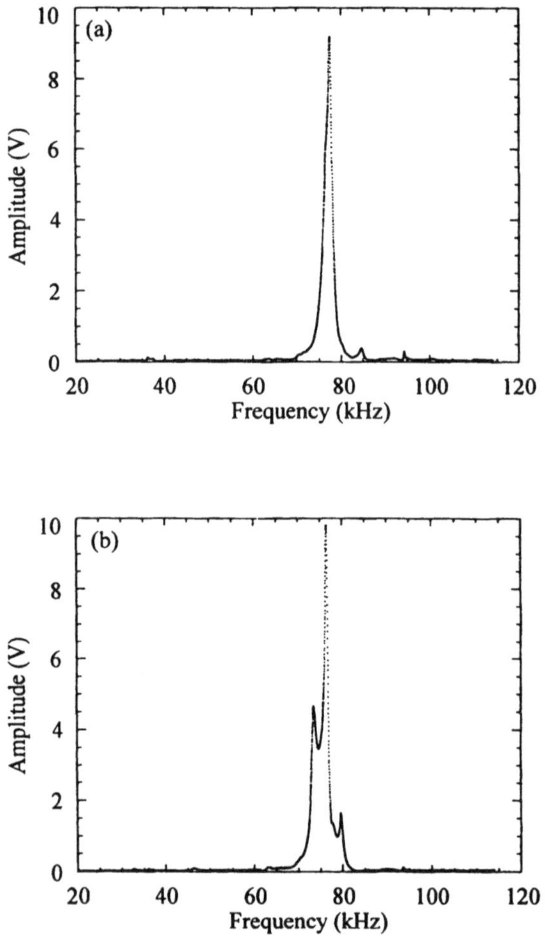

Figure 1. The measured frequency response of a driven cantilever used in this study. The amplitude of the cantilever vibration was measured by the response of a photodiode to a beam reflected off the back of the cantilever. The response of a bare cantilever is given in (a). In (b), the response of a cantilever mounted with an 8 μm diameter polystyrene sphere is shown.

Cantilevers used in this study were Ultralevers (available from Park Scientific Instruments) with specified dimensions of 180, 36, and 2 μm corresponding to the length, width, and thickness, respectively, and having a nominal spring constant of 2.2 N/m. The spring constant was determined for each cantilever from a measurement of the resonance frequency of the free cantilever at room temperature. A representative curve of the frequency response of a bare lever and a lever with a sphere attached is shown in Fig. 1. From the shift in resonance frequency of these data, an added mass of 3.8×10^{-13} kg to the cantilever is inferred. From the known range of densities for polystyrene (1062 ± 8 kg/m^3) and for spheres of 8.5 ± 0.5 μm diameter, we infer a mass for the polystyrene sphere of $3.5 \pm 0.6 \times 10^{-13}$ kg, a result in good agreement with the value inferred from the resonance frequency shift of the cantilever.

From these data, the spring constant of the cantilever can also be calculated using [12]

$$k_{\text{bar}} = 2\pi^3 l^3 w \sqrt{\rho^3/E}\, f_0^3, \tag{1}$$

where l is the length of the bar, w is its width, ρ is the density of the cantilever material (in the case of a silicon bar, $\rho = 2330$ kg/m^3), and E is the elastic modulus ($E_{\text{Si}} = 1.79 \times 10^{11}$ N/m^2). Finally, for a v-shaped cantilever, $k \approx 2k_{\text{bar}}$. Following this approach, k typically varies between 2.7 and 3.2 N/m.

The measurement of k for each cantilever is crucial for accurate measurements. We have measured k for a variety of cantilevers from different wafers. While k for cantilevers from the same wafer may only vary by ~2%, cantilevers from different wafers can vary by as much as ~40%. Also, the measured values have been found to differ from theoretical values by a factor of 1.5. These numbers are representative for 2 μm thick cantilevers. For soft levers with a nominal thickness of 0.6 μm, slight variations in thickness could have even a greater effect.

3. THEORY

The theory required to analyze the data obtained in this experiment is based on the classical analysis of a sphere on a flat plane by Hertz and the well-known extension of the Hertzian model by Johnson, Kendall, and Roberts (JKR) that includes the surface force interaction between the sphere and plane [13].

Because of the surface force interaction, a separation force is required to lift off the sphere from the substrate. The force necessary to effect separation will just occur when the contact radius decreases to $a_0/\sqrt[3]{4}$, where a_0 is the contact radius of the sphere with the substrate at zero load. This will occur for a lift-off force given by

$$P_{\text{lift-off}} = -\frac{3}{2}\pi \Delta\gamma R, \tag{2}$$

where R is the radius of the sphere and $\Delta\gamma = \gamma_1 + \gamma_2 - \gamma_{12}$ (γ_1 and γ_2 are the solid–vacuum interface energies of the sphere and substrate). The term γ_{12} is the interface energy of the sphere–substrate contact. Typically, $\Delta\gamma$ is approximated by $2\sqrt{\gamma_1\gamma_2}$. The values of γ and E appropriate for the materials investigated here are summarized in Table 1.

Table 1.
Input parameters for the surface energies and elastic modulus of the materials investigated in this study

Material	γ (J/m^2)	E (N/m^2)
Mica	0.3 [14]	1.7×10^{11} [18]
HOPG	0.1 [14]	1.7×10^{10} [18]
Glass (SiO_2)	0.35 [15]	6×10^{10} [19]
Sn	0.68 [16]	4.7×10^{10} [20]
Polystyrene	$\sim$ 0.03 [17]	3×10^{9} [21]

4. RESULTS

4.1. Overview

The basic experiment is shown in schematic form in Fig. 2 (a–d). A particle attached to a cantilever is held at a fixed distance above an atomically smooth substrate. As the substrate is moved toward the suspended particle, the interaction between the particle and substrate can be monitored by sensing the deflection of the cantilever. Knowing the spring constant allows a conversion between the cantilever motion and the relevant force. The force exerted on the particle as a function of the substrate displacement is shown in Fig. 2e. The region labeled d is most sensitive to the surface force interactions that are of interest in this study. Atomically smooth substrates are useful here because they serve to isolate the influence of surface asperities on the particle surface. Also, it is useful to study a combination of different spheres and substrates with a range of elastic moduli and electrical conductivities.

4.2. Particle removal

A typical force versus displacement curve for a nominal 8 μm diameter Sn sphere on HOPG is shown in Fig. 3. The characteristics of this curve, which are qualitatively similar to those obtained when studying the adhesion forces between a cantilever tip and a substrate [22, 23], are easy to describe. As the particle, supported by the cantilever, is brought close to the substrate, there is little interaction apparent until the separation distance becomes sufficiently small to allow the forces of attraction to overcome the stiffness of the cantilever. At this point, the cantilever flexes and the particle jumps into contact with the substrate. Upon continued application of a positive load, the force appears to increase linearly with the sphere displacement.

Upon unloading, the cantilever motion initially retraces the loading curve. However, due to adhesion forces along the contact region, the particle may not separate from the substrate at the same point where contact was initially established. Rather, a substantially larger negative load is required before an abrupt separation occurs. The separation force measured from this experiment can be directly compared with the predictions of equation (2). The detailed shape of this region of the curve is sensitive to the details of the contact region and is dominated by the short-range surface forces.

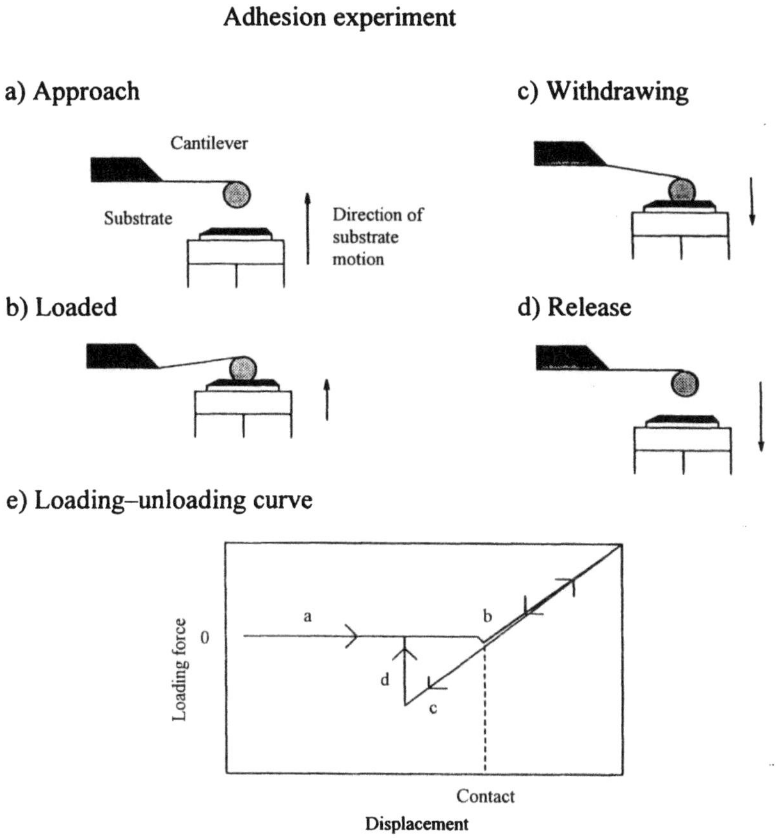

Figure 2. (a–d) A schematic diagram of the procedure used to obtain the force of interaction between a micrometer-size sphere mounted on a cantilever and a substrate material. In (e), the regions of interest on the force curve are labeled to show the approach, jump-to-contact, loading, unloading, and removal regions of the experiment.

While acquiring such data, the total time that the sphere is in contact with the sample is under 10 s. If there were a significant force on the particle due to contact charging, the approach and withdrawal curves would not be superimposed until much greater separation distances.

Systematic measurements on a nominal 8 μm diameter Sn sphere show that the removal force is independent of the applied load, in agreement with the behavior predicted by equation (2). The measured lift-off (removal) force as a function of the applied load for 8 μm diameter glass, tin, and polystyrene spheres is summarized in Fig. 4. Each data point represents the average of three independent measurements of the removal force. Typically, each measure of the removal force was reproducible to within $\pm 2\%$. From these data an average removal force can be estimated for each system.

The use of HOPG as a substrate produced the most reliable data. When the insulating mica substrate was used, charging problems between the sphere and substrate became evident. This was especially the case for glass spheres on mica, where a large

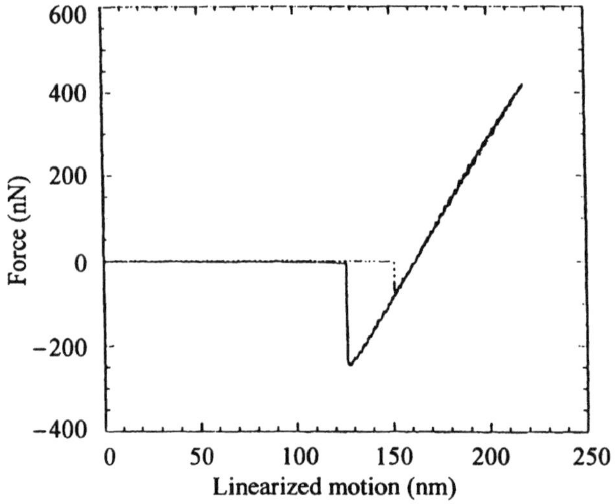

Figure 3. Force data of an 8 μm diameter tin sphere against an atomically flat HOPG substrate. In this experiment, the sphere was externally loaded to a maximum value 450 nN before removal from the substrate. From the data, a lift-off force of ~250 nN is required to separate the tin sphere from the HOPG substrate.

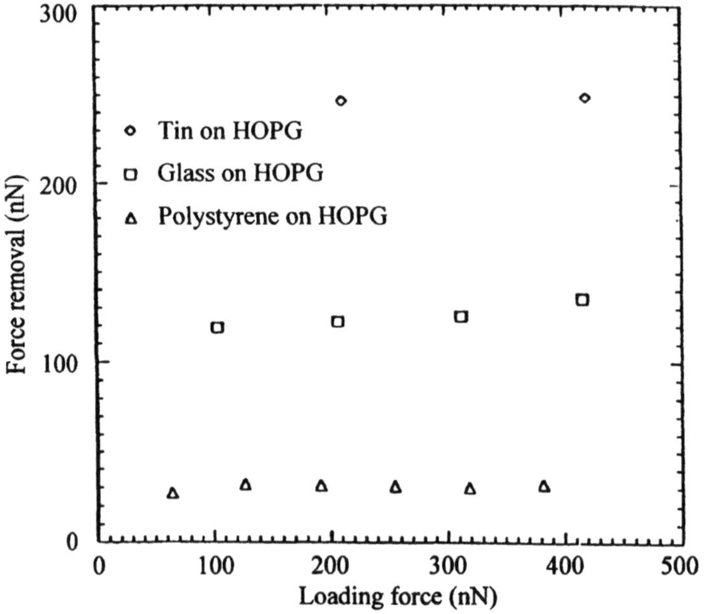

Figure 4. Summary of the lift-off (removal) force as function of the loading force for tin, glass, and polystyrene spheres on an HOPG substrate. The data show little or no dependence of the lift-off force on the externally applied load and indicate that for the range of loads studied, minimal damage to the sphere–substrate interface occurs.

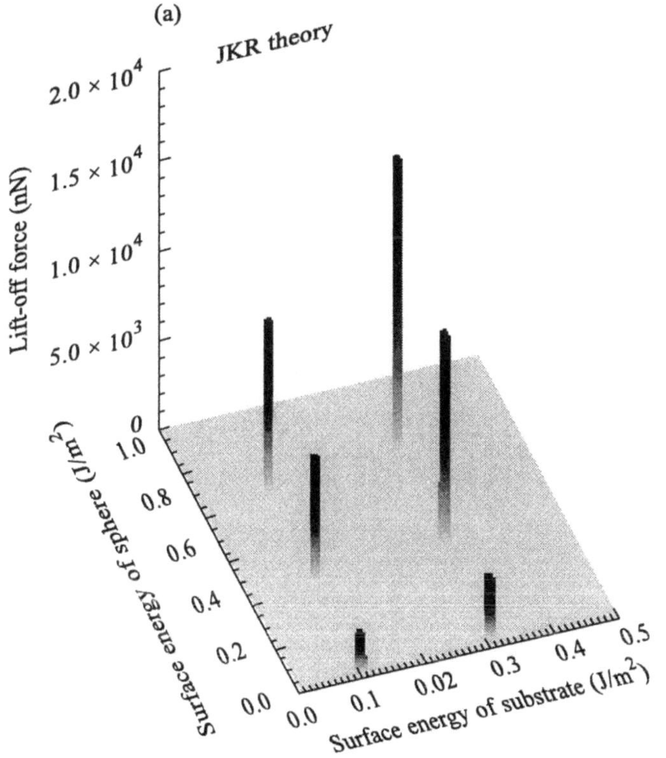

Figure 5. A bar-chart representation of the lift-off force as a function of the substrate and sphere surface energies. In (a), the predictions of the JKR model for $P_{\text{lift-off}}$ are plotted for the materials studied here (see Table 2). In (b), the experimental data are plotted in a similar way. Apart from an overall factor of ~50, the relative values of the data are in reasonable agreement with theory [see equation (2)], indicating that the surface force interaction between the substrate and particle is determining the adhesion.

adhesion force related to charging between the substrate and sphere was occasionally encountered. This effect will be published later. Figure 5 summarizes in bar-chart format the JKR predictions and our experimental results. Table 2 summarizes the numerical results for the different materials. Using equation (2) and the values listed in Table 1, it is possible to calculate the expected lift-off forces. The results are shown under the theory column in Table 2. A quantity r

$$r = \frac{\text{theoretical lift-off force}}{\text{experimental lift-off force}}, \tag{3}$$

defined as the multiplicative factor required to obtain agreement between experiment and theory, is also given in Table 2. In general, the measured lift-off force is smaller than expected by a factor varying between 50 and 175. The origin of this reduction in lift-off force is of considerable interest.

One possibility is that the presence of a contamination layer between the sphere and substrate reduces the lift-off force. Moisture and volatile oils are known to affect

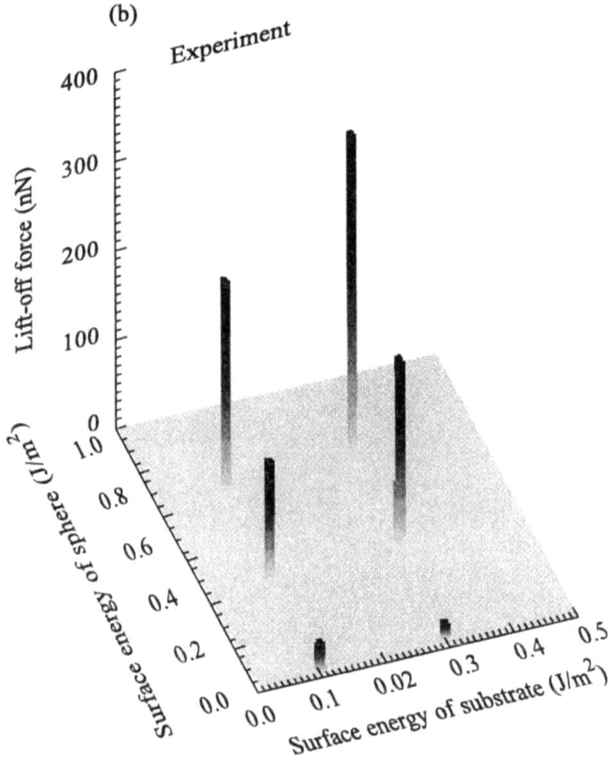

Figure 5. (Continued).

Table 2.
Comparison between the measured and calculated lift-off force (in nN) for 8 μm diameter spheres. The parameter r is the ratio of the calculated to the measured lift-off force

	Glass			Tin			Polystyrene		
Substrate	Exper.	Theory	r	Exper.	Theory	r	Exper.	Theory	r
Mica	219 ± 4	12 250	56	377 ± 2	17 100	45	20 ± 1	3560	178
HOPG	137 ± 18	7 050	51	244 ± 3	9 800	40	30 ± 2	2050	68

adhesion and such contamination effects cannot be discounted in this type of study. Water vapor is known to coat surfaces with many monolayers after a brief exposure to ambient conditions. Although all but the first layer or two are weakly physisorbed, a significant water layer may be present. However, due to the outgassing procedure followed here, only a few layers of a contaminant are expected to be left on the surface of the sphere or substrate. Furthermore, AFM measurements of adhesion forces indicate that the presence of water on a surface tends to increase adhesion [1, 24], an effect opposite to that required here. Also, water vapor should be less of a factor for low surface energy materials (like polystyrene) than for high surface energy materials (like tin). This is also contrary to observation, further suggesting that adsorbed water vapor is not the problem.

Surface contaminants other than water may also play a role, and quantitative estimates of how strongly a thin layer may affect adhesion have been reported in the literature. For instance, model calculations have shown that a thin hydrocarbon contamination layer can reduce the attractive force between two materials by about 50% [25]. A factor of 2 will not explain the discrepancy evident from Table 2.

Also, the influence of an oxide layer on the tin particles must be considered. Careful studies of the effect of an oxide on Ni surfaces have been reported in which the adhesion was studied as a function of the oxide layer thickness [26]. Roughly, a factor of 2 reduction in adhesion is reported for low applied loads. As the load is increased into the 100 μN range, the oxide layer is disturbed and a reduction in adhesion disappears due to disruption of the oxide layer. The decrease in adhesion due to an oxide layer on tin is probably too small to account for the results presented in Table 2.

It is possible that geometrical effects may be the source of the observed discrepancy. For instance, the presence of atomic scale terracing near the contact perimeter may greatly alter the surface force interaction when compared with a smooth profile [27]. While such terracing is expected for brittle particles like ceramics, it is not expected to be a problem in this study where softer materials were used. In addition, AFM images of representative particles used in this study provided no evidence for terracing. Another geometrical effect, that of surface roughness, is discussed in more detail in the next section.

4.3. Surface roughness

It is generally recognized that surface roughness also acts to reduce the adhesion between two solids. The effective radius of the sphere that appears in equation (2) to estimate the lift-off force must be decreased if asperities are present. Further analysis of how asperities reduce adhesion has been given by Fuller and Tabor [28]. Using AFM techniques, it is possible to characterize the surface asperities of micrometer-size particles by scanning the surface of representative spheres to determine the dimensions of the asperities present.

AFM studies showed small asperities, a few tens of nanometers high. Root-mean-square roughness values can be calculated for each sphere, but it is not clear how these statistical values can be easily incorporated into a theory for lift-off. Rather, we have chosen to characterize the asperities by analyzing individual AFM line scans across these asperities as shown in Fig. 6. From these scans, a mean radius of curvature β for the asperity can be estimated. Analysis of the AFM data in this way gives values of β ranging between 0.25 and 0.3 μm. This value is presumably an upper limit for β since the measured radius of the asperity can be increased by tip convolution effects.

Using these values in equation (2) rather than the 4 μm radius of the sphere, the theoretical estimate of the lift-off force is reduced by a factor of between 13 and 16. It should be noted that the asperities on the particles are compressed due to surface

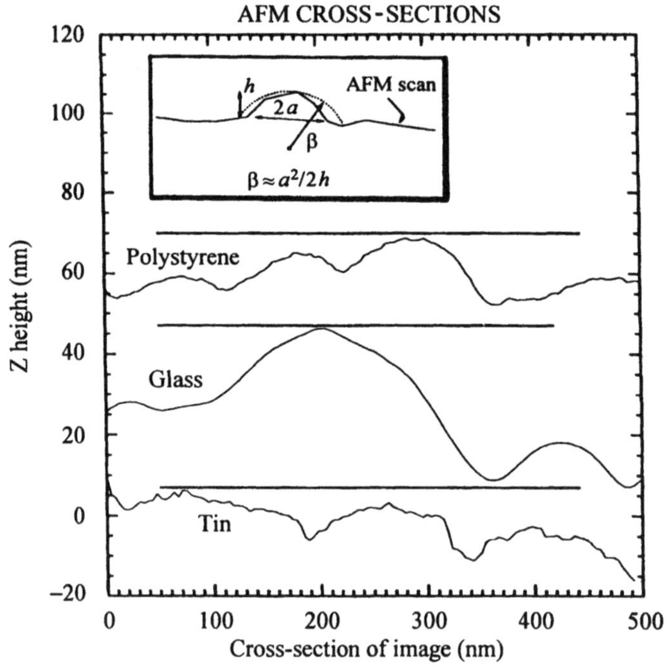

Figure 6. Representative AFM line scans characterizing the surface of typical polystyrene, glass, and tin spheres used in this study. The inset shows the geometry assumed when estimating the radius of the microscopic asperities. The long horizontal line next to each scan represents the expected diameter of the contact zone calculated using the JKR theory for each sphere resting on an HOPG substrate under zero load.

forces. If the compression is sufficient, other asperities will also contact the substrate, thereby increasing the removal force. In this instance, however, the compression calculated using the JKR model and the values of the parameters listed in Table 1 are small compared with the average height of the measured asperities. Therefore, the force of removal can be estimated by substituting the number and radii of the asperities for the radius of the particle.

Following this analysis of surface roughness, it can be concluded that the experimental measure of the lift-off force is roughly three times smaller than the predicted lift-off force (with the exception of polystyrene on mica which is too small, i.e. by a factor of $\sim$15). This factor of 3 discrepancy, while disturbing, might be accounted for by contamination effects and uncertainties in the solid–vacuum interface energies of the spheres and substrates. More accurate ways of estimating these quantities for the actual surfaces under study would be helpful. Also, a time-dependent measurement of the lift-off force might also provide further information about this discrepancy.

5. CONCLUSION

A study of the interaction of individual micrometer-size spheres with atomically flat substrates using atomic force techniques has been reported. These experiments were performed in an inert nitrogen gas environment on particles and substrates that were outgassed in a moderate vacuum. Contact times of the order of 10 s were employed. For these short contact times, non-equilibrium conditions may exist.

The lift-off force for nominal 8 μm diameter particles of tin, glass, and polystyrene was measured from the atomically flat mica and HOPG substrates. The measurements indicate that the *relative* lift-off forces between the various spheres and the substrates are proportional to the JKR model. By utilizing atomically flat substrates, the influence of surface roughness of the particles was isolated and analyzed. We find that a simple method of estimating the radius of the surface asperities reduces the discrepancy between theory and experiment to a factor of $\sim$15. However, after including surface asperities, the measured lift-off forces are still a factor of $\sim$3 smaller than expected from the JKR predictions. It remains to be seen if this remaining discrepancy can be accounted for by better control of surface contamination, by a more rigorous introduction of surface roughness into the theory, or by invoking a time-dependent adhesion process that is not accounted for in the equilibrium JKR model.

Acknowledgement

The work at Purdue University was funded, in part, by a grant the Office Imaging Research and Technology Development Division of Eastman Kodak.

REFERENCES

1. D. M. Schaefer, M. Carpenter, R. Reifenberger, L. P. DeMejo and D. S. Rimai, *J. Adhesion Sci. Technol.* **8**, 197 (1994).
2. H. A. Mizes, K. G. Loh, M. L. Ott and R. J. D. Miller, in: *Particles on Surfaces: Detection, Adhesion, and Removal*, K. L. Mittal (Ed.), pp. 47–59. Marcel Dekker, New York (1994).
3. D. S. Rimai, L. P. DeMejo, W. B. Vreeland, R. C. Bowen, S. R. Gaboury and M. W. Urban, *J. Appl. Phys.* **71**, 2253 (1992).
4. D. S. Rimai, L. P. DeMejo, W. B. Vreeland, R. C. Bowen, S. R. Gaboury and M. W. Urban, *J. Appl. Phys.* **73**, 668 (1993).
5. L. P. DeMejo, D. S. Rimai, J. Chen and R. C. Bowen, *J. Adhesion* **39**, 61 (1992).
6. D. S. Rimai, R. S. Moore, R. C. Bowen, V. K. Smith and P. E. Woodgate, *J. Mater. Res.* **8**, 662 (1993).
7. D. S. Rimai, R. C. Bowen, L. P. DeMejo and J. D. Morris, in: *Particles on Surfaces: Detection, Adhesion, and Removal*, K. L. Mittal (Ed.), pp. 1–31. Marcel Dekker, New York (1994).
8. G. Meyer and N. M. Amer, *Appl. Phys. Lett.* **53**, 1045 (1988).
9. G. Meyer and N. M. Amer, *Appl. Phys. Lett.* **53**, 2400 (1988).
10. S. Alexander, L. Hellemans, O. Marti, J. Schneir, V. Elings, P. K. Hansma, M. Longmire and J. Gurley, *J. Appl. Phys.* **65**, 164 (1989).
11. R. Piner and R. Reifenberger, *Rev. Sci. Instrum.* **60**, 3123 (1989).
12. J. P. Cleveland, S. Manne, D. Bocek and P. K. Hansma, *Rev. Sci. Instrum.* **64**, 403 (1993).
13. K. L. Johnson, K. Kendall and A. D. Roberts, *Proc. R. Soc. London, Ser. A* **324**, 301 (1971).
14. Burnham *et al.*, *Phys. Rev. Lett.* **64**, 1931 (1990).
15. Overbury *et al.*, *Chem. Rev.* **75**, 555 (1975).
16. H. Jones, *Metal Sci. J.* **5**, 16 (1971).
17. A. J. Kinloch, *Adhesion and Adhesives*, p. 26. Chapman and Hall, New York (1987).
18. C. T. Lynch (Ed.), *Handbook of Materials Science III*, p. 160. CRC Press, Cleveland (1975).
19. H. D. Young, *Mechanics and Heat*, p. 485. McGraw-Hill, New York (1964).

20. C. J. Smithells, *Metals Reference Book*, Vol. II, 3rd edn, p. 614. Butterworths, Washington (1962).
21. D. S. Rimai *et al.*, *J. Appl. Phys.* **68**, 6234 (1990).
22. H. A. Mizes, K. G. Loh, R. J. D. Miller, S. K. Ahuja and E. F. Grabowski, *Appl. Phys. Lett.* **59**, 2901 (1991).
23. H. A. Mizes, *J. Adhesion Sci. Technol.* **8**, 937 (1994).
24. T. Thundat, X. Y. Zheng, G. Y. Chen and R. J. Warmack, *Surface Sci.* **294**, L939 (1993).
25. S. R. Cohen, *Ultramicroscopy* **42–44**, 66 (1992).
26. M. D. Pashley, J. B. Pethica and D. Tabor, *Wear* **100**, 7 (1984).
27. J. D. J. Ross, H. M. Pollock and Q. Guo, *Powder Technol.* **65**, 21 (1991).
28. K. N. G. Fuller and D. Tabor, *Proc. R. Soc. London, Ser. A* **345**, 327 (1975).

Fundamentals of Adhesion and Interfaces, pp. 49–60
D. S. Rimai, L. P. DeMejo and K. L. Mittal (Eds)

JKR–DMT transition in the presence of a liquid meniscus

D. MAUGIS[1,*] and B. GAUTHIER-MANUEL[2]
[1] *UMR 113, LCPC/CNRS, 2 Allée Kepler, 77420 Champs, France*
[2] *Electrochimie des Solides, UA 436, Route de Gray, 25030 Besançon Cedex, France*

Revised version received 20 June 1994

Abstract—In a previous paper [*J. Colloid Interface Sci.* **150**, 243 (1992)], one of the authors (D.M.) has shown how, for the contact of elastic spheres, fracture mechanics with a Dugdale model of adhesion forces can lead to a general theory with the Johnson–Kendall–Roberts (JKR) and Derjaguin–Muller–Toporov (DMT) theories as limiting cases. The main equations are briefly recalled and the theory is applied to the case of a liquid meniscus around the contact. It is shown that the results are similar to those of Fogden and White [*J. Colloid Interface Sci.* **138**, 414 (1990)]. Experiments with a surface force apparatus are described showing this transition in the presence of a liquid meniscus. The measured profiles are in good agreement with the theory.

Keywords: Contact mechanics; fracture mechanics; capillarity; adhesion.

1. INTRODUCTION

In the Hertz theory of contacting spheres, only compressive stresses in the contact area are assumed. In the JKR theory [1], tensile stresses in the contact area are allowed, whereas in the DMT theory [2], adhesion forces act around a unilateral contact. In a previous paper [3], one of us has shown how fracture mechanics with a Dugdale model of adhesion forces can lead to a general theory with the JKR and DMT theories as limiting cases. The method used is summarized below and the main equations are recalled without derivation. The reader is referred to the original paper [3] for historical background, calculations, and references.

A liquid meniscus at the edge of a contact is a perfect example of a Dugdale zone and the above theory for the JKR–DMT transition can be used to interpret the experimental results on crossed mica cylinders [4–6] which have shown that a liquid meniscus can change the shape of the profile from a JKR profile to a Hertzian one. This problem was previously studied by Fogden and White [7] and it is shown that the results are similar. In the present work, profiles in the presence of menisci of various sizes have been measured with a high precision and are found to be in good agreement with the theory.

---

*To whom correspondence should be addressed.

2. THE HERTZ RESULTS

Let us consider a rigid sphere of radius R in frictionless contact with an elastic half-space (Young's modulus E, Poisson's ratio ν) under a load P. Only compressive stresses are allowed in the contact area (unilateral problem). The radius of contact a, the penetration δ, the stress distribution $\sigma_y(r,0)$ in the area of contact, and the air gap $[u_y]$ between the sphere and the deformed plane are given by

$$a^3 = \frac{PR}{K}, \tag{1}$$

$$\delta = \frac{a^2}{R}, \tag{2}$$

$$\sigma_y(r,0) = -\frac{3aK}{2\pi R}\sqrt{1-\rho^2}, \qquad \rho < 1, \tag{3}$$

$$[u_y] = \frac{a^2}{\pi R}\left[\sqrt{\rho^2 - 1} + (\rho^2 - 2)\cos^{-1}\frac{1}{\rho}\right], \qquad \rho > 1, \tag{4}$$

where $\rho = r/a$ and K is given by

$$\frac{1}{K} = \frac{3}{4}\left(\frac{1-\nu^2}{E}\right).$$

The stresses vanish at the limit of the contact, ensuring a tangential contact between the sphere and the half-plane. For a given sphere of radius R and a given radius of contact a, the profile is determined and does not depend on any material property.

3. THE JKR RESULTS

The contact between a sphere and a plane can be seen as an external crack in mode I that recedes or advances as the contact area increases or decreases. In the JKR model [1], tensile forces are allowed in the contact area but there is no adhesion force between the crack lips, as in linear elastic fracture mechanics (LEFM), so that the stresses are singular and the profile is parabolic at the edge of the contact. Introducing the stress intensity factor, K_{I},

$$K_{\mathrm{I}} = \frac{\frac{a^3K}{R} - P}{2a\sqrt{\pi a}}, \tag{5}$$

the penetration δ, the stress distribution in the area of contact, and the air gap between the sphere and the deformed plane are given by

$$\delta = \frac{a^2}{3R} + \frac{2P}{3aK}, \tag{6}$$

$$\sigma_y(r,0) = \frac{K_{\rm I}}{\sqrt{\pi a}} \frac{1}{\sqrt{1-\rho^2}} - \frac{3aK}{2\pi R}\sqrt{1-\rho^2}, \tag{7}$$

$$[u_y] = \frac{2(1-\nu^2)}{\pi E} K_{\rm I}\sqrt{\pi a}\cos^{-1}\frac{1}{\rho} + \frac{a^2}{\pi R}\left[\sqrt{\rho^2-1} + (\rho^2-2)\cos^{-1}\frac{1}{\rho}\right]. \tag{8}$$

The equilibrium is given by the Griffith relation:

$$G = w, \tag{9}$$

where $w = \gamma_1 + \gamma_2 - \gamma_{12}$ is the Dupré energy of adhesion (γ_1 and γ_2 are surface energies; γ_{12} is the interfacial energy) and G is the energy release rate, which, in the LEFM approximation, is related to the stress intensity factor by

$$G = \frac{1}{2}\left(\frac{1-\nu^2}{E}K_{\rm I}^2\right) \tag{10}$$

(compared with classical fracture mechanics, the factor 1/2 is introduced as the sphere is rigid, so that energy is released by the half-plane only; we are faced with half a Griffith crack*). For $w = 0$, one returns to the classical Hertz theory.

At equilibrium, equation (8) can be rewritten

$$[u_y] = 4\left(\frac{2aw}{3\pi K}\right)^{1/2}\cos^{-1}\frac{1}{\rho} + \frac{a^2}{\pi R}\left[\sqrt{\rho^2-1} + (\rho^2-2)\cos^{-1}\frac{1}{\rho}\right], \tag{11}$$

which shows that for a given sphere of radius R and a given radius of contact a, the shape of the profile depends only on the material property $\sqrt{w/K}$.

4. THE EXTERNAL CRACK WITH INTERNAL LOADING

Let us assume that the stresses acting between the lips, around the contact, have a constant value σ_0 up to a radius c and let $d = c - a$. (This stress can be the Laplace pressure in a liquid meniscus.) This internal loading leads to a stress intensity factor†:

$$K_m = -\frac{\sigma_0 a}{\sqrt{\pi a}}\left[\sqrt{m^2-1} + m^2\tan^{-1}\sqrt{m^2-1}\right] \tag{12}$$

---

*We will neglect the main problem of interfacial cracks, namely the oscillations of stresses at the crack tip when the Dundurs parameter β is not zero. Such oscillations and their release by a local slip are well known in contact mechanics [8] and are neglected in the Hertz, JKR, and DMT theories, which, more or less implicitly, assume frictionless contacts. (In this field, a contact with infinite coefficient of friction is designated as adhesive contact, which is rather confusing!)

†As explained in [3], the theory of Lowengrub and Sneddon [9] has been used, giving stresses and displacements for an axisymmetric pressurized external crack. The main difficulty was that the stresses in the neck did not equilibrate the pressure in the crack, due to their implicit assumption of no elastic displacement at infinity. (The stresses vanished at infinity but their integral was finite.)

with $m = c/a > 1$. The distibution of stresses in the contact area and the elastic displacement of the plane, due to this loading, are

$$\sigma_y(r,0) = \frac{K_m}{\sqrt{\pi a}} \frac{1}{\sqrt{1-\rho^2}} + \frac{2\sigma_0}{\pi} \tan^{-1} \sqrt{\frac{m^2-1}{1-\rho^2}}, \qquad \rho < 1, \tag{13}$$

$$\begin{aligned} u_y = {} & \frac{2(1-\nu^2)}{\pi E} K_m \sqrt{\pi a} \cos^{-1} \frac{1}{\rho} - \frac{4(1-\nu^2)}{\pi E} \sigma_0 a \\ & \times \left[\sqrt{m^2-1} \left(\sqrt{\rho^2-1} - \cos^{-1} \frac{1}{\rho} \right) - m^2 \int_1^{\min(\rho,m)} \frac{\sqrt{\rho^2-t^2}}{t^2 \sqrt{m^2-t^2}} \, \mathrm{d}t \right]. \end{aligned} \tag{14}$$

5. THE DUGDALE MODEL

Adding the stresses due to external and internal loadings, equations (7) and (13), we see that the non-physical stress singularity disappears if

$$K_{\mathrm{I}} + K_m = 0. \tag{15}$$

This relation gives the ratio c/a as a function of σ_0, P, and a, and ensures the continuity of stresses at the edge of the contact.

The crack opening displacement (COD) δ_t is the air gap at $r = c$ (i.e. $\rho = m$). Adding equations (8) and (14) and using relation (15), the COD is

$$\begin{aligned} \delta_t = {} & \frac{a^2}{\pi R} \left[\sqrt{m^2-1} + (m^2-2) \tan^{-1} \sqrt{m^2-1} \right] \\ & + \frac{4(1-\nu^2)}{\pi E} \sigma_0 a \left[\sqrt{m^2-1} \tan^{-1} \sqrt{m^2-1} - m + 1 \right]. \end{aligned} \tag{16}$$

According to the Rice J-integral, the energy release rate in a Dugdale model is simply

$$G = \sigma_0 \delta_t \tag{17}$$

and the equilibrium is given by the Griffith relation, equation (9). The problem is thus solved. When $m \to 1$, one recovers the JKR results with

$$G = \frac{\left(\frac{a^3 K}{R} - P \right)^2}{6\pi a^3 K}, \tag{18}$$

and when $m \to \infty$, the DMT results are recovered, with

$$G = \frac{\frac{a^3 K}{R} - P}{2\pi R}. \tag{19}$$

Note that in the JKR approximation, the distribution of adhesion forces is reduced to a singular stress at $r = a_+$. The total attraction force outside the contact being zero, the integral of stresses in the contact area is equal to the applied load P and negative applied loads are sustained by the elastic restoring forces inside the contact area.

In the DMT approximation, the adhesion stresses tend towards zero to have a continuity with the stress at $r = a_-$, but their integral is finite and the total attraction force outside the contact is $2\pi wR$. In the area of contact, the distribution of stresses is Hertzian and their integral is $P + 2\pi wR$. Negative applied loads are sustained by adhesion forces outside the contact area.

Introducing the dimensionless parameters

$$A = \frac{a}{\left(\pi w R^2/K\right)^{1/3}},$$

$$\overline{P} = \frac{P}{3\pi w R},$$

$$\Delta = \frac{\delta}{\left(\pi^2 w^2 R/K^2\right)^{1/3}},$$

and

$$\lambda = \frac{2\sigma_0}{\left(\pi w K^2/R\right)^{1/3}},$$

the Griffith relation, equation (9), gives

$$\frac{1}{2}\lambda A^2\left[\sqrt{m^2-1} + (m^2-2)\tan^{-1}\sqrt{m^2-1}\right] + \frac{4}{3}\lambda^2 A\left[\sqrt{m^2-1}\tan^{-1}\sqrt{m^2-1} - m + 1\right] = 1. \tag{20}$$

For a given material parameter λ and a given radius of contact a, one can extract m and obtain the load and the penetration at equilibrium:

$$\overline{P} = A^3 - \lambda A^2\left[\sqrt{m^2-1} + m^2\tan^{-1}\sqrt{m^2-1}\right], \tag{21}$$

$$\Delta = A^2 - \frac{4}{3}\lambda A\sqrt{m^2-1}. \tag{22}$$

The air gap can be written with the elliptic integrals of the first and second kind, $E(\varphi, k)$ and $F(\varphi, k)$. Using the same normalization as that for δ, we have for $r < c$:

$$\overline{[u_y]} = \frac{A^2}{\pi}\left[\sqrt{\rho^2-1} + (\rho^2-2)\cos^{-1}\frac{1}{\rho}\right] + \frac{8\lambda A}{3\pi}\sqrt{m^2-1}\cos^{-1}\frac{1}{\rho} + \frac{8\lambda A}{3\pi}\left[\frac{\sqrt{\rho^2-1}}{\sqrt{m^2-1}} - mE(\varphi, k)\right] \tag{23}$$

with

$$\varphi = \sin^{-1}\left(\frac{m}{\sqrt{m^2-1}}\frac{\sqrt{\rho^2-1}}{\rho}\right)$$

$$k = \frac{\rho}{m} = \frac{r}{c},$$

and for $r > c$:

$$\overline{[u_y]} = \frac{A^2}{\pi}\left[\sqrt{\rho^2-1} + (\rho^2-2)\cos^{-1}\frac{1}{\rho}\right] + \frac{8\lambda A}{3\pi}\sqrt{m^2-1}\cos^{-1}\frac{1}{\rho}$$
$$+ \frac{8\lambda A}{3\pi}\left[\frac{\sqrt{m^2-1}}{\sqrt{\rho^2-1}} - \rho E(\varphi, k) + \left(1-\frac{m^2}{\rho^2}\right)\rho F(\varphi, k)\right] \quad (24)$$

with

$$\varphi = \sin^{-1}\left(\frac{\sqrt{m^2-1}}{m}\frac{\rho}{\sqrt{\rho^2-1}}\right),$$

$$k = \frac{m}{\rho} = \frac{c}{r}.$$

When $\lambda \to \infty$ ($c/a \to 1$), one has the JKR air gap, and when $\lambda \to 0$ ($c/a \to \infty$), one has the DMT (Hertzian) air gap.

Equations (23) and (24) describe the air gap between a rigid sphere and an elastic half-space; they also give the profile of an elastic sphere on a rigid plane or the air gap between two elastic cylinders crossed at right angles.

6. APPLICATION TO A LIQUID MENISCUS

As a liquid meniscus at the edge of a contact is a good example of a Dugdale zone [10], with the constant stress σ_0 equal to the Laplace pressure: $\sigma_0 = \gamma_L/\rho_k$, and a crack opening displacement $\delta_t = 2\rho_k \cos\theta$ (where γ_L is the surface energy of the liquid, ρ_k is the meniscus radius of curvature, and θ is the contact angle), so that $J = \sigma_0\delta_t = 2\gamma_L \cos\theta$ is independent of ρ_k, the above theory can be used to study such contacts. There is difficulty, however, with the solid–solid interactions across the liquid meniscus. In the above theory, the progressive decrease of the adhesion stress outside the contact from the theoretical stress σ_0 to zero was replaced, for convenience, by a constant stress σ_0 acting on an annulus of width d. In the presence of a liquid, the stress must decrease from the theoretical stress σ_{th} at the edge of the contact to the Laplace stress σ_L in the bulk meniscus. This point can be taken into account by the J-integral around the crack tip giving

$$J = w' + 2\gamma_L \cos\theta,$$

where $w' < w$ is the Dupré energy of adhesion across the liquid, but is not taken into account in the calculation of G where no distinction can be made between σ_{th} and σ_L; both are assumed equal and called σ_0.

As stated in [3], the problem of contact in the presence of capillary condensation was previously studied by Fogden and White [7] in 1990. Let us compare these two independent approaches.* For this purpose, their symbols $\mathcal{D}$, W, P, Δp, ρ_k, x, x_k, $2r_k$, must be replaced, respectively, by our $1/K$, P, σ, $-\sigma_0$, c, ρ, m, δ_t. Referring to unpublished papers (White's PhD thesis and a preprint), they summarize the 'physically consistent approximate theory' of contact and write that 'this mixed boundary value formulation can be readily reduced to dual integral equations with Bessel function kernels and solved by the use of Erdelyi–Kober operators'. They obtain very general formulae which, in the case of no interaction forces outside the contact, reduce either to the Hertz results or to the JKR results according to the choice of closure relation. Imposing continuity of pressure leads to the Hertz theory, whereas minimizing the total energy leads to the JKR theory. In the case of a Laplace pressure acting outside the area of contact (solid–solid interactions across the meniscus are also neglected), and choosing the continuity of pressure as the closure relation, they obtained a 'generalized Hertz theory', the incorporation of capillary condensation into the JKR theory of adhesive contact ('generalized JKR theory') being postponed for a paper II never published to our knowledge.

Looking more closely at their 'generalized Hertz theory', a somewhat misleading title, one can see that their closure relation equation [1.19], or [3.1], is our condition $K_I + K_m = 0$, equation (15). Their Laplace equation $\sigma_0 = 2\gamma/\delta_t$ is our Griffith equation $G = w$ for equilibrium. As this last equation results from a minimization of the total energy, there is no need to choose between two closure relations! This is the reason why, in their 'generalized Hertz theory', they obtain a transition from the JKR results to the DMT results† when the size of the meniscus increases. Their equation [3.2] is our equation (21), and their equation [3.3] is our equation (16) for δ_t. Their parameter k is related to our parameter λ by $k^2 = 9/(8\lambda^3)$. It can be shown that upon inserting their closure relation [1.19] in their general formula [2.3] for the profile, one obtains our equations (23) and (24) after calculation of their elliptic integral.

(The great difference in the treatment of the problem is thus a good indication of the correctness of both results.)

7. EXPERIMENTAL

The profile measurements were carried out with the surface force apparatus (SFA) shown schematically in Fig. 1. The principle of the measurement is abundantly described in the literature [11]. We glue (with an epoxy adhesive, thickness 10–20 μm) two freshly cleaved mica plates (about 2 μm thick), back-silvered, onto two cylindrical lenses (radius of curvature 2 cm). With the stepper motor, we slowly approach the mica surfaces, mounted in crossed-cylinder configuration in the SFA, until the van der Waals force attracts the surfaces. Because of the instability of the spring

*The principle of the method followed in [3] has been given in [10].

†Although the DMT theory seemed to be unknown to these authors (not cited in their paper).

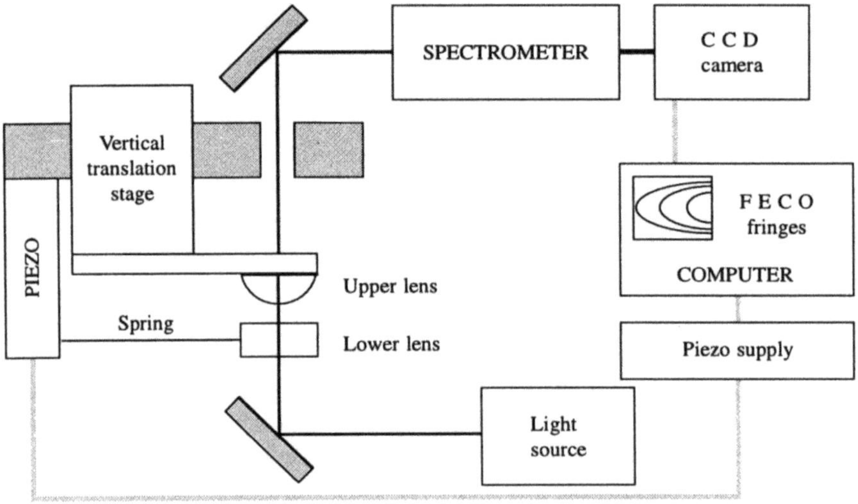

Figure 1. Schematic diagram of the surface force apparatus.

(symmetrical, non-tilting spring), we observe a jump of the surfaces to the molecular contact. A white collimated light beam passes through the interferometer constituted by the two silvered mica plates. A grating spectrometer disperses the emerging light focused on its entry slit and gives fringes of equal chromatic order (FECO) at its exit plane.

A CCD camera, mounted at the exit of the spectrometer, digitizes the fringes and sends the data to a PC computer for subsequent calculations. The CCD array has 4096 pixels, 7 μm width. It measures simultaneously the position of all the fringes in the exit plane of the spectrometer. It can be translated in the vertical direction with the stepper motor, each step 7 μm high, to record the geometric profile of the fringes. With a transverse magnification of 14, the resolution is equal to 0.5 μm in the radial direction, while the interferometric method gives a sensitivity as high as 0.1 nm in the gap direction.

We calculate the gap as a function of the radial distance by using the theoretical result of Hunter and Nabarro [12] and taking into account the non-linear dispersion of the grating. When all the optical elements are correctly aligned, the measured profile is independent of the fringe chosen for the calculation.

At the end of the experiment, to prove the cleanliness of the mica surfaces used, we measure the force required to separate the contacting mica surfaces in air. The surface energy calculated through the relation $F = 3\pi\gamma R$ always gives surface energies γ greater than 300 mJ/m^2.

Due to the uncontrolled thickness of the glue, the radius of curvature R can be different from that of the lenses. It is computed by plotting the gap $[u_y]$ versus r^2 and measuring the slope of the linear part far from the contact, according to the formula

$$[u_y] = \frac{r^2}{2R} - \delta,$$

which is an approximation of equations (4), (8), and (24) at large r. One of the main experimental difficulties is to obtain symmetrical curves; an asymmetric curve leads to different apparent radii of curvature at the left and right sides of the contact. As the optical axis is set perpendicular to the contact area within half a degree, it is not yet clear why such asymmetries sometimes occur.

7.1. Mica–mica contact in dry air

Before bringing the surfaces together for molecular contact, we dehydrate all the apparatus with a flow of dry nitrogen. The profile measured is sketched in Fig. 2. It is a JKR profile, in agreement with equation (11) fitted with $w/K = 2.5 \times 10^{-11}$ m. (Note that this value is similar to the value $w/K = 3 \times 10^{-11}$ m deduced from the JKR profile under zero load given by Tabor [13].) The Hertzian profile is drawn for comparison.

7.2. Mica–mica contact in a large meniscus

After molecular contact is realized, we introduce a drop of pure water (ELGA UHQ) at the edge of the contact. The JKR profile changes to a Hertzian one, as sketched in Fig. 3. For that profile, given by equation (4), no fitting parameter is needed.

7.3. Mica–mica contact in humid air

When molecular contact of two surfaces is produced in humid air, a liquid meniscus of radius c around a contact of radius a is formed. In Fig. 4 we can observe a profile

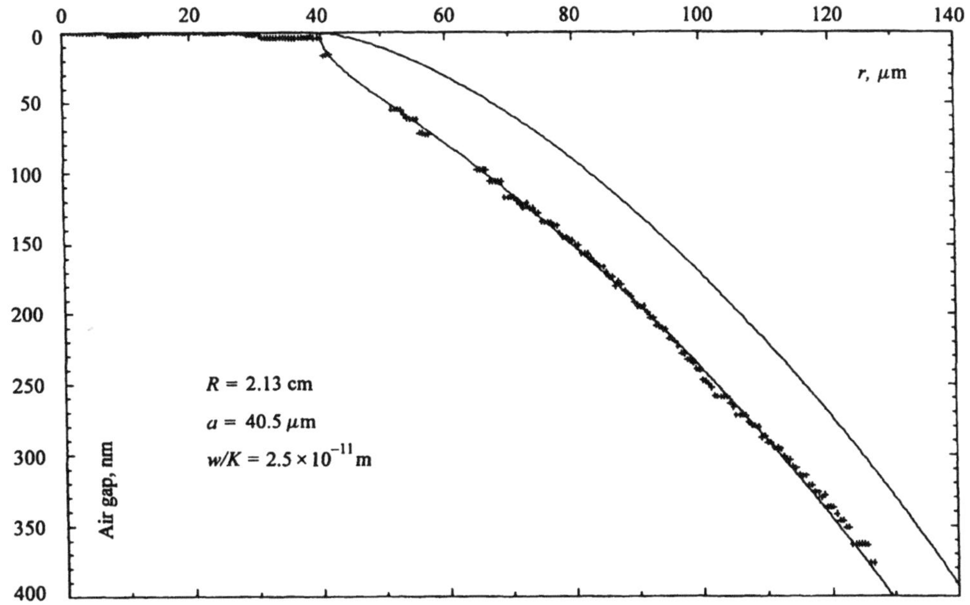

Figure 2. Mica–mica contact in dry air: a JKR profile is observed, in agreement with equation (11) fitted with $w/K = 2.5 \times 10^{-11}$ m. The Hertzian profile is given for comparison.

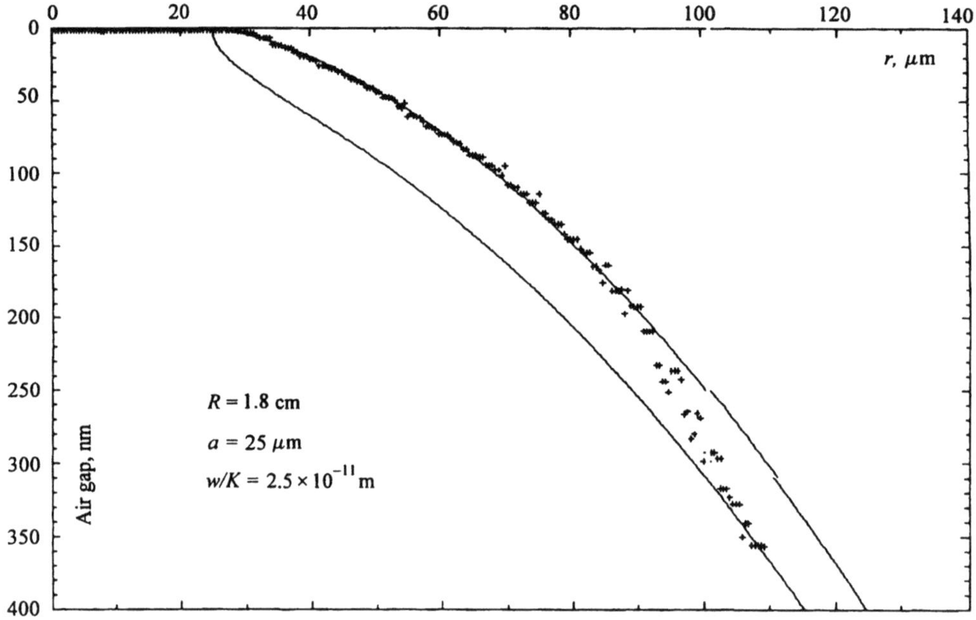

Figure 3. Mica–mica contact with a large water meniscus: a DMT (Hertzian) profile is observed, in agreement with equation (4). A JKR profile (with $w/K = 2.5 \times 10^{-11}$ m) is given for comparison.

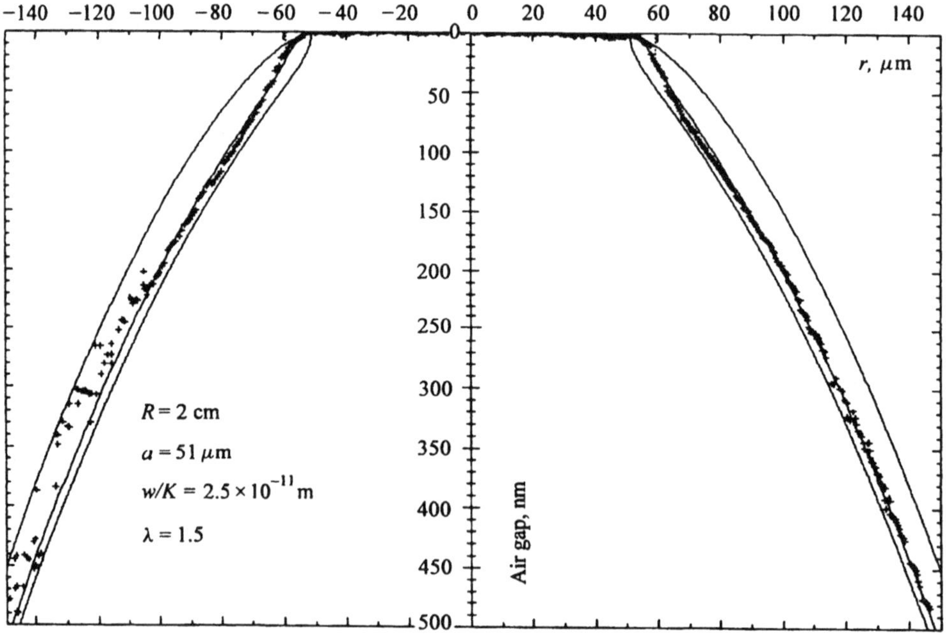

Figure 4. Mica–mica contact in humid air. The Hertzian profile and the JKR profile (with $w/K = 2.5 \times 10^{-11}$ m) corresponding to this radius of contact (51 μm) are drawn for comparison. The observed profile (+) can be fitted by equations (23) and (24) with $\lambda = 1.5$ and $w/K = 2.5 \times 10^{-11}$ m.

intermediate between the JKR and the Hertzian ones, which is in good agreement with equations (23) and (24) using the fitting parameters $w/K = 2.5 \times 10^{-11}$ m and $\lambda = 1.5$.

8. DISCUSSION

The change in profile with the size of the meniscus confirms the earlier results [4–6] and is in agreement with the theory, using one or two adjusting parameters. It remains to be verified that these parameters have reasonable values, consistent with the fact that measurements are made at nearly zero applied load.

First consider the DMT profile for a large meniscus (Fig. 3). From equation (19) the radius of contact under zero load is given by

$$a^3 = \frac{2\pi w R^2}{K}. \tag{25}$$

Inserting $a = 25$ μm and $R = 1.8$ cm leads to $w/K = 7.7 \times 10^{-12}$ m. For such menisci it is known that the pull-out force is essentially due to the Laplace pressure and that the contribution due to solid–solid interactions across the meniscus is negligible [4–6]. One can thus imagine replacing w by $2\gamma_L$, with $\gamma_L = 72$ mJ/m^2 the surface energy of water.* This results in $K = 1.9 \times 10^{10}$ Pa, a value in the range of those measured by Horn *et al.* [14] (3×10^9 Pa $< K < 2 \times 10^{10}$ Pa) for similar materials.

Now consider the JKR profile in dry air (Fig. 2). From equation (18) the radius of contact under zero load is given by

$$a^3 = \frac{6\pi w R^2}{K}.$$

Inserting $a = 40.5$ μm and $R = 2.13$ cm leads to $w/K = 7.8 \times 10^{-12}$ m, i.e. the same value as above, as if a micromeniscus were present. Unfortunately, using this value of w/K in equation (11) leads to a curve which does not fit the experimental values. On the other hand, the value $w/K = 2.5 \times 10^{-11}$ m which fits the observed profile and is in agreement with the large values of the adhesion forces measured in dry air would correspond in the JKR model to negative applied load. To clear up these points, it is necessary to study the profiles along complete $a(P)$ curves. This will be done in the near future.

*However, this procedure is not rigorously correct. For large menisci the separation of the two solids occurs after a point contact, but a liquid bridge remains, the stability of which depends on the stiffness of the spring [15]. The pull-out force $4\pi\gamma_L R \cos\theta$ corresponds to the instability of this liquid bridge at fixed load. It is thus clear that only the surface energy of the liquid appears in the formula. This does not imply that solid–solid interactions do not exist at zero load.

Acknowledgement

This work was undertaken with the support of the GDR 936 of the CNRS 'Mesure des forces de surface en milieu liquide'.

REFERENCES

1. K. L. Johnson, K. Kendall and A. D. Roberts, *Proc. R. Soc. London, Ser. A* **324**, 301–313 (1971).
2. B. V. Derjaguin, V. M. Muller and Yu. P. Toporov, *J. Colloid Interface Sci.* **53**, 314–326 (1975).
3. D. Maugis, *J. Colloid Interface Sci.* **150**, 243–269 (1992).
4. L. R. Fisher and J. N. Israelachvili, *Colloids Surfaces* **3**, 303–319 (1981).
5. H. K. Christenson, *J. Colloid Interface Sci.* **104**, 234–249 (1985).
6. H. K. Christenson, *J. Colloid Interface Sci.* **121**, 170–178 (1988).
7. A. Fogden and L. R. White, *J. Colloid Interface Sci.* **138**, 414–430 (1990).
8. D. A. Spence, *J. Elasticity* **5**, 297–319 (1975).
9. M. Lowengrub and I. N. Sneddon, *Int. J. Eng. Sci.* **3**, 451–460 (1965).
10. D. Maugis, *J. Adhesion Sci. Technol.* **1**, 105–134 (1987).
11. J. N. Israelachvili, *J. Colloid Interface Sci.* **44**, 259–272 (1973).
12. S. C. Hunter and F. R. N. Nabarro, *Philos. Mag.* **43**, 538 (1952).
13. D. Tabor, *J. Colloid Interface Sci.* **58**, 1–13 (1977).
14. R. G. Horn, J. N. Israelachvili and F. Pribac, *J. Colloid Interface Sci.* **115**, 480–492 (1987).
15. D. Maugis, in: *Adhesive Bonding*, L. H. Lee (Ed.), pp. 303–335. Plenum Press, New York (1991).

Fundamentals of Adhesion and Interfaces, pp. 61–71
D. S. Rimai, L. P. DeMejo and K. L. Mittal (Eds)

Adhesion of charged particles

DAN A. HAYS

Wilson Center for Research and Technology, Xerox Corporation, 0114-22D, Webster, NY 14580, USA

Revised version received 3 November 1994

Abstract—The adhesion properties of charged particles are of considerable importance in the electrophotographic process. Measurements on irregularly-shaped, pigmented particles, called toner in the electrophotographic industry, show that adhesion increases with toner charge but that the magnitude is much larger than expected from a simplified electrostatic image force model. An enhanced electrostatic adhesion is also seen in electric field detachment measurements on spherical charged particles. In both cases, this unexpected large adhesion can be attributed to a nonuniform distribution of charge on the surfaces of the particles.

Keywords: Adhesion; electrophotography; electrostatics; particles.

1. INTRODUCTION

The controlled placement of small, charged particles is a key requirement in electrophotographic marking, also known as xerography. These so-called toner particles, $\sim$10 μm in diameter and consisting primarily of a pigment dispersed in a polymer resin, are charged triboelectrically by mixing with larger carrier beads. The charge on toner enables the electrostatic transfer of these particles between surfaces, allowing the development of an electrostatic latent image and subsequent transfer of the developed image to paper [1]. For transfer of a charged toner particle to occur, the applied electrostatic force must overcome the force of adhesion between the particle and substrate. Because of this requirement for toner transfer, the adhesion properties of toner particles are of considerable importance in optimizing the electrophotographic process [1, 2]. A fundamental understanding of toner adhesion becomes even more important when the average particle size is reduced ($\sim$7 μm) to enable high-quality color electrophotographic images [3].

The toner used in most current electrophotographic machines is irregular in shape, since the particles are formed by a grinding process. Many adhesion studies (primarily using centrifuge) on irregularly-shaped toner particles have been reported over the past 30 years [3–9]. A survey of this literature reveals considerable disagreement over the relative importance of non-electrostatic and electrostatic contributions to toner adhesion. It is the purpose of this paper to show that the adhesion of triboelectrically charged particles, *regardless of their shape*, is dominated by an electrostatic force which is enhanced by a nonuniform surface charge distribution.

2. ADHESION OF IRREGULARLY-SHAPED PARTICLES

If it is assumed that an irregularly-shaped toner particle can be approximated as a sphere with the charge uniformly distributed over the surface, the magnitude of the electrostatic image force can be calculated as

$$F_{\mathrm{I}} = \alpha \frac{Q^2}{16\pi \varepsilon_0 R^2}, \tag{1}$$

where Q is the particle charge, R is the average particle radius, ε_0 is the permittivity of free space and α is a correction factor which takes into account the polarization of the dielectric particle. (For a dielectric constant of 4, $\alpha = 1.9$.) For a typical toner particle with a charge-to-mass ratio of 15 mC/kg and an average toner diameter of 10 μm, the particle charge Q is 8 fC. The electrostatic image force as calculated from equation (1) is ~10 nN. A survey of the literature on toner adhesion measurements shows that the typical toner adhesion for this particle charge and size is 5 to 50 times larger [3–9]. It has been argued that the large discrepancy between the adhesion measurements and the simplified electrostatic image model supports the view that the adhesion is dominated by short-range and/or contact forces. If this is true, the adhesion should be independent of toner charge, contrary to observations reported in several studies [3, 6, 9].

Figure 1 illustrates an irregularly-shaped toner particle which has been triboelectrically charged by mixing with larger carrier beads. It is assumed that the protrusions on the insulative particle are charged to a constant charge density σ that depends on the triboelectric charging properties of the toner and carrier bead materials. The total area of the charge patches A_{t} represents a small fraction of the toner surface area. The sum of the charged areas in contact with the conducting substrate is A_{c}. If the extent of the charged area in contact with the substrate is much larger than the average distance between the charged surface and substrate, the electrostatic force of adhesion can be approximated as

$$F_{\mathrm{E}} = \frac{\sigma^2 A_{\mathrm{c}}}{2\varepsilon_0}. \tag{2}$$

We have assumed that the adhesion due to the remaining charge on the toner particle is approximated by equation (1) and can be neglected in comparison to the force given by equation (2).

The total toner adhesion can then be written as

$$F_{\mathrm{A}} = \frac{\sigma^2 A_{\mathrm{c}}}{2\varepsilon_0} + W A_{\mathrm{c}} = Qf\left[\frac{\sigma}{2\varepsilon_0} + \frac{W}{\sigma}\right], \tag{3}$$

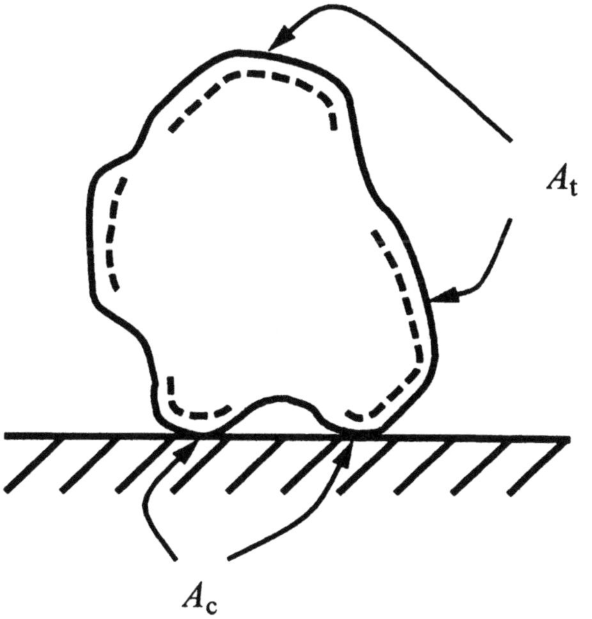

Figure 1. An irregularly-shaped toner particle triboelectrically charged in patches at a charge density of σ. The total charged area is A_t and the charged area in contact with the substrate is A_c.

where $Q = \sigma A_t$, $f \equiv A_c/A_t$ and a non-electrostatic adhesion contribution, WA_c, has been included. (Since the electrostatic force has a longer range than the non-electrostatic force, the actual contact areas will differ. Equation (3) assumes that the smaller area associated with the non-electrostatic force is proportional to the area of the charge patches, A_c, in close proximity to the substrate.) To determine the electrostatic component of adhesion in equation (3), we estimate from geometrical considerations that the average value of f is 0.2 and note from a survey of literature on contact charging that σ typically ranges from 0.5 to 5 mC/m^2 for triboelectrically active materials. For $Q = 8$ fC and $\sigma = 1$ mC/m^2, typical values for xerographic toner, the electrostatic force of adhesion is 100 nN. This value is an order of magnitude larger than that expected from equation (1), and is comparable to the measured values. Furthermore, equation (3) shows that when an electrostatic force QE is applied to a toner layer, the charge-to-mass ratio of the incrementally detached toner is independent of E, as observed [8]. The spread in the electric field required to overcome the force of toner adhesion is attributed to a distribution in the values of f since it is assumed that σ is a constant for a particular set of toner and carrier bead materials. This implies that the electrostatic adhesion of a single particle is not unique but varies widely according to the contact area [8].

3. ADHESION OF SPHERICAL PARTICLES

To obtain insight into the physics of charged particle adhesion, it is desirable to perform experiments with spherical particles in which both the adhesion and charge of a

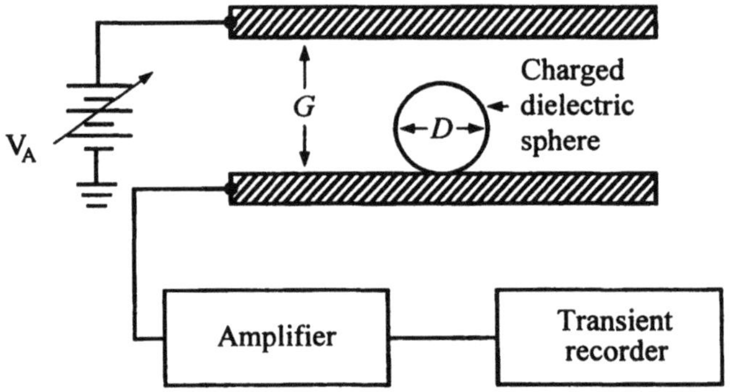

Figure 2. Apparatus for electrostatically detaching a charged particle and detecting the charge and motion of the particle as it bounces against a conducting electrode.

given particle are measured. For this reason, we have reported studies of electric field detachment measurements which show that the adhesion of a spherical, triboelectrically charged particle is much larger than the electrostatic image force calculated if one assumes a uniform charge distribution over the surface [10, 11]. Figure 2 shows the apparatus used to measure both the adhesion and charge of a dielectric sphere repeatedly detached by an electric field between closely-spaced electrodes in a vacuum of 1 mPa [10]. A charge-sensitive amplifier is connected to the bottom electrode for detecting the particle charge. Charge induced in the electrode provides information about the particle charge, position and motion within the gap. A high speed transient recorder captures the induced charge signal. The upper electrode is connected to a power supply which can be ramped through a voltage range of ±2500 volts. When the particle becomes detached, the voltage is held constant until the particle comes to rest on the other electrode. The applied voltage is then reset to zero before the voltage is ramped again with opposite polarity. Repeated detachments are obtained by reversing the voltage polarity after each detachment. The gap between the 5 mm diameter polished (arithmetic surface roughness of 0.03 μm) steel electrodes is varied with a micrometer head.

Adhesion and charge measurements have been obtained on 99 μm diameter dielectric spheres consisting of styrene divinylbenzene. The spheres were triboelectrically charged by mixing with 450 μm polymer coated steel beads. After placing a single particle in the cavity, the system was evacuated and the voltage ramped. Figure 3 shows an induced charge signal due to a 99 μm particle detached from the upper electrode in a 520 μm gap. (The lower base line shift in the signal is due to a poor low frequency response of the amplifier.) The major portion of the induced charge signal is due to translation of the charged particle, Q_T, across the gap. (There is also a component due to particle rotation, Q_R.) Since the gap is much larger than the particle diameter, most of the induced charge signal associated with detachment is due to particle translation.

Figure 4 shows the induced charge signal for the same 99 μm particle detached from the upper electrode for a gap setting of 137 μm. The initial induced charge signal, Q_T, is due to translation of the charged particle across the gap. Since the

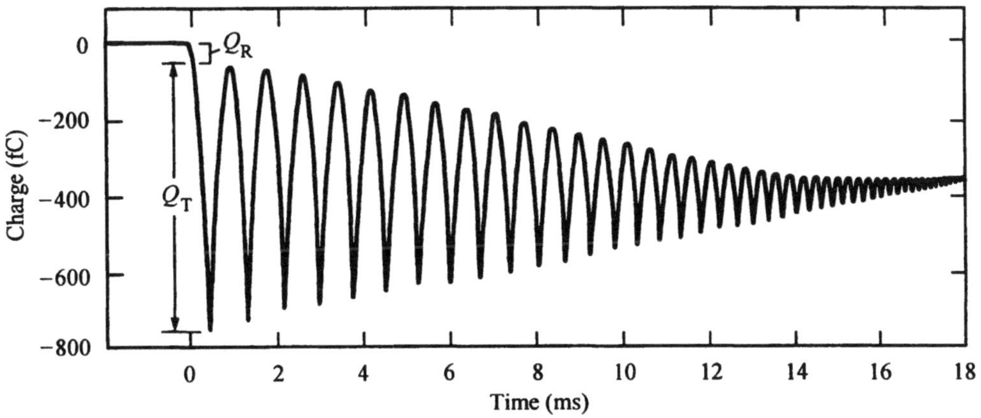

Figure 3. Charge induced in the charge-sensing electrode for a 99 μm bouncing particle detached by −1300 V across a 520 μm gap.

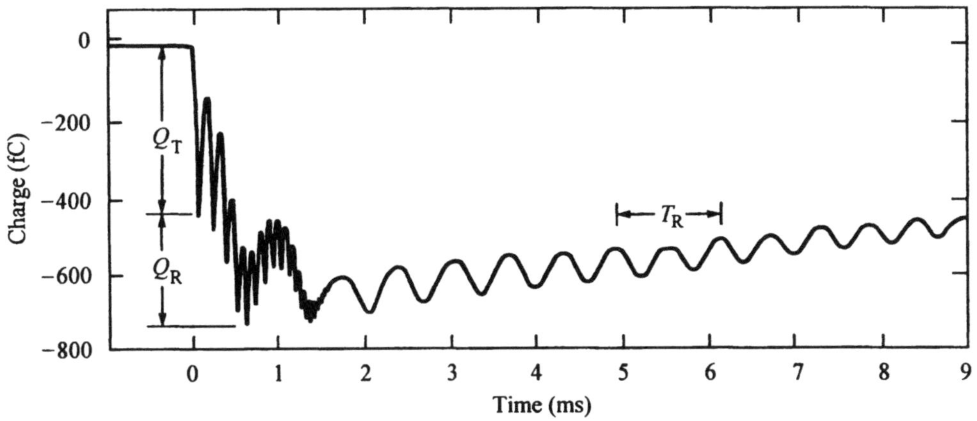

Figure 4. Charge induced in the charge-sensing electrode for a 99 μm bouncing particle detached by −580 V across a 137 μm gap.

center of the charge distribution is not at the center of mass, there is also a signal due to rotation of the particle through an angle of 180 degrees, Q_R. As the particle comes to rest on the electrode, a rocking motion is observed due to the nonuniform distribution of charge on the surface of the sphere. The period of rocking, T_R, is twice the period of the induced charge signal.

The induced charge signal Q_T is less than the particle charge Q, since the translation is limited to the gap G minus the particle diameter D. The particle charge is calculated from the expression $Q = Q_T G/(G - D)$. Figure 5 shows the particle charge for a different 91 μm styrene divinylbenzene sphere as a function of the number of detachments. The figure also shows the arbitrary variation in the gap setting during the sequence of detachments. An increase in the amount of contact charging per detachment for the larger gap settings is due to an increase in the total contact area per detachment, since the kinetic energy of the bouncing particle is larger and the number of bounces is much greater for the larger gaps.

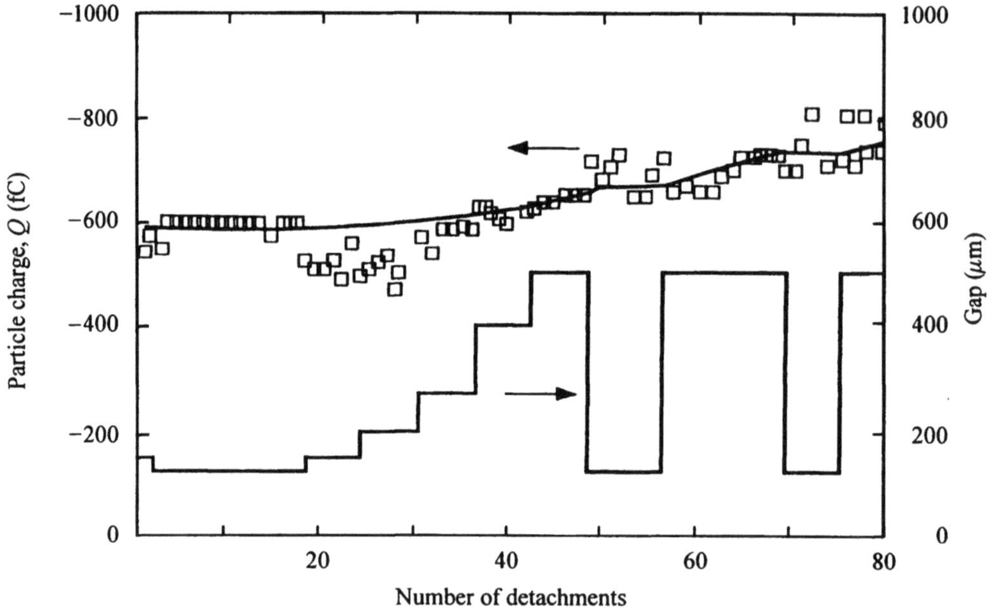

Figure 5. Particle charge and electrode gap as a function of the number of detachments for a 91 μm dielectric sphere.

The charge transferred to a particle upon repeated detachments is not distributed uniformly over the surface. If the center of charge is at a distance δ from the center of mass, the induced charge signal due to a 180 degree particle rotation is $Q_R = 2Q\delta/G$. Normalizing δ to the particle radius R, $\delta/R = Q_R G/(QD)$. The average value for δ/R increases from 0.24 to 0.31 when the particle charge increases by repeated detachments.

4. ANALYSIS

To calculate the electrostatic force and torque acting on a rocking, nonuniformly charged dielectric sphere in an electric field, we assume the surface charge density on the sphere is given by the expression

$$\sigma(\theta) = \sigma_U + \sigma_D \cos\theta + \sigma_Q \frac{(3\cos^2\theta - 1)}{2}, \tag{4}$$

where the coefficients σ_U, σ_D, and σ_Q represent the first three terms of a spherical harmonic expansion for an arbitrary charge distribution with axial symmetry. For the geometry in Fig. 2, the origin is taken at the center of the sphere and $\theta = 0$ occurs when the radial vector is directed at the contact point. The total charge on the sphere is

$$Q = \sigma_U 4\pi R^2 \tag{5}$$

and the center of charge normalized to the sphere radius is

$$\frac{\delta}{R} = \frac{\sigma_{\rm D}}{3\sigma_{\rm U}}. \tag{6}$$

If the sphere is far from electrodes, the electric field outside the surface is equivalent to collinear monopole, dipole and quadrupole moments at the center of the sphere. The monopole moment, $p^{(0)}$, is the total charge Q. The source dipole moment for a sphere of dielectric constant κ is

$$p_{\rm s}^{(1)} = -\frac{\sigma_{\rm D} 4\pi R^3}{\kappa + 2} = -\left(\frac{\sigma_{\rm D}}{\sigma_{\rm U}}\right)\frac{QR}{\kappa + 2} = -\frac{3\delta Q}{\kappa + 2}, \tag{7}$$

where we have used equations (5) and (6) to obtain an expression in terms of δ and Q. The source quadrupole moment is

$$p_{\rm s}^{(2)} = \frac{\sigma_{\rm Q} 4\pi R^4}{2\kappa + 3} = \left(\frac{\sigma_{\rm Q}}{\sigma_{\rm U}}\right)\frac{QR^2}{2\kappa + 3}. \tag{8}$$

In the presence of an applied electric field, the dielectric sphere has an additional polarization. The electric field outside the sphere is given by the applied field, E, plus the field due to a dipole at the center. This additional source dipole moment induced by the applied field and oriented along the field direction is

$$p_{\rm es}^{(1)} = \frac{4\pi\varepsilon_0(\kappa - 1)R^3 E}{\kappa + 2}, \tag{9}$$

where ε_0 is the permittivity of free space.

When a charged sphere in an electric field is resting on one of the electrodes, the source multipole moments induce charges in the conductor. (Induced charges in the counter electrode are presently ignored.) The electric field outside the conductor is equivalent to the field from a set of multipole image charges due to the presence of the conductor. The electric field from the image charges polarizes the dielectric sphere. As shown by Jones [12], the electric field outside a dielectric sphere due to the source multipole moments and the polarization of the sphere by an electric field can be described by an infinite set of linear multipole moments, $p^{(n)}$, at the center of the sphere. The infinite set of multipole moments induces an infinite set of image multipole moments in the adjacent conductor. A methodology for obtaining a self-consistent solution for a finite set of the linear multipole moments $p^{(n)}$ was described by Fowlkes and Robinson [13].

We have used the Fowlkes and Robinson formalism to obtain an explicit equation for the electrostatic force acting on a nonuniformly charged dielectric sphere resting on an electrode in an applied electric field. We assume that the distribution function for the surface charge density is given by equation (4). The total electrostatic force is

$$F_E = -\frac{\alpha Q^2}{16\pi\varepsilon_0 R^2} + \beta QE - \gamma\pi\varepsilon_0 R^2 E^2 - \frac{\eta 3 p_s^{(1)2}}{32\pi\varepsilon_0 R^4} + \frac{\lambda Q p_s^{(1)}}{8\pi\varepsilon_0 R^3} - \frac{\mu p_s^{(1)} E}{R}$$
$$- \frac{\nu 15 p_s^{(2)2}}{128\pi\varepsilon_0 R^6} - \frac{\xi 3 Q p_s^{(2)}}{32\pi\varepsilon_0 R^4} + \frac{\rho p_s^{(2)} E}{R^2} + \frac{\chi 3 p_s^{(1)} p_s^{(2)}}{16\pi\varepsilon_0 R^5}, \tag{10}$$

where the coefficients depend on the dielectric constant κ. For $\kappa = 2.5$, $\alpha = 1.45$, $\beta = 1.29$, $\gamma = 0.298$, $\eta = 1.79$, $\lambda = 1.71$, $\mu = 0.446$, $\nu = 1.96$, $\xi = 2.10$, $\rho = 0.487$, and $\chi = 1.95$. The dipole moment $p_s^{(1)}$ is calculated from equation (7) and the quadrupole moment $p_s^{(2)}$ is obtained by measuring the particle rocking frequency (~800 Hz) when the electrode gap is small, as shown in Fig. 4. (See [10] for details.)

The electric field for particle detachment can be calculated from equation (10) by setting F_E to 0 and solving the quadratic equation for E. The calculated detachment electric field values are within ~20% of the measured values [10]. The difference in the measured and calculated detachment electric fields might be due to a non-electrostatic contact force, but it seems more likely that the neglect of higher order source multipole moments in the expansion of the charge distribution in spherical harmonics could account for the discrepancy.

To check for dwell time effects, a particle was allowed to rest on an electrode for 1 h instead of the normal 2 min before the particle was detached. No increase in the detachment field was measured, implying that the charge on the particle did not migrate towards the contact point. This observation also tends to rule out the significance of a non-electrostatic contact force since dwell time effects attributed to interfacial deformation have been reported [4, 14].

To describe the electrostatic force acting on a charged particle when either the electrode gap is small compared to the particle diameter or the particle is bouncing between the electrodes, it is necessary to consider a series of image multipole moments induced in *both* electrodes [11]. Figure 6 illustrates several sets of image multipole moments when the particle is at a distance S from an electrode. The multipole moments $p^{(n)}$ which describe the electric field external to the particle are given by the source multipoles, equations (5–8), and the multipole moments induced by the

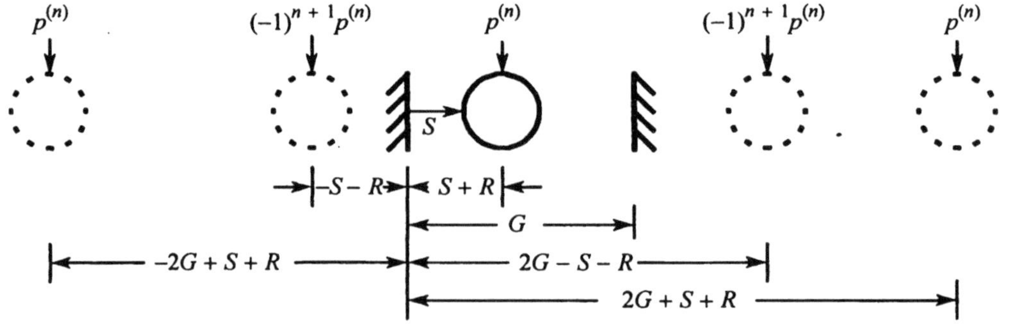

Figure 6. Several sets of image multipole moments and locations in the electrodes.

infinite set of image multipole moments. The induced multipole moments located in the electrode at $-S-R$ are $(-1)^{n+1}p^{(n)}$. The multipole moments at $S+R$ and $-S-R$ induce multipole moments in the counter electrode at the positions $2G-S-R$ and $2G+S+R$, respectively. Likewise, this pair of multipole moments induces another pair of moments in the electrode at $-2G+S+R$ and $-2G-S-R$, and so on. The electrostatic force acting on the charged particle can be calculated by solving for $p^{(n)}$ in a self-consistent manner and then calculating the electric field, E_z. According to Jones [12], the electrostatic force is

$$F_{\mathrm{E}} = \sum_{n=0}^{\infty} \frac{p^{(n)}}{n!} \frac{\partial^n E_z}{\partial z^n}. \tag{11}$$

This equation contains both the adhesion and detachment force terms, and was used to obtain the expression in equation (10).

Since the electrostatic force acting on a charged particle can be calculated for any position between the electrodes, it is of interest to see if this model can describe the bounce height of a particle as it rebounds from an electrode. To determine the bounce height, the electrostatic force is first calculated as a function of S/R and then integrated to obtain the potential energy. The potential energy is set to zero when the

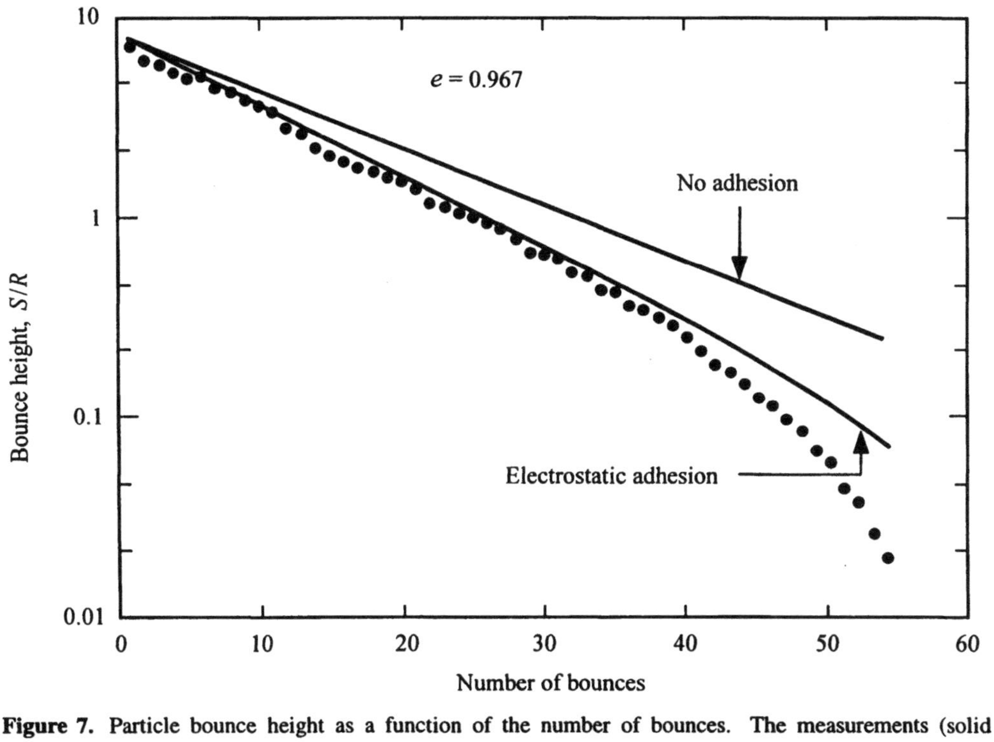

Figure 7. Particle bounce height as a function of the number of bounces. The measurements (solid circles) are compared to a calculated curve (solid line) obtained for a coefficient of restitution *e* of 0.967 and an electrostatic potential well of 0.30 pJ. If there is no adhesion, one would obtain the straight line for the same value of *e*.

particle is in the middle of the gap with no applied electric field. For a 98 μm particle with $G = 513$ μm and $E = 2.11$ V/μm, the calculated potential energy well for the particle at $S/R = 0$ is -0.30 pJ. To calculate the particle bounce height from the potential energy curve, we assume that at $S/R = 0$ the rebound kinetic energy is e^2 times the incident kinetic energy, where e is the coefficient of restitution. Figure 7 compares the measured bounce height of a 98 μm particle with the curve calculated from the potential energy curve for $e = 0.967$. The figure also includes a straight line which is obtained when one assumes no adhesion between the particle and electrode. For bounce heights less than $S/R \simeq 0.2$ or $S \simeq 10$ μm, the calculated curve does not fit the data. A fit can be obtained if the depth of the potential energy well at $S/R = 0$ is increased from -0.30 pJ to -0.44 pJ and e is increased from 0.967 to 0.970. Since this increase in the depth of the potential well cannot be accounted for by an adhesion energy from short-range forces (a force range of several tens of nanometers is much less than the range of electrostatic forces), it is assumed that the increase is either due to contact charge exchange or to an enhanced electrostatic adhesion from a textured surface charge on a scale of micrometers. An atomic force microscope as described by Terris *et al.* [15] would be helpful in measuring any texture in the surface charge.

5. SUMMARY

Electrostatic force calculations which assume a uniform surface charge distribution grossly underestimate the adhesion of charged particles. Measurements and analysis of the electric field detachment of both irregularly-shaped and spherical charged particles show that the most of the observed behavior can be explained by an electrostatic image force due to a nonuniform surface charge distribution. As a charged particle comes to rest on an electrode, the more highly charged regions on the particle will be drawn towards the conducting surface. Although most of the adhesion properties of charged particles can be explained by a nonuniform surface charge distribution, a more detailed understanding of the bouncing motion of spherical, charged particles near a conducting electrode can be obtained by invoking either contact charge exchange or a textured surface charge density on a scale of micrometers.

Acknowledgement

The author would like to thank Dr Elliott Eklund for a critical reading of the manuscript.

REFERENCES

1. D. M. Pai and B. E. Springett, *Rev. Modern Phys.* **65** (1), 163–211 (1993).
2. F. W. Schmidlin, in: *IS&T Proceedings, Eighth Intl. Cong. on Non-Impact Printing*, E. Hanson (Publication Chairman), pp. 25–30. IS&T, Springfield, Virginia (1992).
3. H. Akagi, *SPIE Proc.* **1670**, 138–145 (1992).
4. H. Krupp, *Adv. Colloid Interface Sci.* **1**, 111–239 (1966).
5. D. K. Donald, *J. Appl. Phys.* **40** (7), 3013–3019 (1969).

6. N. S. Goel and P. R. Spencer, in: *Adhesion Science and Technology*, L.-H. Lee (Ed.), Part B, pp. 763–829. Plenum Press, New York (1975).
7. C. J. Mastrangelo, *Photogr. Sci. Eng.* **26** (4), 194–197 (1982).
8. D. A. Hays, in: *Particles on Surfaces 1: Detection, Adhesion and Removal*, K. L. Mittal (Ed.), pp. 351–360. Plenum Press, New York (1988).
9. M. H. Lee and J. Ayala, *J. Imaging Technol.* **11** (6), 279–284 (1985).
10. D. A. Hays and W. H. Wayman, *J. Imaging Sci.* **33** (5), 160–165 (1989).
11. D. A. Hays, in: *Electrostatics 1991*, B. C. O'Neill (Ed.), Institute of Physics Conference Series Number 118, pp. 223–228. The Institute of Physics, Bristol (1991).
12. T. B. Jones, *J. Electrostatics* **18** (1), 55 (1986).
13. W. Fowlkes and K. S. Robinson, in: *Particles on Surfaces 1: Detection, Adhesion and Removal*, K. L. Mittal (Ed.), pp. 143–155. Plenum Press, New York (1988).
14. D. S. Rimai, L. P. DeMejo, R. C. Bowen and J. D. Morris, in: *Particles on Surfaces: Detection, Adhesion and Removal*, K. L. Mittal (Ed.), pp. 1–31. Marcel Dekker, New York (1995).
15. B. D. Terris, J. E. Stern, D. Rugar and H. J. Mamin, *Phys. Rev. Lett.* **63** (24), 2669–2672 (1989).

Fundamentals of Adhesion and Interfaces, pp. 73–94
D. S. Rimai, L. P. DeMejo and K. L. Mittal (Eds)

Adhesion and cohesion mechanisms of lunar dust on the moon's surface

LIENG-HUANG LEE*

*Wilson Center for Research and Technology, Xerox Corporation, Webster, NY 14580, USA***

Revised version received 17 January 1995

Abstract—This review is based on many publications from the US and some from the former Soviet Union. It briefly describes the sources of lunar dust. The problems posed to the past *Apollo* missions are summarized. The physical properties of lunar soils, such as cohesion, angle of repose, angle of friction, and bearing capacity, as reported by scientists from these two nations, are compared. The composition of lunar soils is described in terms of acidic and basic components. Since lunar soils are mainly silicates, we are the first to point out the importance of the Lewis acid–base interaction to the cohesion and adhesion of lunar dust. The adhesion of lunar dust is similar to that of a xerographic developer. For the adhesion between lunar dust and dissimilar materials, both van der Waals and electrostatic interactions can be the driving force, depending on the size of the dust particles. For particles smaller than 50 μm, van der Waals interaction predominates, while for those larger than 50 μm, electrostatic interaction controls. For the electrostatic interaction, lunar dust should charge positively. If lunar dust is pre-charged, e.g. by solar wind, the electrostatic interaction should predominate regardless of the size of the particles. It is important to note that for the electrostatic interaction, both electronic and ionic mechanisms can influence charge exchange. Thus, the Lewis type acid–base interaction can also affect the charge exchange. Experimental work with a simulated lunar dust is recommended for the elucidation of the adhesion and cohesion mechanisms. Before that, all methods of control of lunar dust are, at best, speculative.

Keywords: Acid–base interaction; *Apollo*; lunar dust; moon surface; particle adhesion; van der Waals interaction; electrostatic interaction.

1. INTRODUCTION

In 1990, a Lunar Particle Adhesion Program was initiated at NASA Lewis Research Center with the purpose of setting up a permanent station, e.g. Lunar Observer [1], on the moon in the future. At that time, the main objective of the program was to prevent lunar dust from adhering (or clinging) to the components of power systems, such as radiator and photovoltaic devices.

---

*Honorary Professor, Chinese Academy of Sciences.

**New address: Consultant on Adhesion, 796 John Glenn Blvd., Webster, NY 14580, USA.

Twenty-four years after the first exploration of the moon, on 25 January 1994 an unmanned spacecraft, *Clementine I*, started a 7-month journey to photograph the moon [2]. The early returns of the photographs of the moon from this spacecraft have been of excellent quality. It is hopeful that this may be the beginning of the next phase of lunar exploration. It appears that this is the time for us to review some earlier problems related to the prevention of lunar dust from contaminating the surfaces of optical and solar devices prior to the return of human exploration to the moon.

In this brief review [3], I will describe the composition of the lunar atmosphere and the sources and properties of lunar dust and soils. Several previous experiments carried out by others on lunar soils will be used to lead to the discussion of the adhesion and cohesion mechanisms of lunar dust. In some instances, I will draw an analogy between the adhesion of a xerographic developer and that of lunar dust because I have some firsthand experience with the former, but not with the latter.

2. LUNAR ATMOSPHERE

The atmospheric pressure on the moon [4] is between 10^{-12} and 10^{-16} Torr. Interestingly, although the moon is under ultrahigh vacuum, it does have a *changing* atmosphere. During the day, with a daily high of 400 K at noon, the atmosphere is filled very thinly with only a trace of helium, ^4He (2×10^3 molecules/cm^3), while at night with a midnight low of 100 K, the atmosphere is refilled with H_2, ^4He, ^{20}Ne, ^{36}Ar, and ^{40}Ar, as well as traces of O_2 and CO_2. The rare gases, except ^{40}Ar, originate, presumably, from the solar wind, and the presence of these gas molecules at night may lower the vacuum somewhat. In general, there may be 1–10 million molecules/cm^3 in the atmosphere during the lunar day, down to 10^5 molecules/cm^3 at night.

We have always been led to believe that the lunar surface is ultraclean after 3.5–4.0 Ae (1 Aeon = 10^9 years) of desorption. Although there were some occasional, cosmic impactions with foreign matter, the moon's surface should have been cleansed well under ultrahigh vacuum. However, at night the rare gases can adsorb or condense on the surface of lunar soils. Consequently, any adsorption of foreign matter such as rare gases on the surface may reduce the adhesion or cohesion of lunar soils slightly. Besides adsorption, the presence of rare gases through entrapment or chemical bonding in lunar rocks [5] has been identified by many scientists with modern instrumental techniques, e.g. Auger, SIMS, ESCA, etc.

The effect of gas adsorption on lunar soil particles is well known [5]. Once exposed to a terrestrial atmosphere, the particles show significantly greater fluidity, i.e. adsorbed gases act as a lubricant. Astronauts had much greater difficulty than expected in obtaining core samples. It was observed that after sitting in dry N_2 for some time, the core samples appeared to 'shrink'. Actually, what was observed was settling. The effect is real and while water vapor may be more effective than dry N_2, the latter also has an effect.

3. SOURCES OF LUNAR DUST

Lunar dust (approximately < 60 μm in diameter) [6, 7] is presumably formed by cosmic processes: hypervelocity impacts of meteors and micro-meteorites, exposure to solar cosmic rays and galactic cosmic rays, and implantation of solar wind particles. Lunar dust is a part of lunar regolith found on the surface of the moon, and the bulk of surface materials is lunar soil containing particles less than 1 mm in diameter. All of the impacts in the last 3 billion years have created a regolith blanket 3–20 m thick. Lunar dust particles are generally accreted with smaller particles adhering to large particles through thermal fusion to form agglutinates. By analogy, the surface of the fused particles in appearance somewhat resembles that of carriers impacted by toners in a xerographic machine (Fig. 1). However, the actual mechanisms of impactions are not the same. Actually sample #74220 was a fire fountaining (volcanic) soil. All particles were spheres or parts of spheres. They were *not* agglutinates consisting

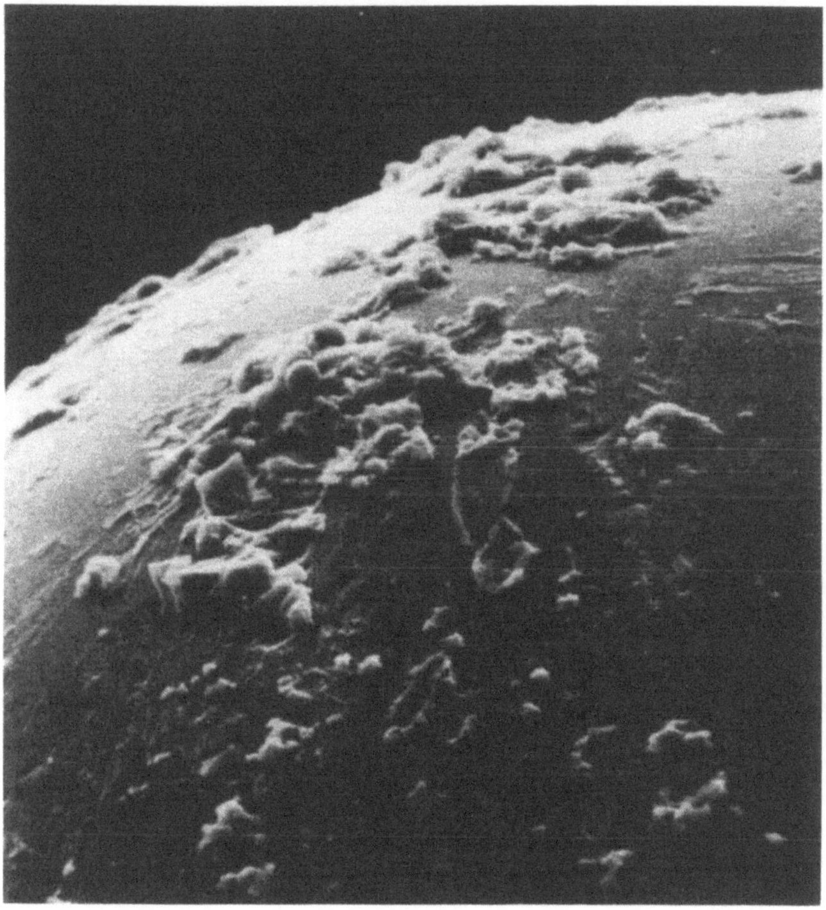

Figure 1. The surface of an orange glass spherule (sample #74220) from *Apollo 17*, coated with spatter. Magnification ×5000; photographed by D. S. McKay (NASA S.7317687) [8].

of part-molten and part-crystalline particles. The 'bumps' on the surface of #74220 samples were caused by the collision of small spherical (molten) particles with larger ones. Agglutinates are impact-generated. Sample #74420 was one of two such fire fountaining soils; the other was the *Apollo 15* green glassy soil. Thus, in fact, the collision mechanism does not resemble that in xerography.

Agglutinates (Fig. 2) are a composite of comminuted rock and mineral fused together with glass fragments. Figure 2 represents a part of a glassy agglutinate surface and it does not readily convey an idea of what an agglutinate is. Lunar soils and fragmentary rocks including agglutinates form the top layer, usually several meters thick, of regolith covering the entire moon. The major components of lunar soils and rocks [10, 11] are clinopyroxenes, $(CaMgFe)_2Si_2O_6$; plagioclase feldspar, $CaAl_2Si_2O_8-NaAlSi_3O_8$; ilmenite, $FeTiO_3$; and olivine, $Mg_2SiO_4-Fe_2SiO_4$. Miner-

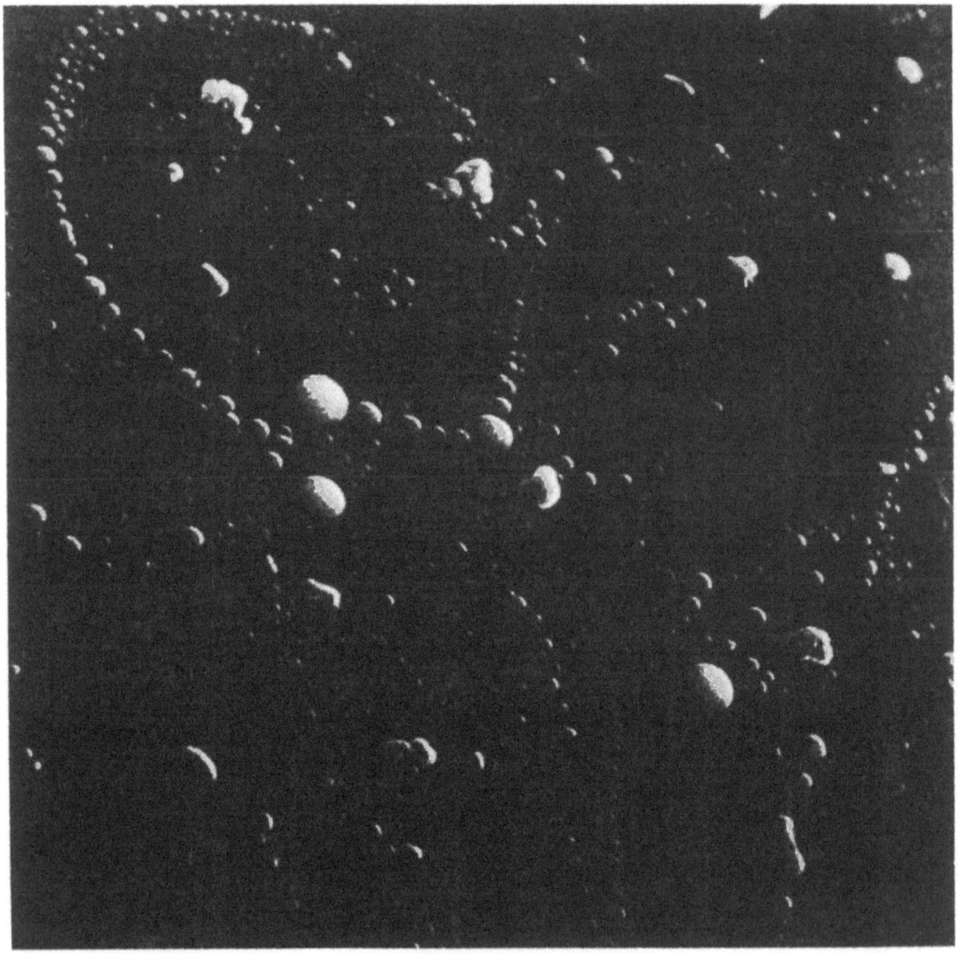

Figure 2. SEM photograph of a glassy agglutinate surface; the beads are formed by small iron droplets. Iron sources include both the impaction projectile and the reduction of target material. The photograph is of an area 4.5 μm^2, photographed by D. S. McKay (NASA S72-53166) [9].

als found on the highlands of the moon also include orthopyroxenes, $MgSiO_3-FeSiO_3$. Other important accessory minerals are *armalcolite,** $(FeMg)Ti_2O_5$; *pyroxferroite,** $CaFe_6(SiO_3)_7$; *transquillityite,** $Fe_8(Zr,Y)_2Ti_3Si_3O_{24}$; apatite, $Ca_5(PO_4)_3(F,Cl)$; chromite, $Fe_2Cr_2O_4$; christobalite or tridymite, SiO_2; triolite, FeS; iron–nickel, Fe–Ni; etc. Approximately 2000 minerals have been identified on the earth, while on the moon, the number is only about 100. Most of them have been found by scientists since the successful launch of the *Apollo* missions, 25 years ago.

The compositions and particle size distributions of lunar soil samples brought back by the *Apollo 11, 12, 14, 15, 16,* and *17* missions have all been reported in *Science* in great detail by the Preliminary Teams [12–17]. In Table 1, the composition of lunar

Table 1.
Typical lunar soil composition — determined by the former Soviet Union and US scientists [18]

Major oxides	Soviet *Luna-16* (wt%)		US *Apollo 12* (wt%)	
	Basaltic rock	Fine fraction	Basaltic rock	Fine fraction
SiO_2	43.8	41.7	40	42
TiO_2	4.9	3.39	3.7	3.1
Al_2O_3	13.65	15.32	11.2	14
FeO	19.35	16.8	21.3	17
MgO	7.05	8.73	11.7	12
CaO	10.4	12.2	10.7	10
Na_2O	0.33	0.37	0.45	0.40
K_2O	0.15	0.10	0.065	0.18
MnO	0.2	0.21	0.26	0.25
Cr_2O_3	0.28	0.31	0.55	0.41
ZrO_2	0.04	0.015	0.023	0.009

Table 2.
Typical lunar soil composition — Maria vs. Highlands [19]

Major oxides	Maria[a] (wt%)	Highlands[b] (wt%)
SiO_2	40.09	44.2
TiO_2	9.23	0.26
Al_2O_3	10.70	29.4
Cr_2O_3	0.49	—
FeO	17.85	2.96
MnO	0.24	0.6
MgO	9.92	16.30
CaO	10.59	0.73
Na_2O	0.36	0.11
K_2O	0.08	—

[a] Soil 71060, *Apollo 17*.
[b] Soil 67710, *Apollo 16*.

*Highlands only; species unique to the moon are in italics.

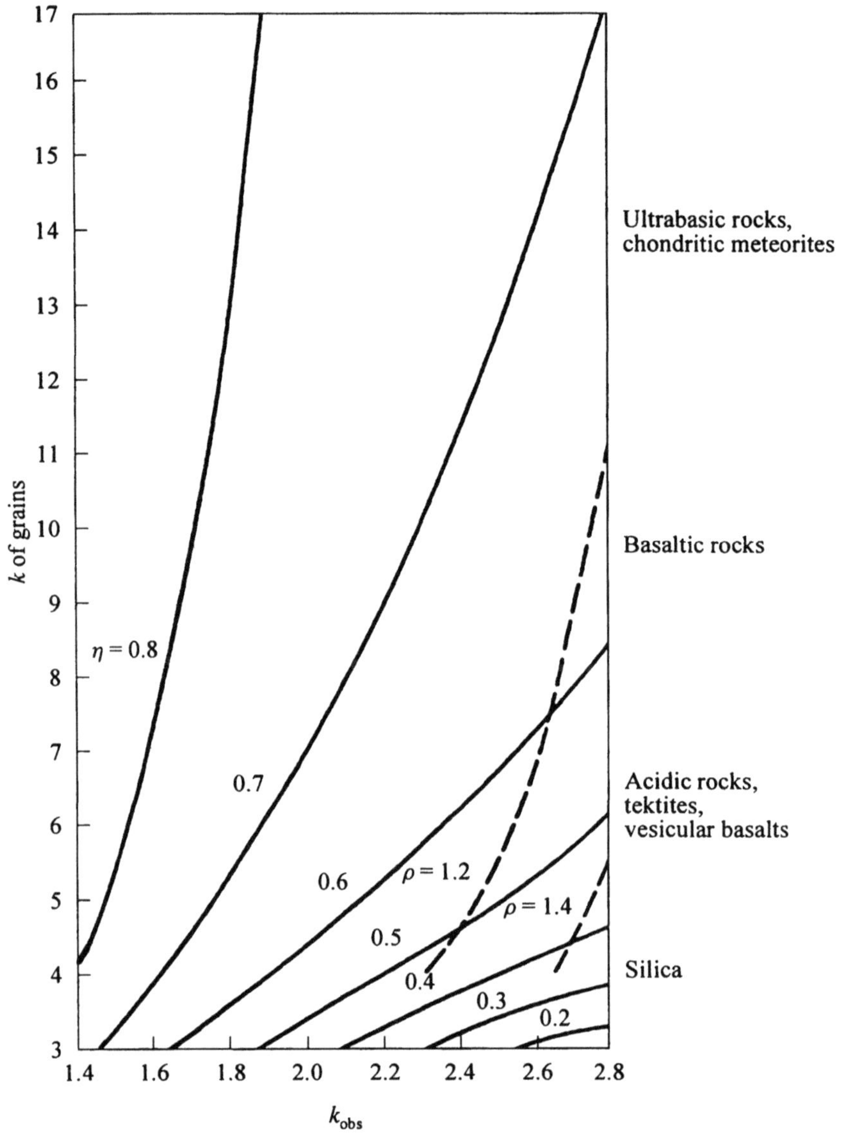

Figure 3. Relation between the dielectric constant of the grains, k, and the bulk dielectric constant, k_{obs}, according to Betner's formula; η is the porosity and ρ is the density [20].

samples brought back to earth by Americans has been verified by the former Soviet scientists [18] on the basis of lunar soil samples obtained by their unmanned flights. Interestingly, the differences are very small and some of the differences are obviously due to different sampling sites.

In Table 2, the chemical composition of lunar soils obtained from Maria and from Highlands is compared in terms of oxides [19]. The richer components are SiO_2, FeO, CaO, and Al_2O_3. The soil from Highlands contains more SiO_2, MgO, and Al_2O_3, but less FeO and TiO_2. Although most of these oxides exist only as bound components of lunar soils, the dust may be monopolar, i.e. acid or base, or bipolar, i.e. both acid and base. It is clear that some of the oxides, such as SiO_2, are acidic while others, like FeO, are basic. Of course, there are oxides, like Al_2O_3, which are amphoteric. Some simulated lunar soil close to the lunar soil from Maria, obtained from a quarry in Duluth, Minnesota, is called Minnesota Lunar Simulant (MLS-1). MLS-1 has been used by others for studies of some problems related to the adhesion and cohesion of lunar dust, described in a later section.

The distinction between acids and bases is more obvious among rocks. It should be noted that terrestrial rocks have been classified as acidic, basic, and ultrabasic. Figure 3 is a plot for various rocks [20] showing the relation between the bulk dielectric constant, k_{obs}, and that of the grains, k, according to Betner's formula, in which η is the porosity, i.e. the ratio of intergranular void space to total volume:

$$\frac{k_{obs} - 1}{3k_{obs}} = \frac{k - 1}{k + 2k_{obs}}(1 - \eta). \tag{1}$$

The minerals from the Highlands of the moon (Table 2) appear more acidic than those from Maria. The Maria samples are typically basaltic (see [21]) and are thought to arise via seepage through cracks in (impact-generated) crators (see also [22]). The minerals with a high SiO_2 content resemble the acidic pumice on earth. This important identification of acid and base may be one of the clues to understanding the adhesion and cohesion mechanisms of lunar dust.

A typical size distribution of lunar dust [23] is shown in Fig. 4. The median particle diameter is 40–130 μm and 10–25% of the soil is smaller than 20 μm. In the dust samples, there was a depletion of particles smaller than 1 μm.

McKay once proposed a steady-state situation and particle size distribution of the lunar regolith. This idea may explain the breakdown of the bedrock to small particles and the fusion of small particles to form agglutinates. In this sense, the formation of regolith does resemble the agglomeration of particles in xerography.

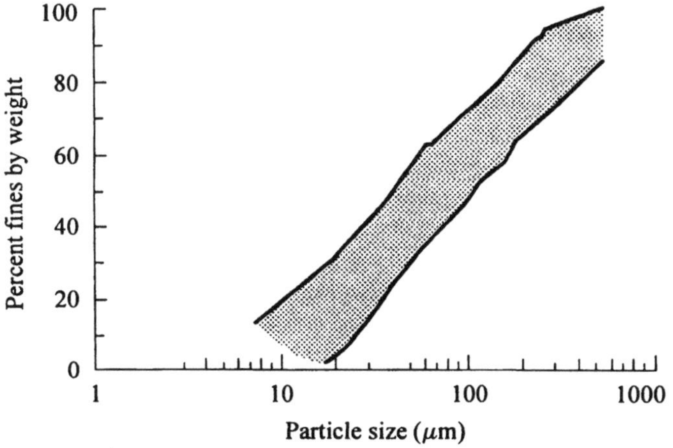

Figure 4. Grain size distribution for lunar soil. Samples from *Apollo 11, 12, 14* and *15* all fall within this band [23].

4. DETRIMENTAL EFFECT OF LUNAR DUST

Every *Apollo* mission had both hardware and operational problems [7] caused by contamination with lunar dust. During the *Apollo 11* mission, dust that collected in the pulleys of the lunar equipment conveyor resulted in jamming. The television cable, which was coated with lunar dust, could not be readily sighted. The dust problem was particularly bad for *Apollo 12*. Exhaust gases from land rockets would blow off the dust approximately 52.5 m above the surface, and the ground was completely obscured by 15 m. The *Apollo* crews complained that after 3 days their upper arm bearings and wrist disconnects were notably more difficult to move because of the dust penetration. An *Apollo* EMU life-support connector could not be properly engaged due to dust accumulation at the connector part. Generally, dust adhered readily to the astronaut's suits. Some dust on the surface could be brushed off, but dust was already 'rubbed in' the fabric. Thus, brushing the sticky dust off became a routine ritual before returning to the Lunar Module. Once the dust got into the Module, it became a nightmarish, respiratory irritant.

Batteries and radiators on the Lunar Roving Vehicle (LRV) had to be periodically cleaned. During the *Apollo 15* mission, dust accumulation on the LRV was not a serious problem; however, small particulate matter did accumulate over a long period of time. Large particles were controlled very well by fenders. The majority of the dust accumulation occurred on the lower horizontal surfaces, such as floor boards, seatpans, and the rear wheel area.

The effects of lunar dust accumulation on the performance of photovoltaic arrays have been studied by Katzan *et al.* [24]. For future lunar base operations, during rocket launch and landing, lunar dust will be an expected major problem. According to their calculation, at 2 km from the source, there will be 0.28 mg/cm^2 of dust accumulated

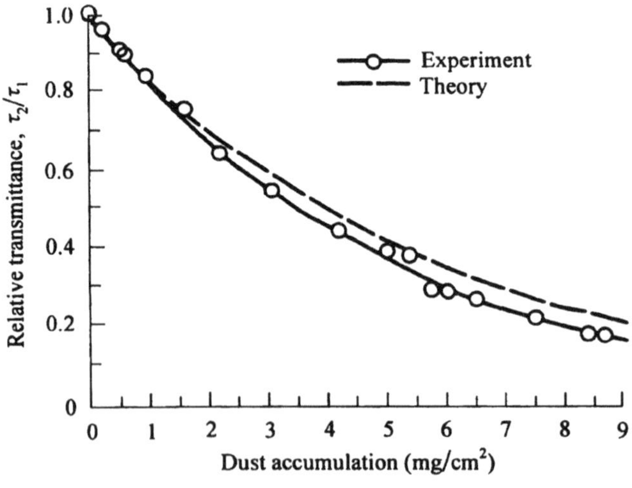

Figure 5. Relative transmittance of a glass coverslip with varying amounts of dust. Theory and experiment are compared for a specific particle size and shape. The dust material used was MLS-1, with an average thickness of 18 μm; length, 78 μm; width, 56 μm; relative transmittance, 0.45; and density, 3.0 g/cm^3 [24].

after ten surface missions with a 26 000 N excursion vehicle. The attenuation of light to photovoltaic arrays, due to the dust accumulation [25], may be in excess of 60% at 1 km and more than 10% at 2 km from the source after ten surface missions. Two models were developed by Katzan and Edwards [26] to estimate the effect of dust accumulation on the performance of high-temperature radiator surfaces and photovoltaic arrays. Both models predict an exponential decay in the performance of the arrays with increasing accumulation (Fig. 5).

However, for high-temperature radiators, very thin dust layers may actually enhance performance because of the high emissivity of lunar dust when the first layer is in intimate contact with a radiator. When lunar dust is thicker than 11 μm, the dissipation of heat would be limited because of its poor thermal conductivity. Katzan and Edwards' study indicates that the performance of photovoltaic arrays on a lunar base is met with insurmountable problems of dust accumulation.

Aside from natural sources of suspended dust, there are several human activities which can disturb significant amounts of lunar dust. Walking, operating surface vehicles, construction and mining, and, most importantly, rocket launch and landing can all create dust problems. From the components returned from *Surveyor III*, it was estimated that dust accumulation had occurred at a distance of 155 m from an *Apollo* lunar landing site. From shadowing and pitting, the average particle velocity was estimated to be between 40 and 100 m/s, arriving at an angle of alomost 0° during touchdown.

5. PROPERTIES OF LUNAR SOILS

Harkins [27] introduced the term cohesional work, or the work of cohesion W_c, in dynes/cm^2 to denote the work required to separate a column of a liquid 1 cm^2 in cross-section into two. Since the result of this operation is to produce 2 cm^2 of liquid surface where no interface was before, the cohesional work of a liquid is $2\gamma_l$, where γ_l is the liquid surface tension. The same definition can be applied to a solid with a γ_s, where γ_s is the solid surface energy. Now let us define the cohesion of soil, which is very different on the moon. The cohesion C, which is a force in dynes, of the earth and lunar soils (Fig. 6) can be expressed by the following equation [28]:

$$C = mg(\sin\alpha - \mu\cos\alpha), \tag{2}$$

where α is the angle of repose, μ is the coefficient of friction, m is the mass in grams, and g is gravity. When $C = 0$ for cohesionless soils, the angle of repose is the same on the earth and the moon, and $\mu = \tan\alpha$. Since the gravity of the moon is less than one-sixth of the gravity of the earth, the angle of repose is generally steeper for the same value of C when $C > 0$ for cohesive soils, but flatter, when $C < 0$, when there is some repulsion between the particles. C is also a function of particle size (Fig. 7). For the same size particle, C on the moon is greater than that on the earth.

It is noteworthy that the angle of the internal friction, ϕ, is expressed by a more general form of the Coulomb equation [29]:

$$S = N\tan\phi + C, \tag{3}$$

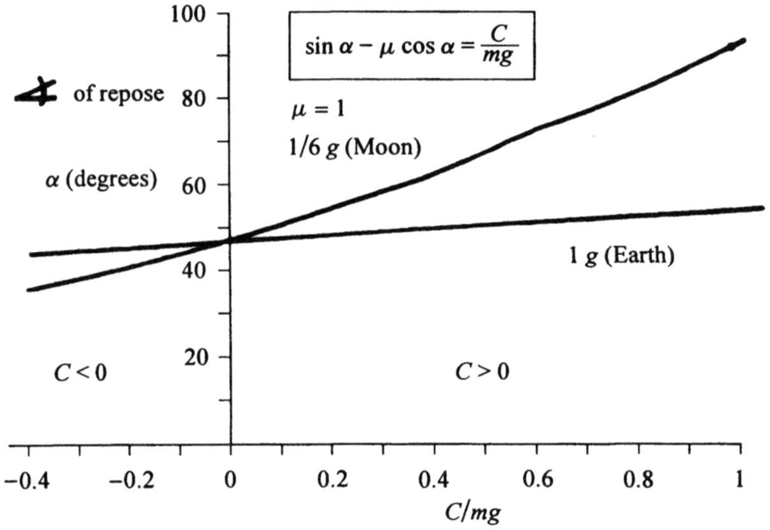

Figure 6. The angle of repose for the earth and the moon soils [28].

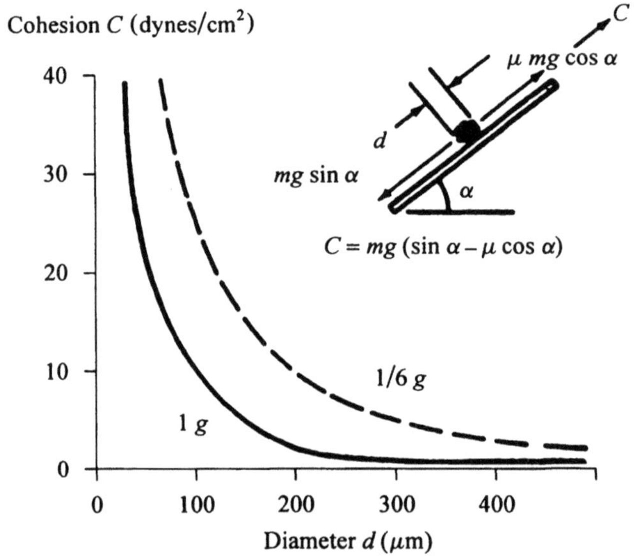

Figure 7. Cohesion in dry 'natural' quartz particles [28].

where S is the shear strength and N is the load normal to the shear plane. The shear strength of lunar soil is relatively high. The angle of the internal friction is a function of the coefficient of friction between individual grains, the shape and distribution of particles, and the void ratio (volume of void space/volume of solids) or packaging factor of the aggregates. When $C = 0$, $\phi = \alpha$, and when $C > 0$, $\phi < \alpha$. Thus, in general, the angle of repose is not equal to that of the internal friction.

Table 3.
Physical and mechanical properties of lunar soils from the *Surveyor* stations [30]

Soil properties	*Surveyor-1*	*Surveyor-3*	*Surveyor-5*	*Surveyor-6*	*Surveyor-7*
Grain size (mm)	Mostly < 1	Mostly < 0.06	Largely 0.002–0.06	Similar to *Surveyors-1* and *-3*	Similar to *Surveyors-1*, *-3*, and *-6* with stones
Angle of internal friction, ϕ (degrees)	30–40	35–40	37	37	37–39
Cohesion (kg/cm^2)	0.0013–0.004	0.0015–0.0071	—	0.005–0.012 (jet)	0.035–0.071 (manipulator)
Density (g/cm^3)	1.5 (mean)	1.5 (mean)	1.1 (mean)	0.7–1.2 in the upper mm; 1.6 in the upper cm	1.2–1.5 (mean)
Bearing capacity (kg/cm^2)	0.36	0.20–0.51	0.28	0.1 in the upper mm; 0.18 in the upper cm; 0.34 (mean)	0.34
Maximum depth of investigated soil layer (cm)	8 (support)	17.5 (manipulator)	12 (support)	9 (support)	15 (manipulator)

The lunar soils on all five of *Surveyor*'s landing sites [30] possess similar properties (see Table 3). They consist of a finely grained, weakly cohesive material with most of the grains ranging from 0.002 to 0.06 mm and fine earth averaging around 0.01 mm. The soil density also increases with depth. The topmost layer, a few millimeters thick, is looser, with a density of 1.0 g/cm^3. On the other hand, the mean soil density in the top 5 cm layer is 1.5 g/cm^3. The cohesion, C, is 0.005 kg/cm^2 down to a depth of several centimeters, but it is an order of magnitude less in the topmost layer. The grains tend to become aggregated. The angle of internal friction is approximately 35°. The static bearing capacity is $\cong$ 0.01 kg/cm^2 at a depth of a few millimeters and $\cong$ 0.15 kg/cm^2 at a depth of approximately 5 cm. The maximum depth investigated is 18 cm. The total thickness of granular deposits on the moon's surface is estimated at 1–20 m in Maria, and 1–20 cm, but not over 100 cm, in mountainous regions. These results are rather similar to those measured at the former Soviet *Luna 13* station [18].

6. EXPERIMENTAL RESULTS ON THE ADHESION AND COHESION OF LUNAR SOILS

Lunar soils resemble moist, silty soils [1]. The soil particles tend to stick to each other and form lumps or pseudolinks. During several *Apollo* missions, the footprints left behind by astronauts [30] (Fig. 8) remained clearly undisturbed and uncut, and so were the walls of the dug trenches (Fig. 9). For example, the cohesion of the lunar soil at the trench site was shown and reported [32] to be rather strong, in the neighborhood of 3.5–7 kN/m^2 (or 0.004–0.007 kg/cm^2), throughout the *Apollo 11* and *12* missions. However, for *Apollo 14* [14], the cohesion of the soil at a similar kind of trench site, but of course not at the same spot, was approximately only 10% of the earlier values. This decrease in cohesion could be due to the high glass content of the soil and/or the presence of lithic (or stony) fragments. The presence of glassy beads could transform the soils into a graphite-like consistency with low cohesion, and the presence of lithic fragments could reduce the contact areas between particles. This finding strongly indicates that the morphology of the soils can seriously affect cohesion. The ultrahigh vacuum environment on the moon surface definitely enhances the cohesion of the soils. Vertical walls produced by the indentation of the lunar soil sample persisted in vacuum, in dry N_2, but not in air. The air exposure seemed to reduce the cohesiveness with time. Experimentally, lunar soil sample #10084-93 from *Apollo 11* (sealed under N_2) [33] was evacuated to 5×10^{-9} Torr, and then exposed

Figure 8. Undisturbed lunar surface after having been stepped on by an astronaut (after Costes, NASA Marshall Space Flight Center, Alabama) [31].

to O_2, and $O_2 + 3.5\%$ H_2O at 2 Torr, 500 Torr, and 760 Torr, and at 27 °C and 200 °C. Loss of cohesion after exposure to air was observed in all these cases. It should be noted that these results were obtained with the assumption that the *Apollo* samples stored in a high vacuum were uncontaminated.

Earlier work by Ryan [34, 35] on the adhesion of silicate glass under high vacuum indicated the possible role of electrostatic interaction. The later work by Grossman *et al.* [33] also showed that there might be some electrostatic attraction between the fractured surface and particles of the lunar sample. However, they could not conclude whether the electrostatic attraction was the cause of cohesion and adhesion in the beginning prior to fracture or whether it operated on the lunar surface. In fact, it is rather common to find electrostatic charges on a fractured surface, as in the case of the cleaved mica crystal. However, the fact that all lunar dust lost stickiness upon

Figure 9. A trench dug in the lunar soil by a manipulator scoop [18].

exposure to air indicates that the electrostatic mechanism is not the major contribution to cohesion. A similar observation was obtained by Salisbury *et al.* [36] on the adhesion behavior of silicate powders in ultrahigh vacuum. Loss of cohesion for silicates in the presence of air was also observed. It is important to note that cohesion was restored after the system was evacuated.

Besides cohesion between particles, lunar dust has been reported to stick to anything in sight on the surface of the moon and to the clothing and tools of astronauts. In vacuum, lunar dust sticks to all kinds of surfaces, whether metallic, mineral, or organic (plastics). Photographic analysis of astronaut boot prints revealed that the value of the cohesion of lunar soil was higher than the soil–boot adhesion. The particle size distribution for the samples brought back by *Apollo 11*, *12*, *14*, and *15* ranged between 0.008 and 0.5 mm, and 50% by weight of the particles were less than 0.07 mm. In general, small dust particles less than 0.01 mm ($< 100\ \mu$m) travel the farthest distance and exhibit the strongest adhesion to various surfaces. Now, the questions remaining are: What causes the adhesion and cohesion of lunar dust and why does adhesion disappear rapidly after exposure to the atmosphere?

7. ADHESION AND COHESION MECHANISMS OF LUNAR DUST

There are at least six mechanisms proposed for the adhesion of solids [37]:

(1) diffusion;

(2) mechanical interlocking;

(3) chemical (or atomic) bonding;

(4) adsorption based on van der Waals (vdW) forces;

(5) electronic (or Coulombic) interaction [38]; and

(6) molecular interaction [39, 40] based on donor–acceptor (or acid–base) interaction.

In the case of lunar soil adhesion, the first mechanism, diffusion, is not relevant. However, on the basis of the above briefly mentioned results, we cannot rule out any of the last five. As in the case of particle adhesion, the total forces [41] may consist of Lifshitz–van der Waals forces, F_{LW}; the electrostatic induced image forces, F_{im}; the chemical forces, F_{ch} (such as the force for the acid–base interaction, F_{ab}); and the electrical double layer force, F_{ed}:

$$F = F_{\mathrm{LW}} + F_{\mathrm{im}} + F_{\mathrm{ch}} + F_{\mathrm{ed}}. \tag{4}$$

According to Krupp [42], in general F_{LW} dominates among spherical particles, and in the case of a sphere and a plane, the first term representing the vdW forces is

$$F_{\mathrm{LW}} = \frac{\hbar\omega R}{8\pi Z^2}, \tag{5}$$

where R is the radius of the sphere, Z is the distance between the sphere and the plane, and $\hbar\omega$ is the Lifshitz–van der Waals constant. The second term, the image force F_{im}, is

$$F_{im} = K_{im}\left(\frac{q^2}{r^2}\right), \tag{6}$$

where q is the point charge, r is the distance between point charges, and K_{im} is a constant which depends on the roughness of the spherical particle and the density of the particle layer [43].

The third term, F_{ab}, can be significant if the donor–acceptor (or acid–base) interaction is involved. Fowkes [44] studied the role of the acid–base interaction especially for polymers, and he introduced the enthalpy of the acid–base interaction to be a function of Drago's parameters [45]:

$$-\Delta H^{ab} = E_a E_b + C_a C_b, \tag{7}$$

where E_a and E_b are the susceptibilities of the acid (a) and base (b), respectively, to undergo the electrostatic interaction, and C_a and C_b are those to undergo covalent interaction. The work of adhesion W_A^{ab} of the acid–base interaction is then expressed in terms of the enthalpy of the interaction:

$$W_A^{ab} = -f n^{ab} \Delta H^{ab}, \tag{8}$$

where n^{ab} is the fraction of surface involved in the acid–base interaction and f is a correction factor, which is smaller than unity [46]. From the work of adhesion, one can calculate the force of the acid–base interaction, F_{ab}.

Finally, the fourth term, F_{ed}, is the electrical double layer (EDL) force proposed by Deryagin *et al.* [47] and is expressed as

$$F_{ed} = 2\pi\sigma S_d, \tag{9}$$

where σ is the surface charge density and S_d is the area. Thus, even in the case of particle adhesion, the EDL is only part of the total forces of attraction. Since Uber *et al.* [48] report that the particle adhesion is not affected by charge transfer, the EDL force, F_{ed}, in the fourth term may be insignificant or equal to zero.

The dominating role of the Lifshitz–van der Waals component in the adhesion of smaller particles ($< 50\ \mu$m) has also been shown on a semiconductor surface [49]. These vdW forces can increase with time due to particle and/or surface deformation, which increases the contact area. Micrometer size particles can be held to surfaces by forces exceeding 100 dynes, which corresponds to a pressure of 10^9 dynes/cm^2 or more. The total forces of adhesion of micrometer size particles exceed the gravitational force on the particle by factors greater than 10^6. Generally, electrostatic forces only become predominant for particles larger than 50 μm.

In the case of *pre-charged* particles, e.g. xerographic toners, the electrostatic image (Coulombic) component is more dominant [50–54] than the vdW component. The contact charging and adhesion of pre-charged particles are easier to understand than those of neutral particles.

So far, the evidence suggests that the sixth nechanism, i.e. the acid–base interaction, may be the most important for the lunar dust cohesion for the following reasons:

- The lunar surface is very clean and there are supposed to be no other adsorbed materials except a trace of rare gases. The relative cleanliness [55] has a very pronounced influence on strength because of a dramatic increase in interparticle forces, which may be chiefly vdW type. However, the holding of lunar dust on a steep slope suggests that the major force is greater than the vdW forces.
- Under ultrahigh vacuum and at a very high level of cleanliness, some degree of cold-welding of lunar soils [25] can take place, partly because the intermolecular distance is very small. If the distance is less than 4 Å, interaction between two molecules can be realized through the overlapping of molecular orbitals. Molecular interaction involves the electrostatic interaction and charge transfer as primary driving forces, and dispersion, polarization, and exchange repulsion as secondary factors. Molecular interaction is generally stronger than vdW interaction in magnitude and this may explain satisfactorily why lunar soils could be held on a steep slope as observed.
- Since molecular interaction requires molecular contacts, the morphology of the soils becomes very important. Poor morphology results in smaller contact areas and a lower interaction energy. This may be what happened to *Apollo 14* soil samples. On the other hand, irregular morphology may also lead to some degree of mechanical interlocking.
- One of the reasons why lunar soil [39, 40] is attracted together may be the Lewis acid–base (or donor–acceptor) interaction through molecular contacts of bipolar structures. It is well known that the Lewis acid–base interaction can take place between insulators or between an insulator and a conductor. The major requirement for this type of acid–base interaction is an intermolecular distance which must be smaller than 4 Å. Good contacts result in larger contact areas and high cohesion or good adhesion because acids and bases can interact fully to produce the maximum interaction energy. This may also explain why lunar dust can form pseudolinks in vacuum. These pseudolinks may consist of loose chains of ab bipolar lunar dust.
- The particle size of lunar soils should be an important factor in determining the adhesion and cohesion strengths. The most troublesome particles are small lunar dust particles (< 100 μm). Assuming that all other things are equal, electrostatic interaction predominates for particles greater than 50 μm, while vdW interaction predominates for particles smaller than 50 μm. In fact, the smaller the particles, the greater the vdW interaction.

Although we should consider the significance of the molecular interaction mechanism [39], we cannot overlook the other four mechanisms. If the intermolecular distance is larger than 4 Å, we should be thinking about the adsorption mechanism on the basis of vdW and Coulombic interactions [56]. If the distance is larger than

Table 4.
Estimated electric fields on the surface of the moon [57]

Dominant component of the environment	Electron density $(cm^3)^{-1}$	Plasma temp. (eV)	Debye length (m)	Field (V/m)
Solar wind (day side)	10	10	7	−0.01 to −2
Lunar ionosphere	10^2	10	2	−0.5 to −5
Photoelectrons	50	4	2	+0.5 to +5
Shocked solar wind	35	250	20	~ −5
Geomagnetic tail plasma (day side)	1	10–1000	20–500	±0.5 to +5

100 Å, we should be considering only the Coulombic interaction, which operates both at short range and at long range.

Another reason why we cannot rule out the electrostatic interaction is that there are also electric fields [57] on the moon due to solar wind (day side), lunar ionosphere, photoelectrons, shocked solar wind, and geomagnetic tail plasma (day side) (Table 4).

If lunar soils, which are primarily silicates, are pre-charged by any of these events or by foreign matter imported from the earth, such as Teflon gloves and plastic space suits, etc., we may have to consider the electrostatic mechanism on account of contact charging alone. For example, most silicates [58], like nylon, would charge positive, and Teflon negative. A recent study by Lee [59] indicates that the electrostatic contact charging is a dual mechanism in nature. Both electronic and ionic factors are involved. Previously, Gallo and Lama [60] also showed that the dielectric constant K correlates well with the work function ϕ of dielectrics:

$$\phi = \left(\frac{e^2}{4X_0}\right)\left(\frac{K+7}{K+1}\right), \tag{10}$$

where X_0 is the maximum excursion distance of an electron from the surface and e is the electronic charge.

In Fig. 3, we have seen the classification of terrestrial rocks by the observed bulk dielectric constants between 1.4 and 2.8 into acidic, basic, and ultrabasic. From equation (10), this spread can correspond to the spread of work functions of the rocks. By either the ionic or the electronic mechanism of charge exchange, triboelectrification between lunar dust and a dissimilar material is by all means certain.

Since lunar dust is mainly silicates, the triboelectric charging between lunar dust particles can be more ionic, as in the case of silicates, rather than chiefly covalent, as in organic polymers. With the presence of both acidic and basic constituents in lunar soils, the ionic mechanism may be dominant. This could serve to explain the strong cohesion of lunar soils in vacuum and the loss of cohesion after exposure to air. This may indicate that the cohesion of lunar soil is not due to electrostatic, but chiefly due to molecular or acid–base interaction. Indeed, there is no other way to explain the reversal nature of cohesion. In general, molecular interaction is stronger than the electrostatic interaction in magnitude, and this may explain the strong cohesion of lunar dust on a steep slope on the moon.

On the other hand, the adhesion between lunar dust and a dissimilar material under high vacuum can be due to electrostatic interaction, which is somewhat weaker than the cohesion of lunar dust, and this may be why the adhesion between the dust and a dissimilar material was observed to be weaker than the cohesion of dust particles. Whether it is electronic or ionic depends on the nature of the foreign matter contacted by the lunar dust. By analogy, the non-thermal cohesion between lunar dust particles resembles the cohesion of uniformly sized carriers in a xerographic machine, while the adhesion between lunar dust and a dissimilar material resembles that between a toner and a carrier. The former is controlled by the vdW or acid–base interaction, and the latter involves predominantly the electrostatic interaction.

8. SPECULATIVE PREVENTION OF LUNAR DUST PROBLEMS

The lunar environments and soil properties are difficult to modify; thus, lunar dust problems are inherently insoluble. However, if some of the above mechanisms are correct, there may be some preventive measures that one can take during the design and planning stage. Considering the triboelectric events, most lunar dust, like silicates, charges positively [58]. Then the desirable foreign matter surface should be charged positively to result in repulsion, instead of attraction. Since this is beyond control, we should be thinking about preventive mechanical barriers or dust screens to be installed in the vicinity of sensitive equipment [24]. Such screens could be activated by a signal from a flag made of piezoelectric film, such as polyvinylidene fluoride (J. Graf, NASA Johnson Space Center, Houston, TX, 1990). An electrostatic fence might be another method for diverting these particles from photovoltaic arrays or radiators. Such a fence on the basis of an alternating potential has been suggested by Katzan *et al.* [24].

Some of the decontamination attempts might have been achieved previously. For example, we found that the *Apollo 16* solar wind composite foil (Fig. 10), which was deployed at the Descartes site, appeared to be very dust-resistant (NASA AS-16-117-18849) [6]. The foil also incorporated platinum strips, which facilitated an acid washing technique, in the aluminum foil. It is important to find out how this solar wind foil was constructed and how it worked to achieve the dust-repelling effect. For design, in some cases, improved fenders for lunar roving vehicles, more flexible space suits, or other restricted access to sensitive equipment may be helpful in reducing the dust problem. On a lunar base, power components could be strategically placed as far as possible from rocket activities. Launch pads may be specially glazed and temporary tracks or compact roads may be constructed for regular traffic routes [7].

In brief, there may be at least two measures [7] to minimize the detrimental effect of dust: isolation and surface preparation. Particles follow ballistic trajectories and accumulation of dust drops off very quickly from sources. It may be possible to prevent dust accumulation on panels from most energetic mechanisms or activities. For example, mining operations that can generate dust should be isolated.

Figure 10. The *Apollo 16* solar wind composite foil deployed at the Descartes site. It appears that the difficulties of lunar dust contamination were overcome by the use of layered foils and by incorporating platinum strips, which facilitated acid washing techniques, in the aluminum foil (NASA AS-16-117-18849).

9. CONCLUSION

Lunar dust is mysterious. The cohesion of lunar particles can be explained by the difference in gravity and by the Lewis acid–base interaction mechanism, but the adhesion of lunar dust is not well understood. Since lunar particles are irregular in shape, there may be some degree of mechanical interlocking. So far, experimental results obtained by others indicate that van der Waals and molecular (or acid–base) interactions seem to play major roles because the cohesion between particles diminishes after exposure to the atmosphere and reappears after the ultrahigh vacuum is restored. This review is the first attempt to point out the importance of the acid–base interaction under high vacuum, because lunar dust particles are mostly ionic silicates. Strong Lewis (or Brønsted) acid-and-base bipolar lunar dust can adhere through molecular interaction to form pseudolinks under ultrahigh vacuum.

In the case of the adhesion between lunar dust and dissimilar materials, besides the acid–base interaction, both vdW and electrostatic interactions can be involved, depending on the size of the particles. The vdW interaction predominates when the particles are smaller than 50 μm, while the electrostatic interaction dominates when the particles are larger than 50 μm. Electrostatic interaction is definitely involved if the lunar environment is pre-charged, e.g. by solar wind. Furthermore, it has recently been shown that electrostatic interaction is actually a dual mechanism: electronic and ionic. Thus, the ionic nature of lunar soil can lead to electrostatic interaction without pre-charging. However, further verification of the adhesion mechanisms is needed through study of the simulated lunar dust before we know how to control the real lunar dust problem.

Acknowledgements

This paper is, in part, based on an invited presentation to the NASA Lewis Research Center on 18 May 1990. I would like to thank Ms Cynthia M. Katzan and Mr Curtis R. Stidham of Sverdrup Technology, Inc., Lewis Research Center Group for providing reprints of their publications for this review. I especially thank Mr Stidham and a reviewer for their comments on the original manuscript.

REFERENCES

1. J. Eberhart, *Sci. News* **137**, 264 (28 April 1990).
2. *Chem. Eng. News*, 31 (14 Feb. 1994).
3. L. H. Lee, Lecture on 'Adhesion and cohesion mechanisms of lunar dust on the surface of the moon', Presented at the NASA Lewis Research Center, Cleveland, OH (18 May 1990).
4. J. H. Hoffman, R. R. Hodges, F. S. Johnson and D. E. Evans, *Proceedings of the 4th Lunar Science Conference*, Vol. 3, p. 2865. Pergamon Press, New York (1973).
5. J. H. Reynold, C. M. Hohenberg, R. S. Lewis, P. K. Davis and W. A. Kaiser, *Science* **167**, 545 (1970).
6. H. S. F. Cooper, Jr, *Moon Rocks*, p. 92. Dial Press, New York (1970).
7. J. Graf, Private communication to C. Katzan (9 Sept. 1989).
8. S. R. Taylor, *Planetary Science: A Lunar Perspective*, p. 298. Lunar and Planetary Institute, Houston, TX (1982).
9. S. R. Taylor, *Lunar Science: A Post-Apollo View*, p. 74. Pergamon Press, New York (1975).
10. A. A. Levinson and S. R. Taylor, *Moon Rocks and Minerals*. Pergamon Press, New York (1971).
11. S. R. Taylor, in: *Lunar Science: A Post-Apollo View*, p. 132. Pergamon Press, New York (1975).
12. Preliminary examination of *Apollo 11* lunar samples, *Science* **165**, 1211 (1969).
13. Preliminary examination of *Apollo 12* lunar samples, *Science* **167**, 1325 (1970).
14. Preliminary examination of *Apollo 14* lunar samples, *Science* **173**, 681 (1971).
15. *Apollo 15* lunar samples, a preliminary examination, *Science* **175**, 363 (1972).
16. The *Apollo 16* lunar samples, petrographic and chemical description, *Science* **179**, 23 (1973).
17. *Apollo 17* lunar samples, chemical and petrographic description, *Science* **182**, 639 (1973).
18. I. I. Cherkasov and V. V. Shvarev, in: *Lunar Soil Science, Physicomechanical Properties of Lunar Soil*, A. Yu. Ishlinskii (Ed.), p. 24 (translated from Russian). Israel Program for Scientific Translations, Jerusalem (1975).

19. R. V. Morris, R. Score, C. Dardano and G. Heiken, *Handbook of Lunar Soils, Part I: Apollo 11–15* and *Part II: Apollo 16–17*, JSC-19069. NASA Johnson Space Center (July 1983).
20. *Surveyor III — A Preliminary Report*, p. 156, NASA SP-146 (1968).
21. G. J. Taylor, *Sci. Am.* **271**, 40 (July 1994).
22. *Origins of Mare Basalts and their Implications for Lunar Evolution*. Lunar Science Institute, Houston, TX (15–19 Nov. 1975).
23. R. E. Smith and G. S. West (compilers), *Space and Planetary Environment Criteria Guidelines for Use in Space Vehicle Development*, Vol. I, 1982 revision, NASA TM-82478 (1983).
24. C. M. Katzan, D. J. Brinker and R. Kress, Paper presented at the Space Photovoltaic Research and Technology Conference (SPRAT XI), held at NASA Lewis Research Center, Cleveland, OH (7–9 May 1991).
25. C. M. Katzan and C. R. Stidham, 'Lunar dust interactions with photovoltaic arrays', Paper presented at IEEE Photovoltaic Specialists Conference, Las Vegas, Nevada (7–11 Oct. 1991).
26. C. M. Katzan and J. L. Edwards, *Lunar Dust Transport and Potential Interactions with Power Systems Components*, NASA Contractor Report 4404 (Nov. 1991).
27. W. D. Harkins, *The Physical Chemistry of Surface Films*. Rheinhold, New York (1952).
28. J. D. Halajian, in: *The Lunar Surface Layer — Materials and Characterization*, J. W. Salisbury and P. E. Glaser (Eds), p. 67. Academic Press, New York (1964).
29. G. D. Sjaastad, in: *The Lunar Surface Layer — Materials and Characterization*, J. W. Salisbury and P. E. Glaser (Eds), p. 23. Academic Press, New York (1964).
30. L. D. Jaffe, 'Results of the *Surveyor* Lunar Landings', Paper presented at the Symposium on the Physics of the Moon and Planets, in Sbornik 'Fizika Luny i Planet'. Izdatel'stvo 'Nauka' Moskva, (1969).
31. A. A. Levinson and S. R. Taylor, in: *Moon Rocks and Minerals*, p. 34. Pergamon Press, New York (1971).
32. N. C. Costes and J. K. Mitchell, in: *Proceedings of the Apollo 11 Lunar Science Conference*, A. A. Levinson (Ed.), Vol. 3, p. 2025. Pergamon Press, New York (1970).
33. J. J. Grossman, J. A. Ryan, N. R. Mukerjee and M. W. Wegner, in: *Proceedings of the Apollo 11 Lunar Science Conference*, A. A. Levinson (Ed.), Vol. 3, p. 2171. Pergamon Press, New York (1970).
34. J. A. Ryan, *J. Geophys. Res.* **71**, 4413 (1966).
35. J. A. Ryan, J. J. Grossman and W. M. Hansen, *J. Geophys. Res.* **73**, 6061 (1968).
36. J. W. Salisbury, P. E. Glaser, B. A. Stein and B. Vonnegut, *J. Geophys. Res.* **69**, 235 (1964).
37. L. H. Lee, in: *Adhesive Bonding*, L. H. Lee (Ed.), p. 1. Plenum Press, New York (1991).
38. D. Hays, in: *Fundamentals of Adhesion*, L. H. Lee (Ed.), p. 249. Plenum Press, New York (1991).
39. L. H. Lee, *J. Adhesion Sci. Technol.* **5**, 71 (1991).
40. L. H. Lee, *J. Adhesion* **37**, 187 (1992).
41. A. D. Zimon, *Adhesion of Dust and Powder* (Engl. edn). Plenum Press, New York (1969).
42. H. Krupp, *Adv. Colloid Interface Sci.* **1**, 111 (1967).
43. N. S. Goel and P. R. Spencer, in: *Adhesion Science and Technology*, L. H. Lee (Ed.), Part B, p. 763. Plenum Press, New York (1975).
44. F. M. Fowkes, *J. Adhesion Sci. Technol.* **1**, 7 (1987).
45. R. S. Drago, G. C. Vogel and T. E. Needham, *J. Am. Chem. Soc.* **93**, 6014 (1971).
46. J. M. D. Vrbanac and J. C. Berg, in: *Acid–Base Interactions: Relevance to Adhesion Science and Technology*, K. L. Mittal and H. R. Anderson, Jr (Eds), p. 67. VSP, Zeist, The Netherlands (1991).
47. B. V. Deryagin, I. N. Aleinikova and Yu. P. Toporov, *Powder Technol.* **2**, 154 (1968/69).
48. A. E. Uber, J. F. Hoburg and G. Penney, *IEEE Trans. Ind. Appl.* **20**, 148 (1984).
49. R. Allen Bowling, *J. Electrochem. Soc.* **132**, 2208 (1985).
50. D. K. Donald and P. K. Watson, *Photogr. Sci. Eng.* **14**, 36 (1970).
51. D. K. Donald, in: *Recent Advances in Adhesion*, L. H. Lee (Ed.), p. 129. Gordon and Breach, New York (1973).
52. M. H. Lee and J. Ayola, *J. Imaging Technol.* **11**, 279 (1985).
53. L. H. Lee, *Photogr. Sci. Eng.* **22**, 228 (1978).
54. A. R. Akande and J. Lowell, *J. Phys. D* **20**, 565 (1987).

55. J. D. Nelson and E. Vey, *J. Geophys. Res.* **73**, 3471 (1968).
56. L. H. Lee, in: *Fundamentals of Adhesion*, L. H. Lee (Ed.), p. 1. Plenum Press, New York (1991).
57. H. R. Anderson and R. H. Manka, in: *Electromagnetic Exploration of the Moon*, W. I. Linlor (Ed.), p. 117. Mono Book, Baltimore (1970).
58. J. Henniker, *Nature* **196**, 474 (1962).
59. L. H. Lee, *J. Electrostatics* **32**, 1 (1994).
60. C. F. Gallo and W. L. Lama, *IEEE Trans.* **1A-12**, 7 (1976).

Fundamentals of Adhesion and Interfaces, pp. 95–108
D. S. Rimai, L. P. DeMejo and K. L. Mittal (Eds)

The adhesion-induced deformation and the removal of submicrometer particles

S. KRISHNAN,[1] A. A. BUSNAINA,[2] D. S. RIMAI[3,*] and L. P. DEMEJO[3]

[1] *Department of Chemical Engineering, Clarkson University, Potsdam, NY 13699, USA*
[2] *Department of Mechanical Engineering, Clarkson University, Potsdam, NY 13699, USA*
[3] *Eastman Kodak Company, Rochester, NY 14653-6402, USA*

Revised version received 28 April 1994

Abstract—Adhesion-induced deformations of submicrometer polystyrene particles on silicon substrates were observed as a function of time using scanning electron microscopy. The contact area between the particle and the substrate was found to increase with time for a period of approximately 72 hours before reaching a constant value. The ratio of the final contact radius to the particle radius was ≈ 0.4. The time dependence of this deformation appears similar to the creep phenomenon in bulk polymers. These results are related to the studies of particle removal conducted for different time periods, using hydrodynamic and centrifugal removal forces. The removal efficiency was found to decrease with time. This correlates well with the increase in the adhesion force on the particles with time as observed from the SEM measurements. The effect of the particle diameter on the removal efficiency and the correlation between the time dependent adhesion-induced deformation and particle removal efficiency is discussed.

Keywords: Adhesion; particles; adhesion-induced deformation.

1. INTRODUCTION

The adhesion of particles to surfaces and their removal from the surface is important in the semiconductor industry, xerography, pharmacology, etc. For example, in the photocopying process, the adhesion and deformation of the toner particles is crucial to the quality of the image transfer [1]. Similarly, submicrometer particle contamination adversely affects the yields of several processes in the semiconductor industry.

The adhesion of fine particles to surfaces is a complex phenomenon. The surface forces in play can cause deformation of the particle and/or the substrate. Depending on the material properties of the substrate and the particle and the magnitude of the surface-force induced stresses, either elastic or plastic deformation can occur. These deformations result in an increase in the contact area between the particle and the substrate, thereby, further increasing the adhesion force on the particle.

*To whom correspondence should be addressed.

Many studies have been conducted using various methods to detach particles from surfaces [2–4]. Efficient removal of the particles from surfaces requires knowledge of the magnitude of the forces involved and the most effective way of applying them. Experimental studies on the removal of particles and a detailed review of particle/surface interactions have been widely reported [5–11]. Visser [12] investigated the removal of cellulose particles adhered to a surface using centrifugal forces and proposed the criterion of 50% particle removal as a measure of the adhesion force. Hubbe [5] used hydrodynamic forces to remove particles (hydrous oxide) from flat substrates (cellulose and glass). Kurz [13] used a rotating disk (silicon wafer) to generate the hydrodynamic force required to remove 1 μm and larger diameter particles. Removal rates above 90% were reported for particles larger than 2 μm. Taylor, Busnaina and coworkers [13–15] measured the removal force for submicrometer particles on silicon substrates and correlated it with the theoretical van der Waals adhesion force. However, very few studies have tried to experimentally quantify the removal force and to relate this to the particle–substrate contact area. Moreover, the correlation between the removal forces and particle deformation as a function of time has not been studied.

This paper presents the observations of time dependent adhesion-induced deformation of polystyrene latex particles on silicon substrates for periods ranging from approximately 1 hour to 2 weeks. In addition, the removal force for the particles from a silicon substrate was measured and the removal efficiency for several particle diameters was evaluated. The effect of adhesion-induced deformation on the removal efficiency is reported and a qualitative comparison is made between the results from the particle deformation experiments and those from the particle removal experiments.

2. THEORY

2.1. Adhesion and deformation forces

The adhesion forces can be categorized as follows [16]:

a) The adhesion forces that prevail in the region of contact and in the peripheral area such as the van der Waals and electrostatic forces.

b) The adhesion forces that result from chemical bonding as well as intermediate energy bonding such as hydrogen bonding; and

c) The adhesion forces caused by interfacial reactions such as diffusion, condensation and diffusive mixing.

The primary force of adhesion for small ($d < 50$ μm) uncharged particles on uncharged substrates is frequently attributed to van der Waals (vdW) interactions [17]. For the case of an undeformed sphere and rigid plane, this force is proportional to the radius of the sphere [18]. The magnitude of the force of adhesion between a particle and a substrate depends on the contact area between them. Deformations of the contacting materials can, therefore, increase the adhesion forces [16, 17]. According to Bowling [17], when a sphere and a flat substrate come into contact with each other, the attractive force (F_{vdW}) deforms the interface and a circular adhesion area is formed between the adherents. The total adhesion force consists of two additive components, namely, the force acting between the adherents before deformation at the instant of first contact, $F(\text{vdW})$, and the force acting on the contact area due to the deformation, $F(\text{vdW deform.})$:

$$F(\text{vdW}_{\text{total}}) = F(\text{vdW}) + F(\text{vdW}_{\text{deform}}). \tag{1}$$

It has been shown that significantly large deformations can occur as a result of the surface forces [19].

The theoretical formulations for the deformation of particles and/or substrates has been well documented [20–23]. Derjaguin calculated the contact area assuming van der Waals force of interaction between the two surfaces, and assuming the particle to be a Hertzian indentor [21]. Accordingly, for a Hertzian indentor of initial radius R, the contact radius, a, under an applied load P^0 is given by

$$a^3 = \frac{3}{4} P^0 \frac{(1-\nu^2)}{E} R, \tag{2}$$

where E and ν are the Young's modulus and Poisson's ratio of the substrate, respectively. Assuming that the van der Waals interaction is the only load imposed on the system it can be shown that [16]

$$P^0 = \frac{\overline{h\omega}}{8\pi z_0^2} R, \tag{3}$$

where $\overline{h\omega}$ is the Lifshitz–van der Waals constant and z_0 is the separation between the particle and the substrate (typically 4 Å for van der Waals bonded crystals). Hence, the Derjaguin model predicts that,

$$a^3 = \frac{3\overline{h\omega}}{32\pi z_0^2} R^2. \tag{4}$$

Krupp [16] modified the Derjaguin model to account for large adhesion-induced stresses that exceed the elastic limits of at least one of the contacting materials. He postulated that the contact area could be divided into two coaxial regions of stress. The inner zone being subjected to greater stresses would undergo plastic deformation whereas the outer zone with the lower stress would deform elastically. According to this model the radius of the inner zone a_0, and radius of the total contact area a_1, are related to the Young's modulus, E, and a time dependent hardness, $H(t)$, by

$$\pi a_0^2 = \frac{P^0}{H(t)} + \frac{1}{3}\left[\frac{\pi^{3/2}(1-\nu^2)RH(t)}{2E}\right]^2, \tag{5}$$

$$\pi a_1^2 = \frac{P^0}{H(t)} - \frac{2}{3}\left[\frac{\pi^{3/2}(1-\nu^2)RH(t)}{2E}\right]^2. \tag{6}$$

The hardness, $H(t)$, was assumed to reach a constant value H, within a period of 30 minutes with a steady state value of H being approximately of the order of $10^{-3}\,E$ [16]. This ensures that the contact area reaches a constant value.

Johnson, Kendall and Roberts (JKR), using a thermodynamic approach, incorporated the contributions of the tensile interactions to predict contact areas of spherical particles on planar substrates [22]. Accordingly, the radius of the contact area is given by

$$a^3 = \frac{6w_A \pi R^2}{K}, \tag{7}$$

where

$$K = \frac{4}{3\pi(k_1 + k_2)}, \qquad \text{where} \quad k_i = \frac{1 - \nu_i^2}{\pi E_i}. \tag{8}$$

The Young's modulus and Poisson's ratio are given by E_i and ν_i, respectively, of the i_{th} material. The work of adhesion w_A is given by

$$w_A = \gamma_1 + \gamma_2 - \gamma_{12}, \tag{9}$$

where γ_1, γ_2, γ_{12} are the surface energies of the two materials and their interfacial energy, respectively. Maugis and Pollock [23] extended the JKR model to include plastic deformation of the contacting materials. They postulated that the contact radius is related to the externally applied load, P, the hardness H, of the deforming material, the work of adhesion, w_A, and the particle radius R, by,

$$P + 2\pi w_A R = \pi a^2 H. \tag{10}$$

DeMejo *et al.* [24–26] and Rimai *et al.* [27–31] directly observed the adhesion-induced deformation behavior of several particle–substrate systems using scanning electron microscopy (SEM). For a system of polystyrene particles on silicon substrates, Rimai *et al.* [28] found the contact radius to vary as the particle radius to the 0.42 ± 0.13 power. This power law dependence was consistent with the plastic deformation of the particles based on the magnitude of the surface forces generated in this system. Bowen *et al.* [30] studied the surface force induced deformation between glass particles and polyurethane substrates and found that any time dependence for the deformation in those studies would have to occur within a few minutes.

All the models discussed above assume that the particle is in thermodynamic equilibrium with the substrate. In order to ensure equilibrium, Rimai and coworkers [31] would, typically, allow the particles to remain in contact for approximately 2 weeks before measuring the contact radius. It many cases, however, the adhesion of particles to surfaces after short times may be important.

2.2. *Removal forces*

The particle removal studies were conducted by rotating a silicon wafer contaminated with particles about its axis while a water jet simultaneously impinged on the wafer

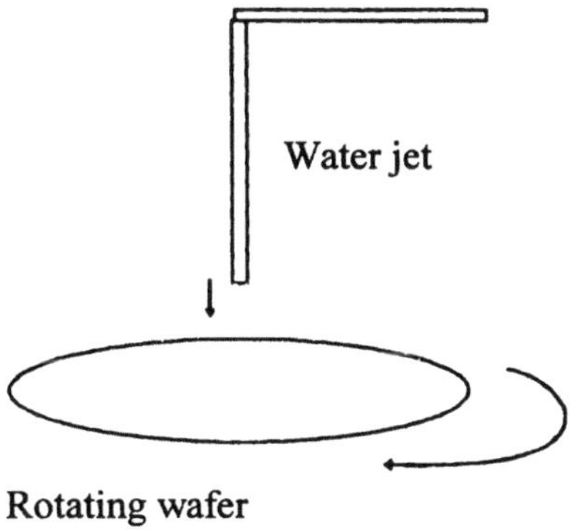

Figure 1. Schematic of the experimental setup for the particle removal experiments.

surface. The schematic for the experimental setup is shown in Fig. 1. As the fluid flows across the spherical particle, the difference in pressure across the surface of the particle gives rise to the Stokes drag force (F_D) which is given as shown below.

For laminar flow [32]

$$F_D = 3\pi\mu d V. \tag{11}$$

For turbulent flow [33]

$$F_D = 8\rho\nu^2\left(\frac{dU}{\nu}\right)^2, \tag{12}$$

where d is the particle diameter, μ and ρ are the viscosity and density of the fluid, respectively, and V is the relative velocity between the particle and the fluid.

The local velocity and the drag coefficient [34, 35] are used to determine the local drag force on a spherical particle. Summation of these local drag forces over the sphere results in the total drag force. Shear stresses in the flow give rise to the hydrodynamic lift force or Saffman lift force [11]. The hydrodynamic lift force is determined based on the shear velocity in the boundary sublayer [11, 13] and is given as follows:

For laminar flow [32]

$$F_L = 1.62\mu d^2\left(\frac{1}{\nu}X\frac{du}{dx}\right)\frac{1}{2}V. \tag{13}$$

For turbulent flow [33]

$$F_L = 10\rho\nu^2\left(\frac{dU}{\nu}\right)^3, \tag{14}$$

where U is the shear velocity of the flow.

Finally, as the wafer is rotated about its axis, a particle on the surface experiences a centrifugal force due to the angular movement caused by the rotation. Thus, the drag force, the hydrodynamic lift force, and the centrifugal force constitute the total removal force used in particle removal by rotational rinsing.

2.2.1. Removal percentage. In this study the adhesion force generated contact area and the particle removal force as a function of time for periods ranging between 1 hour and 360 hours were measured. The removal percentage at a given rotational speed can be used to calculate the average force needed to separate particles of a given size and material from a surface. Quantitatively, this force is related to the particle adhesion by [36]:

$$\text{Removal\%} = \frac{n_{\text{before}} - n_{\text{after}}}{n_{\text{before}}} \times 100 = 53.4X\frac{F_{\text{removal}}}{F_{\text{adhesion}}} + 40.23, \tag{15}$$

where n is the number of particles.

3. EXPERIMENTAL PROCEDURE

3.1. Adhesion-induced deformation

Polystyrene latex (PSL) spherical particles having nominal radii of 0.5 μm, obtained from Duke Scientific, Inc. were used in this study. The substrate comprised of 1 cm by 1 cm pieces of a polished silicon wafer. Prior to deposition of the particles, the Si substrate was cleaned with a 1:1:5 solution (SC1) of ammonium hydroxide, hydrogen peroxide and DI water, to remove undesirable Si particles, generated while cutting the wafer, and organic contamination. The Si substrates were then rinsed with distilled water (18MΩ) and then taken to a Class 10 cleanroom. In the cleanroom, the silicon substrates were rinsed with isopropyl alcohol to speed up the drying of their surfaces and to remove any traces of water. A few drops of the highly concentrated (10^{10}/ml) PSL [polystyrene-latex] suspension were diluted with $\sim$ 6 ml of isopropyl alcohol (IPA) to form a dilute solution. The particles were deposited onto the substrate by atomizing the particle suspension using a nebulizer. The particle deposition was carried out at 1 atm pressure for a period of 2 minutes at 45% relative humidity. It was found that under these conditions the IPA evaporated prior to the deposition of the individual particles and that particle agglomeration was minimal. After the deposition the samples were kept in uncovered Petri dishes in the clean room for the desired time interval before the contact radii measurements.

The contact radii between the polystyrene particles and the silicon substrate were measured using a scanning electron microscope (SEM) (JEOL JSM 6300) as shown in Fig. 2. To avoid the surface charging of the PSL particles by the scanning electron beam, they were coated with gold ($\approx$ 40 Å) immediately prior to viewing. The gold coating was sputtered using argon at pressure of 0.1 Torr and at a constant current of 20 mA for 3 minutes.

In addition, the contact radii were also determined, using low voltage SEM operating near the unity point, on uncoated samples to ensure that any effects of the coating on the apparent contact radius could be taken into account.

After sputtering, the sample was mounted vertically on a sample holder and placed in the SEM. Observations of the contact region were made with the normal to the

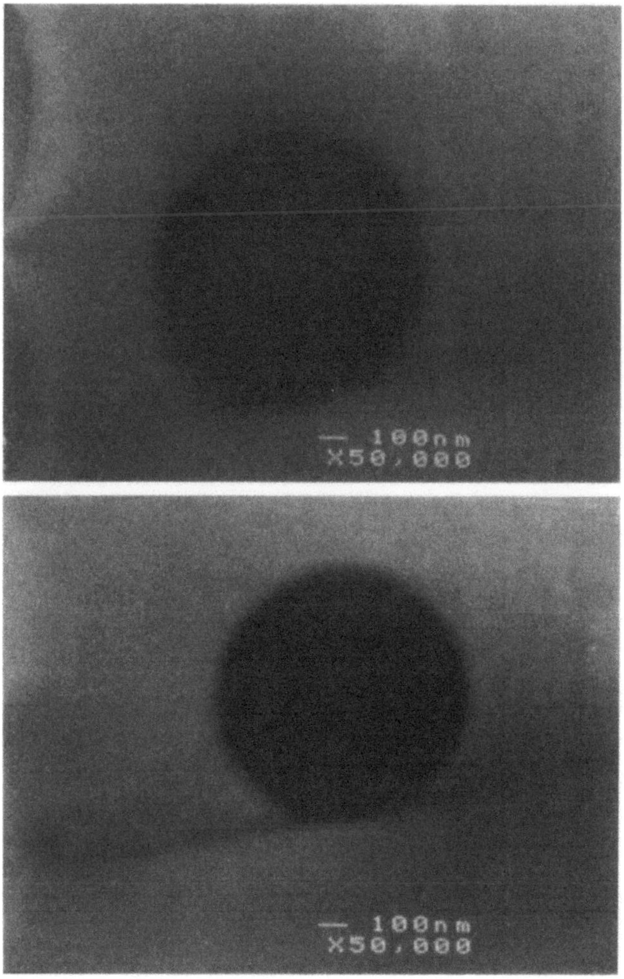

Figure 2. SEM micrographs of polystyrene spheres (r = 0.5 μm) on silicon substrates observed at different times after deposition, a) 1 hour, b) 72 hours, c) 360 hours.

plane of the sample substrate tilted at an angle of 86°–88° from the beam. A beam voltage of 15 kV was used to observe the PSL particles. The total elapsed time between the deposition of the particles on the substrate and initial observations was 60 minutes. The observation times of the individual particles were less than 2 minutes to ensure minimal occurrence of artifacts due to beam induced Joule heating. The contact diameter of the PSL particles on the Si substrates was measured directly from the micrographs of these particles. The contact areas of several spheres in contact with the substrate for comparable periods of time were averaged. The results are presented in Table 1. Figure 3 shows the dependence of the contact area as a function of time for periods ranging from 1 hour to 360 hours.

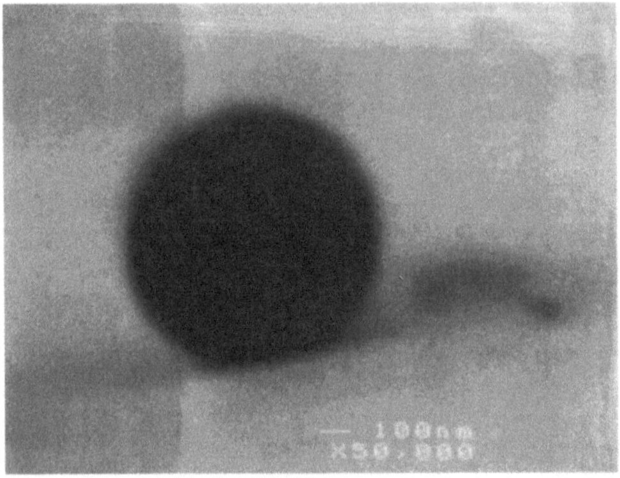

Figure 2. (Continued).

Table 1.
Contact area measurements for the spherical particles ($r = \sim 0.5\ \mu m$)

Time (hours)	No. of spheres	Contact area (μm^2)	Standard deviation
1	7	0.06	0.0090
20	6	0.08	0.0081
72	6	0.11	0.0230
96	4	0.13	0.0095
168	7	0.12	0.0150
360	7	0.12	0.0220

As previously indicated, a few of the particles were not sputter coated with gold in order to assess the effect of the coating on the apparent contact radius. The average contact diameters determined for the gold coated particles and the uncoated particles were approximately the same for any given contact time as shown in Table 2. Therefore, it was not necessary to correct the measurements of the apparent contact radius for coating effects, as had to be done in several previous studies [31].

3.2. *Particle removal*

Particles (PSL spheres) of 0.5 μm radius were used in this particle removal study. Prior to the particle deposition, a clean silicon wafer was inspected to detect surface defects and possible surface contamination using a laser surface scanner (Particle Measuring Systems) [36]. The particles (PSL spheres) were deposited on the silicon wafer and the wafer was scanned to obtain the particle count before the hydrodynamic force was applied. Using a spin coater the wafer was rotated at a known speed while a jet of deionized water gently impinged on the center maintaining a water film of

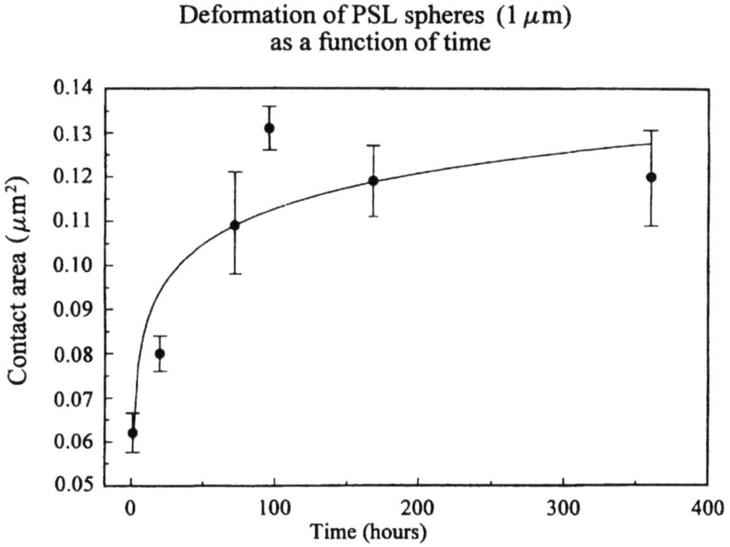

Figure 3. The contact area between the polystyrene spheres ($r = 0.5$ μm) and the silicon substrate as a function of time.

Table 2.
Effect of the gold coating on the contact diameter between the polystyrene particles and the silicon substrate

Time (hours)	No. of particles	Average contact diameter without gold coating (μm)	No. of particles	Average contact diameter with gold coating (μm)
20	2	0.31	2	0.33
72	4	0.37	2	0.36
360	3	0.39	3	0.39

constant thickness. The wafer was dried using the spin coater by rotating at a high speed. The wafer was then scanned to obtain the particle count and distribution. For this set of experiments, particles were deposited on silicon wafers and stored in the clean room. Particle removal experiments were carried out as described above on these samples at different time periods, to study the effect of the adhesion-induced deformation of the particles on the removal efficiency.

4. RESULTS AND DISCUSSION

4.1. Particle deformation

The contact area between the sphere and the silicon substrate was found to increase with time logarithmically as seen from Fig. 3. The relationship between the contact area and time is given by:

$$\text{Contact area} = 0.059 + 0.012 \ln(t) \quad (16)$$

with a time constant, Y, equal to 148 hours (6.2 days). The ratio of the final radius of the contact area to the particle radius was ~ 0.4.

Assuming that in view of [31], the deformation of the polystyrene sphere is plastic in nature, the work of adhesion, w_A, was calculated from the Maugis–Pollock theory using the equation (10). For the value of the contact radius within one hour of deposition, the work of adhesion, w_A is found to be 0.72 J/m^2, utilizing the mechanical properties of a similar polystyrene–silicon system [28].

4.2. Particle removal

The force needed to remove particles from substrates has been determined in a manner similar to that described earlier in this paper for previous studies [5–12]. In those studies the authors typically assumed that at the point where 50% of the particles were removed the adhesion force equalled the removal force [11, 12]. However, as discussed by Busnaina *et al.* [36], it would be expected that, if the removal force is greater than the adhesion force, nearly all the particles should be removed (assuming that the particles are monodisperse particles having the same separation distance from the substrate).

The effect of time on the removal efficiency of 0.5 μm radius particles is shown in Fig. 4. As can be seen, removal efficiency decreases with time according to the relation:

$$\text{Removal\%} = 87.27 - 11.06 \ln(t). \quad (17)$$

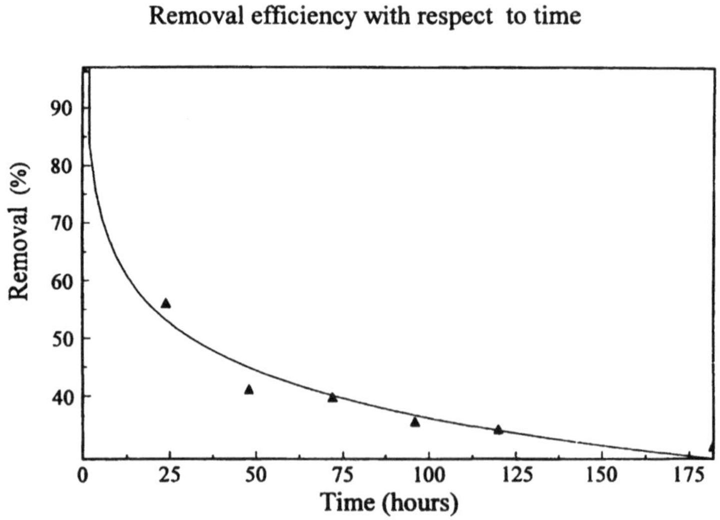

Figure 4. The removal efficiency for the 0.5 μm radius polystyrene spheres from silicon substrate as a function of time.

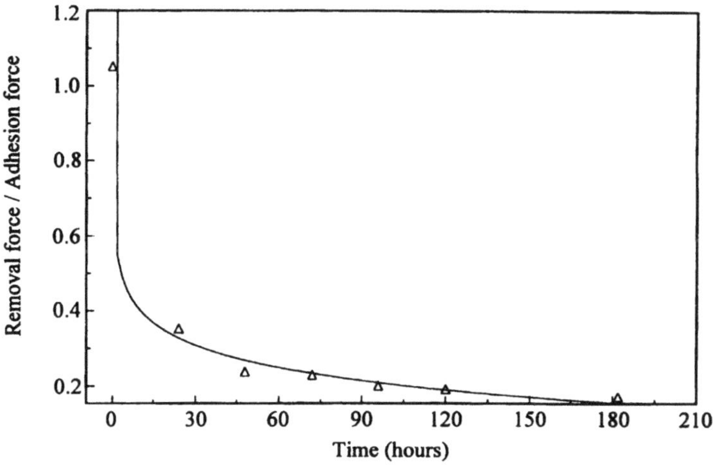

Figure 5. The force ratio ($F_{\text{removal}}/F_{\text{adhesion}}$) as a function of time.

In this instance the characteristic time constant was found to be 164 hours (6.8 days), which compares favorably with the time constant found for the rate of deformation.

From the similarity observed in the shape of the two curves and the agreement with the values of the time constants, it would appear that the decrease in removal efficiency with time correlated to the increasing amount of particle deformation induced by the adhesion force.

The ratio of the removal force to the adhesion force, i.e. the force ratio, was calculated from equation (15). The force ratio plotted with respect to time for 0.5 μm radii particles is shown in Fig. 5. The decrease in the force ratio with time implies an increase in the adhesion force (the removal force is constant) with time. This is consistent with Bowling's analysis that particle deformation causes an increase in the adhesion and consequently the removal forces [17].

The pressure exerted by the silicon on the polystyrene particles due to the surface forces can be calculated following the analysis by Tabor [6]. Assuming van der Waals interactions, the initial attractive force of the undeformed materials is related to the work of adhesion by [6]

$$F_0 = 2\pi R w_{\text{A}}. \tag{18}$$

The increase in the surface force due to the finite contact area can be estimated as reported by Bowling [17]. Accordingly, the force of attraction of the deformed particle F_{d} is related to F_0 by

$$F_{\text{d}} = \frac{F_0 a^2}{z_0 R}, \tag{19}$$

where z_0 is the separation distance for a van der Waals bonded solid (≈ 4 Å). Upon dividing equation (19) by the contact area, the average pressure, p, on the Si wafer is

$$p = \frac{2w_{\text{A}}}{z_0}. \tag{20}$$

Substituting the experimentally obtained value for the work of adhesion, the pressure p is found to be 3.6×10^9 Pa. This is close to the value of Young's modulus for crosslinked polystyrene (2.55×10^9 Pa [28]) and is in excess of its yield point. Hence, the observed plastic response of the particles is consistent with the magnitude of the stresses involved.

Finally, the total adhesion force for the 0.5 μm radius particles, calculated using the contact areas measured from the SEM photographs, is plotted as a function of time in Fig. 6. Figures 5 and 6 show that the change in the adhesion force (calculated using the SEM contact area) agrees with the trend in the measured force ratio with time. The actual magnitude of the adhesion force on the particles will be different for each of the systems involved. For the SEM observations, the particles and substrates constitute a dry system, whereas the particle removal experiments involved an immersed system. The van der Waals interactions (van der Waals–Lifshitz constants) are different for the dry and immersed systems and the corresponding adhesion forces for an immersed system are smaller than those for a dry system [17]. Hence, a qualitative comparison of the results, at this juncture, from these two different systems indicates a good agreement. The Maugis and Pollock model [23] was utilized to predict the contact areas of the PSL spheres on the silicon substrates. The initial contact areas of the particle (within 60 minutes of deposition), incorporating the necessary parameters from a similar polystyrene–silicon system [28], were predicted reasonably well by this model. However, the final contact areas for the time period (15 days) reported in this study could not be predicted. For such a long time period, material creep should have begun to occur for the polystyrene spheres and this phenomenon is not accounted for in the theoretical models [20–23]. The radius of the contact area after 24 hours, for the 0.5 μm radius spherical particles adhered to silicon substrates, is approximately 0.16 μm. This value is reasonably well predicted by the dependence of the radius of the contact area on the particle radius, determined approximately 24 hours after the deposition, by Rimai *et al.* [28] for the polystyrene–silicon system.

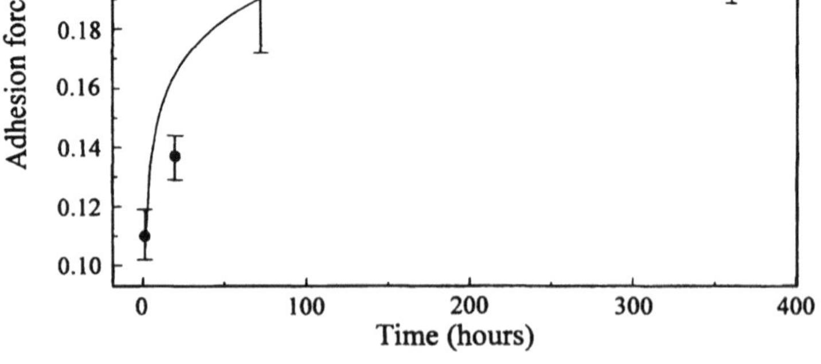

Figure 6. The van der Waals adhesion force as a function of time for the 0.5 μm radius polystyrene particles.

5. CONCLUSIONS

The contact radius of nominal 0.5 μm latex polystyrene spheres in contact with a silicon wafer substrate was found to increase logarithmically with time with a time constant of 148 hours. This result is consistent with plastic flow of the particles due to the forces of adhesion. The work of adhesion, calculated assuming the validity of the Maugis–Pollock model of adhesion, was found to be 0.72 J/m^2. The shape of the time dependence of the contact radius and its time constant were found to be similar to that of the particle removal efficiency, suggesting that the increase in the force necessary to remove these particles from the substrate may be due to the increasing contact area.

Acknowledgements

The work on the adhesion-induced deformation of particles was funded by the Eastman Kodak Company, Rochester, NY. The authors would like to thank Mr Bill Plunkett for his assistance in taking the SEM photographs. The work on particle removal was initially funded by the IBM-General Technology Division, Essex Junction, Vermont and the Center for Particulate Control in Process Equipment at Clarkson University.

REFERENCES

1. C. J. Mastrangelo, *Photogr. Sci. Eng.* **26**, 194 (1982).
2. K. L. Mittal (Ed.), *Particles on Surfaces 1: Detection, Adhesion, and Removal.* Plenum Press, New York (1988).
3. K. L. Mittal (Ed.), *Particles on Surfaces 2: Detection, Adhesion, and Removal.* Plenum Press, New York (1989).
4. K. L. Mittal (Ed.), *Particles on Surfaces 3: Detection, Adhesion, and Removal.* Plenum Press, New York (1991).
5. M. Hubbe, Detachment of Colloidal Hydrous Oxide Spheres from Flat Surfaces Exposed to Flow, PhD Dissertation. Clarkson University (1983).
6. D. Tabor, *J. Colloid Interface Sci.* **58**, 2 (1977).
7. A. D. Zimon, *Adhesion of Dust and Powder.* Plenum Press, New York (1969).
8. M. B. Ranade, *Aerosol Sci. Technol.* **7**, 161–176 (1987).
9. W. C. Hinds, *Aerosol Technology.* John Wiley, New York (1982).
10. R. P. Musselman and T. W. Yarbrough, *J. Environmental Sci.*, 51–56 (1987).
11. P. G. Saffman, *J. Fluid Mech.* **22**, 385–400 (1965).
12. J. Visser, in: *Surface and Colloid Science*, E. Matijevic (Ed.), Vol. 8. John Wiley and Sons, New York (1976).
13. M. Kurz, Measurement of Detachment Forces of Particles on a Silicon Substarte, MS Dissertation. Clarkson University (1988).
14. J. Taylor, Measurement of Detachment Forces for Submicron Particles on a Silicon Wafer, MS Dissertation. Clarkson University (1990).
15. A. Busnaina, F. Kern and R. Kunesh, in: *Proceedings, Inst. Environ. Sci., 36th Annual Technical Meeting*, pp. 422–426. New Orleans, Louisiana, April 23–27 (1990).
16. H. Krupp, *Adv. Colloid Interface Sci.* **1**, 111 (1967).
17. R. A. Bowling, in: *Particles on Surfaces 1: Detection, Adhesion and Removal*, K. L. Mittal (Ed.), pp. 129–142. Plenum Press, New York (1988).
18. H. C. Hamaker, *Physica* **4**, 1058 (1937).
19. D. S. Rimai, L. P. DeMejo and R. C. Bowen, *J. Appl. Phys.* **66**, 3574 (1989).

20. R. S. Bradley, *Philos Mag.* **13**, 853 (1932).
21. B. V. Derjaguin, *Kolloid Z.* **69**, 155 (1934).
22. L. Johnson, K. Kendall and A. D. Roberts, *Proc. R. Soc. London* **A324**, 301 (1971).
23. D. Maugis and H. M. Pollock, *Acta Metall.* **32**, 1323 (1984).
24. L. P. DeMejo, D. S. Rimai and R. C. Bowen, *J. Adhesion Sci. Technol.* **5**, 959 (1991).
25. L. P. DeMejo, D. S. Rimai, J. Chen and R. C. Bowen, *J. Adhesion* **39**, 61 (1992).
26. L. P. DeMejo, D. S. Rimai and R. C. Bowen, *J. Adhesion Sci. Technol.* **2**, 331 (1988).
27. D. S. Rimai, L. P. DeMejo, W. B. Vreeland, R. C. Bowen, S. R. Gaboury and M. W. Urban, *J. Appl. Phys.* **73**, 668 (1993).
28. D. S. Rimai, R. S. Moore, R. C. Bowen, V. K. Smith and P. E. Woodgate, *J. Mater. Res.* **8**, 662 (1993).
29. D. S. Rimai, L. P. DeMejo and R. C. Bowen, *J. Appl. Phys.* **65**, 755 (1989).
30. R. C. Bowen, L. P. DeMejo, D. S. Rimai and W. B. Vreeland, *J. Appl. Phys.* **70**, 3360 (1991).
31. D. S. Rimai, L. P. DeMejo and R. C. Bowen, *J. Appl. Phys.* **68**, 6234 (1990).
32. M. A. Rizk and S. E. Elgobashi, *Phys. Fluids* **28** (1985).
33. J. Cleaver and B. Yates, *J. Colloid Interface Sci.* **44**, 464 (1973).
34. C. W. Oseen, Über die Stokes'sche Formel und über eine verwandte Aufgabe in der Hydrodynamik. Ark. f. Math. Astro. och. Fys. **6**, 29 (1910).
35. T. von Karman, N.A.C.A. TM No. 1092, Washington, DC (1946). Translated from 'Über laminare und turbulente Reibung,' *ZFAMM*, **1**, 233 (1921).
36. A. Busnaina, J. Taylor and I. Kashkoush, *J. Adhesion Sci. Technol.* **7**, 441 (1993).

Fundamentals of Adhesion and Interfaces, pp. 109–124
D. S. Rimai, L. P. DeMejo and K. L. Mittal (Eds)

X-ray evanescent wave-induced fluorescence study of adsorption of a sulfonated polystyrene ionomer from dimethyl sulfoxide to the solution/vapor interface

WEN PING WANG,[1] DENNIS G. PEIFFER[2] and SCOTT W. BARTON[3,*]

[1]*Department of Chemical Engineering, University of Massachusetts at Amherst, Amherst, MA 01003, USA*
[2]*Exxon Research and Engineering Co., Rt. 21 E, Annandale, NJ 08801, USA*
[3]*Polymer Science and Engineering Department, University of Massachusetts at Amherst, Amherst, MA 01003, USA*

Revised version received 9 January 1995

Abstract—We have studied the adsorption of an ionomer, poly(styrene-*ran*-manganese styrene sulfonate), from dimethyl sulfoxide solution to the solution/vapor interface at 20°C. The polymer sulfonation level was 15 mol% and the solution concentration studied was 1.00% (wt/wt). The adsorption was followed over time (4 days) by X-ray evanescent wave-induced fluorescence (XEWIF), which is a non-destructive probe of the polymer adsorption profile. We found that a state of apparent equilibrium was reached after a 4-day period and that the amount of polymer adsorbed was rather larger than anticipated from surface tension measurements. This anomaly is explained by the influence of image charge forces, which increase the surface energy and must be balanced with the surface energy reduction of placing the non-polar repeat segments at the surface. We suggest that this effect may be generalized to aqueous polyelectrolytes.

Keywords: Polymer adsorption; ionomer; X-ray; surface tension.

1. INTRODUCTION

A detailed understanding of polymer adsorption from solution is essential in applications such as coatings, colloid stabilization, surface passivation and modification, adhesion promotion, etc. Ideally, one would like to have complete control over the thickness, density, and morphology of an adsorbed polymer layer and this control may be achieved by (1) modification of the substrate surface, (2) functionalization of the polymer, or (3) changing the thermodynamic state of the bulk solution (temperature, composition, pressure). To determine the effectiveness of the different control strategies, one requires a quantitative measure of the adsorption phenomenon;

*To whom correspondence should be addressed.

this should preferably be *in situ*, so that the surface/solution equilibrium remains undisturbed. The polymer adsorption profile, which describes the local concentration of polymer as a function of the distance from the adsorbing surface, provides the most complete quantitative description. Our research goals are to understand the response of the adsorption profile to the surface/polymer/bulk modifications listed above.

The experimental determination of the adsorption profile, however, presents some difficulties; for example, conventional electron and ion probe methods are precluded due to the presence of solvent liquid or vapor. And while important progress has been made with neutron reflectivity to determine the polymer adsorption profile at the free solution/vapor interface [1] and the solid/solution interface [2], a large facility neutron source is required and deuterium labelling must be used to provide suitable scattering contrast between the polymer and the solvent. Here we describe a different approach to the problem, in which X-ray evanescent wave-induced fluorescence (XEWIF) [3–7] is used to measure the density profile of a poly(styrene-*ran*-manganese styrene sulfonate) ionomer (PSS-Mn) adsorbed to the solution/vapor interface from dimethyl sulfoxide (DMSO) (see Fig. 1). We choose to study polymer adsorption to the solution/vapor interface for several reasons. First, we avoid the presence of physical or chemical heterogeneities that can exist on a solid surface. Second, the physics of adsorption to the free liquid surface is no different than adsorption to a dielectric surface in the absence of specific interactions between the polymer and the surface. Lastly, one can measure the surface tension which provides an important complement to microscopic probes like XEWIF or neutron [8] and X-ray reflectivity [8].

In the XEWIF technique, one measures the X-ray fluorescence of heavy atoms (atomic number $Z \geqslant 14$) as a function of the penetration depth of an excitation X-ray source below an interface. The system for this study was chosen so that we could follow the profile for the manganese ($Z = 25$) associated with the polymer and sulfur ($Z = 16$) found predominantly in the solvent. The adsorption was studied at 20°C for a concentration of 1.00 wt% of polymer which had a sulfonation level of 15.0 mol%. At this sulfonation level, the polymer exhibits polyelectrolyte behavior in DMSO [9] (the dielectric constant of DMSO is 45 [10]). From surface tension measurements and physical arguments based on image charge repulsion, we expected *a priori* that little, if any, of the polymer would be adsorbed from this bulk concentration. We found,

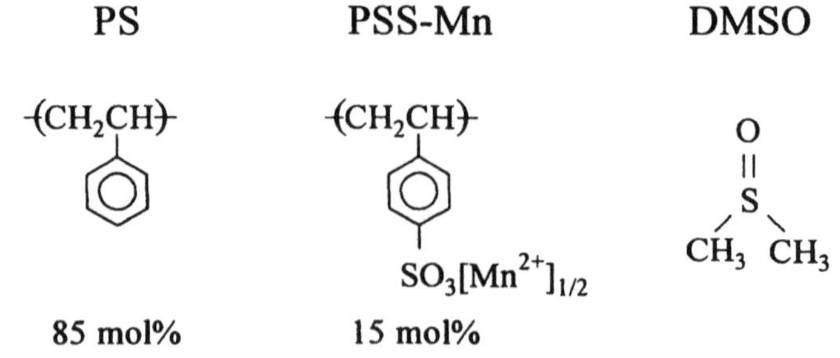

Figure 1. Chemical structure of PS, PSS-Mn, and DMSO.

however, that a rather large amount of polymer adsorbed over a period of several days. Based on our original expectation, the paradox which our results present is an important feature of this study and is discussed later.

Let us first introduce a pictorial model for polymer adsorption with the understanding that one cannot define a 'monolayer' of a polymer in the traditional sense that we use for small molecules. In the cases of interest to us, only a fraction of the total number of chain segments of an adsorbed polymer chain are actually in contact with the surface, with the remaining portions of the chain extending into the bulk solution as 'loops and tails' [11] (this is shown in Fig. 2). A characteristic thickness of the adsorbed layer is given intuitively by a surface correlation length, ξ_{surf}, over which the polymer density decays to the bulk value; in the case of a weakly adsorbed layer from dilute solution, the characteristic thickness is of the order of the polymer radius of gyration, R_g. This qualitative description may be translated into the adsorption profile, where we plot the polymer concentration (or volume fraction) as a function of the distance from the interface (Fig. 3). From the adsorption profile, it is clear

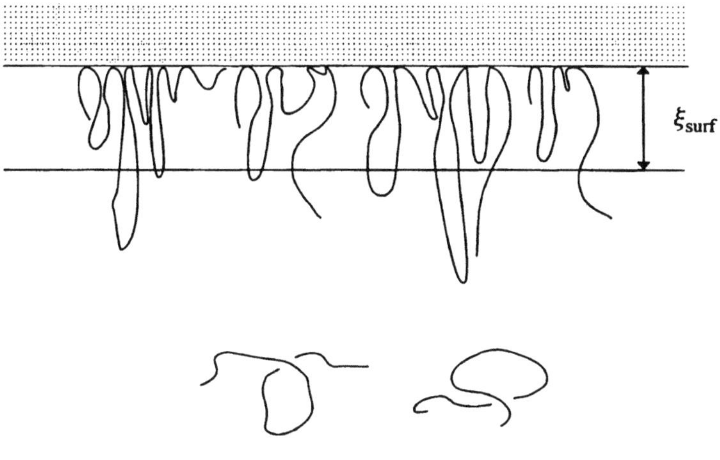

Figure 2. Schematic diagram of adsorbed polymer chains.

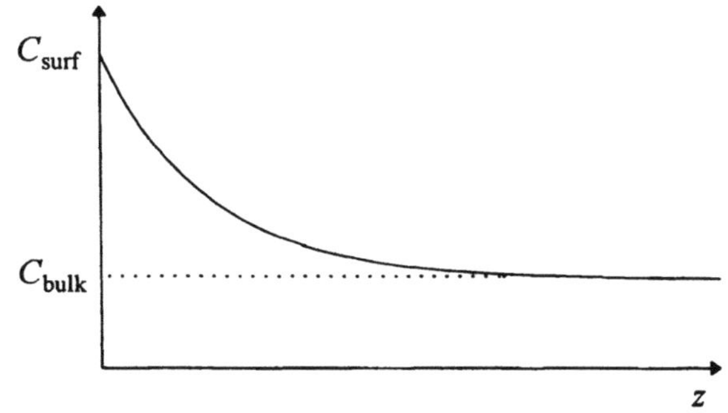

Figure 3. Example of the polymer adsorption profile.

that one can determine the number of segments in direct contact with the surface as well as the total surface excess (the latter is given by the integral over the profile). In this paper, we will concentrate our attention on the total excess as measured by the integral over a simple model adsorption profile.

2. CAHN–DE GENNES THEORY

The mean-field theoretical framework for describing polymer adsorption is based on the Cahn theory for adsorption form solution [12]; further modifications unique to polymer systems are due to de Gennes [13]. The interfacial energy for a polymer adsorbed from a single-phase solution to an external surface is given as

$$\gamma = \gamma_{\text{surf}} + \int_0^\infty \left[\Delta A + \kappa[\nabla\phi]^2\right] \mathrm{d}z, \tag{1}$$

where γ_{surf} is the enthalpic contribution which arises from favorable interactions between the polymer segments and the surface. For van der Waals-type interactions, the segment/surface potential is assumed to be a δ-function where only those segments in direct contact with the surface contribute to γ_{surf}, and this energy is usually expanded in terms of c_{surf}, the polymer concentration in the surface plane. ΔA is the cost in free energy required to create a concentration different from that of the bulk, while the polymer is kept at the same chemical potential. The gradient-squared volume fraction, $[\nabla\phi]^2$, prevents a rapid change in the polymer density away from the surface. For an adsorbed polymer, the stiffness factor, κ, is argued to be inversely dependent on the local concentration; this additional feature takes into account the entropy loss associated with stretching a chain in the vicinity of the surface [13].

While modeling studies [14] have shown that the δ-function potential for γ_s is sufficient for systems where the polymer and surface interact through van der Waals interactions, the inclusion of a long-range potential for stronger interactions (such as electrostatic or dipole–dipole) may be necessary. In particular, our results suggest the importance of these long-range interactions in strongly interacting systems, as well as the effect of image charges at the surface when one deals with charged polymers.

3. X-RAY EVANESCENT WAVE-INDUCED FLUORESCENCE (XEWIF)

To appreciate better the experimental methods described in this study, a brief introduction to X-ray optics is useful [15]. In Fig. 4, we consider an X-ray beam incident, from the vapor side, on a solution surface. The incident beam has a wavevector, **k**, of magnitude $|\mathbf{k}| = 2\pi/\lambda$, where λ is the X-ray wavelength, and θ_I is the grazing incidence angle. The behavior of the incident beam at the interfacial boundary depends on the variation in the refractive index across the interface. In the semi-classical Drüde model, the index of refraction is written as [15]

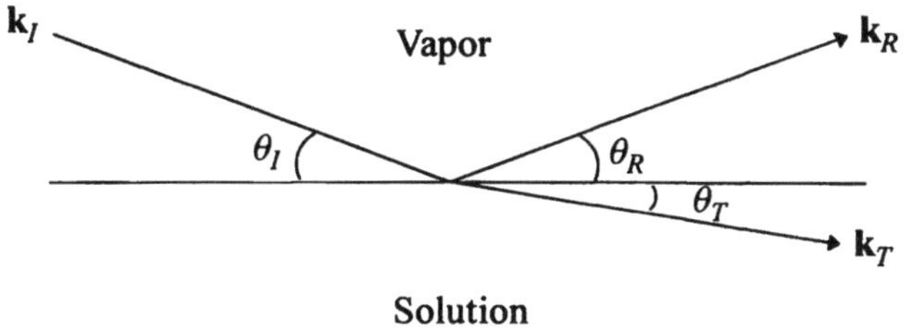

Figure 4. Geometry for XEWIF experiments.

$$n = 1 + \sum_i \frac{A_i}{E_i^2 - E_0^2}, \tag{2}$$

where the A_is are proportional to the electronic polarizability, E_0 is the incident beam energy, and the E_is are the characteristic energies (corresponding to the electronic absorption energies) of the system. The validity of this model for the X-ray refractive index has repeatedly been demonstrated experimentally [8, 16] and can be justified by virtue of the incident beam energy being much greater than the characteristic energies in the material. Since the incident beam energy is greater than greater than the characteristic electron binding energies of the system, the index will be less than 1, and the X-ray refractive index may be written as (including absorption) [15]

$$n = 1 - \delta - i\beta,$$

$$\delta = 2.7019 \times 10^{10} \lambda^2 \sum_i \frac{\rho_i (Z_i + f_i)}{M_i}, \tag{3}$$

$$\beta = \sum_i \frac{\mu_i \lambda}{4\pi},$$

where λ is the wavelength in cm, ρ_i is the partial density in g/cm^3 of the ith component, Z_i is the atomic number of the ith component, M_i is the atomic weight, and f_i is the anomalous correction term [15] for atom i. The imaginary part of the index accounts for absorption, where μ_i is the linear absorption coefficient and may be found in standard tables [17]. For organic materials, both δ and β are small corrections, of the order of 10^{-6}–10^{-7}, and this accounts for the transparency of lighter materials to X-rays.

As a consequence of the refractive index being less than unity, the X-rays transmitted into the solution from the vapor will be refracted towards the interface. If we consider the case where the solution is completely homogeneous up to a sharp interface, a critical angle for total external reflection exists at $\theta_c \approx (2\delta)^{1/2}$ [18], which is of the order of several milliradians. Under conditions of total external reflection, where $\theta \leqslant \theta_c$, the normal component of the transmitted beam becomes purely imaginary and leads to a strongly damped *evanescent* wave beneath the surface [18, 19]. At grazing incidence above the critical angle, the transmitted beam intensity decays

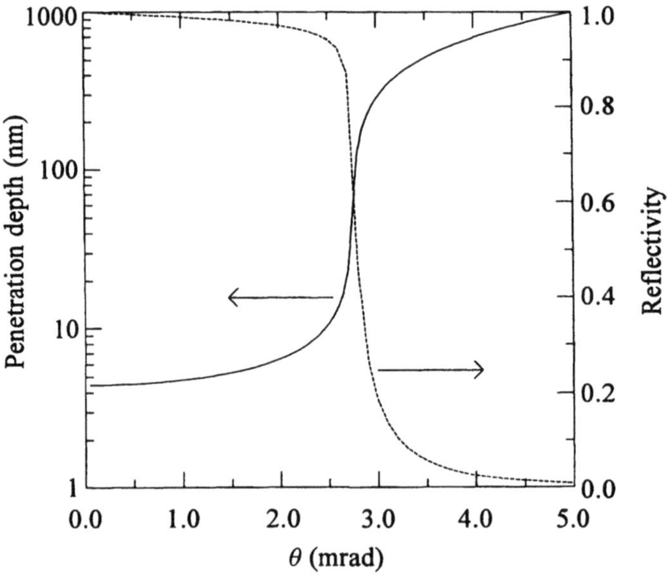

Figure 5. The penetration depth and reflectivity vs. the grazing incident angle for DMSO. The wavelength used for the calculation was 0.154 nm.

slowly due to conventional linear absorption. Hence the penetration depth, defined as the e-folding decay length for the evanescent wave, can be adjusted continuously from microscopic (nm) to macroscopic (μm) length scales by varying the incident grazing angle. For a typical organic liquid, the penetration depth of the evanescent wave can be continuously varied from about 4 nm at angles $\theta \ll \theta_c$ to effectively infinity at $\theta \gg \theta_c$.

X-ray evanescent wave-induced fluorescence (XEWIF) [3–7] exploits the depth selectivity of the transmitted X-rays as a function of the incident grazing angle. By labeling the polymer chain, solvent, or both with a heavy atom ($Z \geqslant 14$), atomic fluorescence can be excited by the transmitted beam at different penetration depths; in this way, the density profile of the adsorbing atom can be determined. In Fig. 5, we show the calculated penetration depth and X-ray reflectivity as a function of the grazing incidence angle for dimethyl sulfoxide using a wavelength of 0.154 nm (Cu $K_{\alpha 1}$). Note the limiting penetration depth of 4.4 nm. One can see the rapid change in penetration depth that occurs at the critical angle of 2.7 mrad. (Due to X-ray absorption, the reflectivity is less than 1 below the critical angle.) The analysis of the fluorescence intensity vs. angle is discussed briefly in Section 4.4.

4. EXPERIMENTAL

4.1. Materials

Anionically polymerized poly(styrene) was purchased from Polymer Laboratories, Ltd. The weight-averaged molecular weight of the polymer before sulfonation was

100 k with $M_w/M_n = 1.04$ as determined by gel permeation chromatography. The sulfonation procedure followed the description in [9]. The free acid was neutralized with manganese acetate, followed by recovery of the polymer by steam stripping. The polymer precipitate was then vacuum-dried for 36 h at 100 °C, and the weight percent sulfur was determined by Dietert analysis. Spectrophotometric grade DMSO was purchased from Aldrich and then dried over molecular sieves before use. All the solutions were prepared in a dry N_2 glove bag.

4.2. Surface tensiometry

The surface tension of the solutions was measured using either a filter paper Wilhelmy plate [20] or the du Nouy ring [20, 21]; the results were identical for both methods. A viscoelastic response [22] of the surface was observed using the du Nouy ring and this will be discussed in a separate communication. The measurements were made in an enclosed cell to ensure liquid/vapor equilibrium and the temperature of the liquid and vapor was controlled to ±0.1 °C.

4.3. X-ray spectrometer

There are several important requirements for an XEWIF spectrometer. Unlike X-ray scattering where one can allow harmonic or K_β radiation contamination, the different energies will have different penetration depths in the sample and would render the fluorescence results uninterpretable. And since most of the important surface information is contained in the fluorescence collected with the incident angle below the critical angle (2.7 mrad), the incident beam must be collimated to better than 0.05 mrad to preclude extensive resolution desmearing. Most importantly, one must have a way to steer the beam down to the liquid sample surface at the desired angle of incidence. We constructed an apparatus which satisfies these requirements, and this is shown in Fig. 6. Here we provide a brief description of the set-up; a more complete description will be provided elsewhere. The X-rays were emitted from a standard 1.5 kW fine focus X-ray tube with a Cu target anode. The beam was then totally reflected from a horizontal, Ni-coated mirror (10 mm × 50 mm) which was slightly bent to focus the X-rays on the sample, over a distance of 1.5 m. In addition to preserving as much intensity as possible, the mirror behaved as a low-pass energy filter to prevent the higher-energy harmonics from being passed by a monochromator, located downstream of the mirror. The monochromator was a channel cut, double-bounce Ge(111), which preserved the forward direction of the X-ray beam. The Cu $K_{\alpha 1}$ component (1.544 nm) of the source was selected. The dispersed $K_{\alpha 2}$ peak was eliminated by the slit in front of the beam monitor, the latter consisted of a 12 μm thick Kapton foil oriented at 45° to the incident beam, which scattered a small fraction of the incident beam into a NaI scintillation detector. The optical arrangement consisting of the mirror, monochromator, slits, and beam monitor was mounted firmly on a granite bench. To adjust the angle of incidence, we varied the relative heights of two vertical translation stages which supported the granite bench at either end along its length. The precision with which the angle could be varied was at least 0.012 mrad. When the desired tilt angle was set, the absolute heights of the bench supports were adjusted

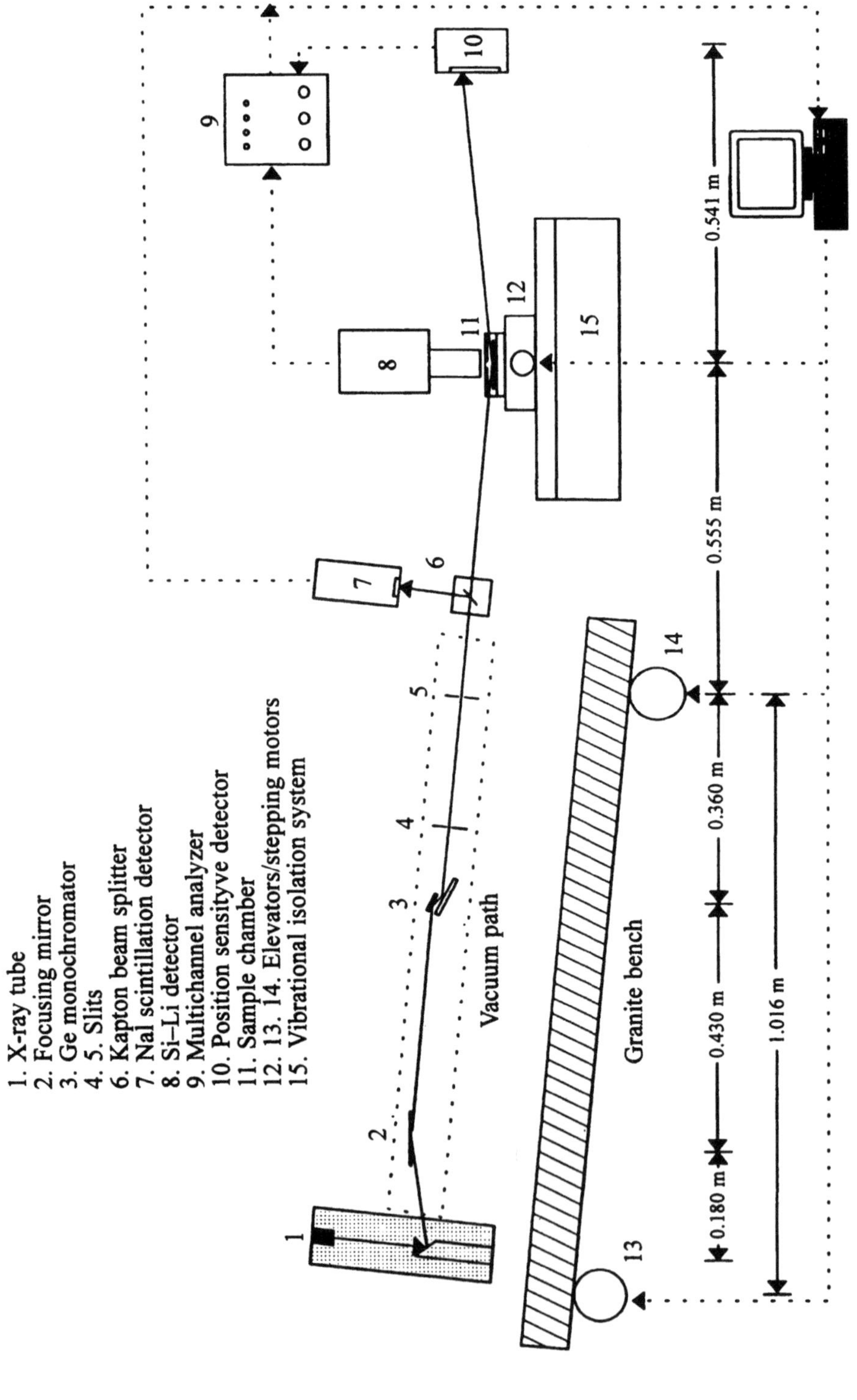

Figure 6. X-ray surface spectrometer. The incident beam monitor is composed of units 6 and 7.

so that the center of the X-ray beam would hit the same position on the liquid sample surface (underneath the energy dispersive X-ray detector) at every grazing incidence angle. A commercial Si(Li) energy dispersive detector (EDD) (EG&G ORTEC, Oak Ridge, TN) was used to collect the fluorescence intensities. The detector energy resolution was less than 180 eV at 5.9 keV and sufficient to discriminate the Mn K_α from K_β; however, it was not sufficient to separate the K_α and K_β components of the sulfur fluorescence which yielded one peak. The sample chamber and detector were kept inside a plastic glove bag purged with dry N_2 to prevent the absorption of water by the hygroscopic solution. Within the protective environment of the glove bag, 8.0 ml of solution was delivered to a Teflon sample trough (100 mm × 45 mm × 1.5 mm) inside the sample chamber. The Teflon trough sat on a water-jacketed plate used to control the temperature of the sample. The chamber was then covered with an O-ring sealed plate which had a thin Mylar window (30 mm diameter, 12 μm thick) in the center to allow the fluorescent X-rays to pass into the EDD above. At either end of the sample cell, Kapton windows allowed transmission of the incident and reflected beams, the latter of which was collected on a linear position sensitive detector (PSD) (Innovative Technology, Newburyport, MA). When collecting the fluorescence intensity at a desired angle, the sample was moved vertically through a distance of about 90 μm in 18 μm steps and the fluorescence intensity was collected at each step height. The maximum fluorescence intensity (for both sulfur and the manganese) at a fixed angle corresponded to the optimal height position for the sample and was taken as the recorded fluorescence intensity. This was done to correct for any slight misalignments at the shallow grazing incidence angles and to ensure that the maximum beam intensity was incident on the sample beneath the detector. Due to the sensitivity of the PSD efficiency to small changes in room temperature, the PSD was used primarily for alignment purposes.

4.4. Data analysis

The full data analysis required to model the adsorption profile using the observed fluorescence will be described elsewhere and we will only outline the procedure in this article. The fluorescence is given as [3]

$$I_{\text{fluor}}(\theta) = K(\lambda_{\text{excit}}, \lambda_{\text{fluor}})\, G(\theta) \int_0^\infty T(z,\theta) c_{\text{fluor}}(z)\,\mathrm{d}z, \qquad (4)$$

where $K(\lambda_{\text{excit}}, \lambda_{\text{fluor}})$ is the efficiency factor which takes into account the fluorescence yield for the species under consideration and the detector efficiency at the fluorescence wavelength, $G(\theta)$ is a geometry factor which will depend on the angle of incidence, $T(\theta, z)$ is the intensity of the transmitted beam at a depth z for a grazing incident angle θ, and $c_{\text{fluor}}(z)$ is the concentration of the fluorescent label as a function of the depth from the interface. In principle, one can measure $I_{\text{fluor}}(\theta)$ and invert the integral equation to determine $c(z)$, using a measured or calculated $K(\lambda_{\text{excit}}, \lambda_{\text{fluor}})G(\theta)$ and calculated $T(z,\theta)$. The problem is, however, ill-posed and further exacerbated by noisy data; so rather than attempting an inversion, we fit parameterized model profiles for $c(z)$ to the experimental fluorescence intensity curves. Furthermore, to eliminate the problem of determining a reliable $G(\theta)$, the data are analyzed from the Mn/S fluorescence ratio, $I_{\text{fluor}}^{\text{Mn}}/I_{\text{fluor}}^{\text{S}}$. In this way, the geometry corrections are divided out. One must still know the detector efficiency and fluorescence rate ratio for the Mn and S characteristic radiations. These were determined experimentally from the Mn/S fluorescence ratios for different mole ratios of Mn to S, using standard solutions of manganese acetate in dimethyl sulfoxide.

We note parenthetically that for the rare case in which there is no variation of the refractive index as a function of depth (when the polymer and solvent are index-matched), the transmission function may be written simply as

$$T(\theta, z) = T(\theta, 0)\,\mathrm{e}^{-z/\Delta}, \tag{5}$$

where $T(\theta, 0)$ is the Fresnel transmission intensity [18] and Δ is the penetration depth for the evanescent wave [3, 4]. In a more realistic case, the index of refraction changes with the composition as a function of depth and this must be considered in the data analysis.

To calculate the fluorescence intensity from a given profile, the interfacial region was divided into thin slabs (thickness $\leqslant$ 0.2 nm), with a constant refractive index inside each slab determined by the polymer concentration at the depth corresponding to the center of the thin slab. We then solved the Fresnel equations to calculate the reflected and transmitted X-ray amplitudes at each slab boundary. Next, the calculated amplitudes were normalized to the incident amplitude, summed within a given layer, and then squared to determine the intensity within a slab. These intensities were used in the integral of equation (4), which was solved numerically. A more complete description may be found in [3].

5. RESULTS AND DISCUSSION

Before presenting our results, we first note that this ionomer system is believed to exhibit polyelectrolyte behavior in DMSO [9].

The integrated S and Mn fluorescence was determined at angles from 1.8 to 5 mrad. The fluorescence ratio $I_{\text{fluor}}^{\text{Mn}}/I_{\text{fluor}}^{\text{S}}$ is shown in Fig. 7 for PSS-Mn (15 mol% S) in DMSO (1.00% wt/wt) at 20°C. Next to the data sets is the time in hours that the scan was made after introducing the sample into the solution chamber. Each scan took about 3–4 h. A long (4 days) time dependence was observed before the experiment was terminated due to a failure in the energy dispersive detector. (We have seen adsorption saturation for other concentrations and sulfonation levels around 4 days.)

Before discussing the fitting results, one can clearly see that the polymer is adsorbed to the solution/vapor interface. At high incident angles, the fluorescence ratio is a probe of the bulk composition, while at angles $\theta \ll \theta_{\text{crit}}$, the ratio reflects the surface composition (within the limiting penetration depth). Since the ratio of Mn to S is much higher at the lowest angles, the surface composition is enhanced in manganese, hence also polymer, over that of the bulk phase.

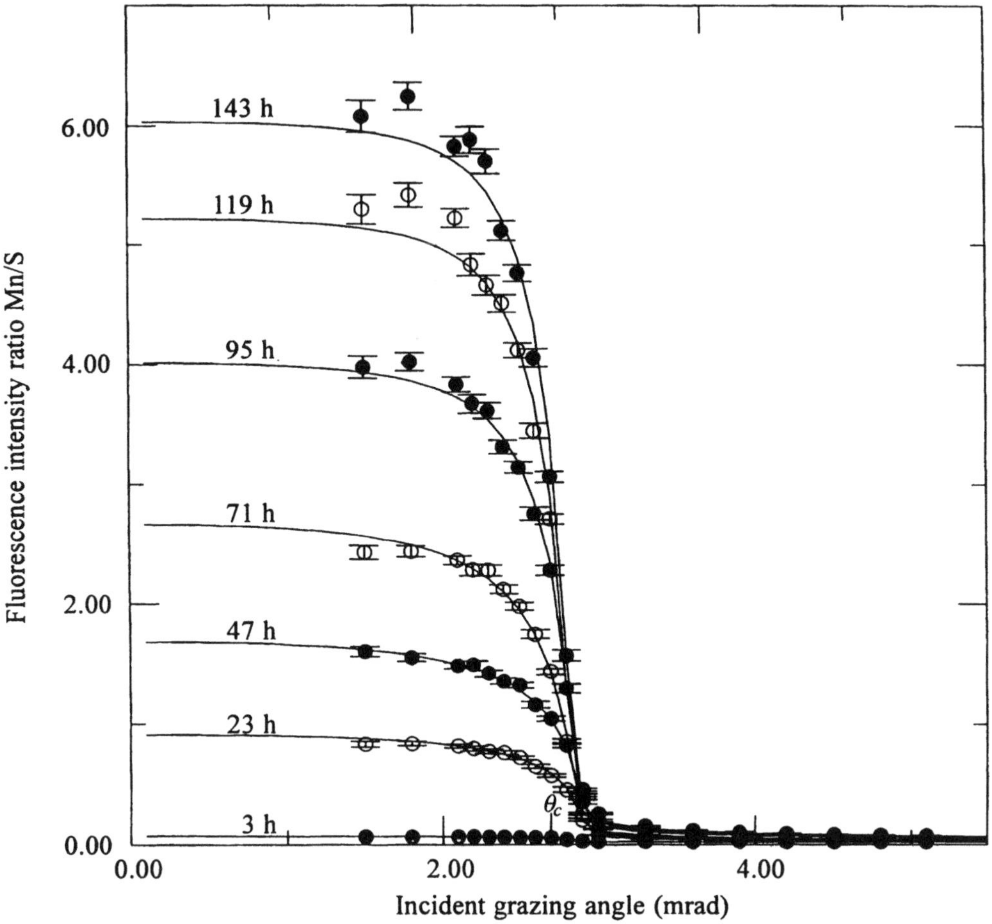

Figure 7. The ratio of manganese to sulfur fluorescence as a function of the incident grazing angle. The time in hours is the time after the sample was delivered into the sample trough. The solid lines correspond to a step profile fitting with the parameters listed in Table 1.

As mentioned earlier, the fluorescence ratios must be fit to a model polymer concentration profile. The selection of a particular form for the profile was difficult in this case, where the polymer is presumed to demonstrate polyelectrolyte behavior. For this reason, we present the fitting results for a step profile, which, while somewhat unphysical, captures the important features of the adsorption process in the most parsimonious fashion. (We will discuss other adsorption profiles elsewhere when we complete the analysis of other experiments at different concentrations and sulfonation levels.) The solid lines in Fig. 7 are the results of a non-linear least squares fit to the data for a model profile in which the width and height of the adsorption step were varied. The resulting surface excesses in mg/m^2 are shown in Fig. 8, and the table of adsorption widths and heights for the different times are given in Table 1.

To complement our XEWIF results, we performed surface tension measurements on the PSS-Mn/DMSO system. Before discussing those results, it is first important

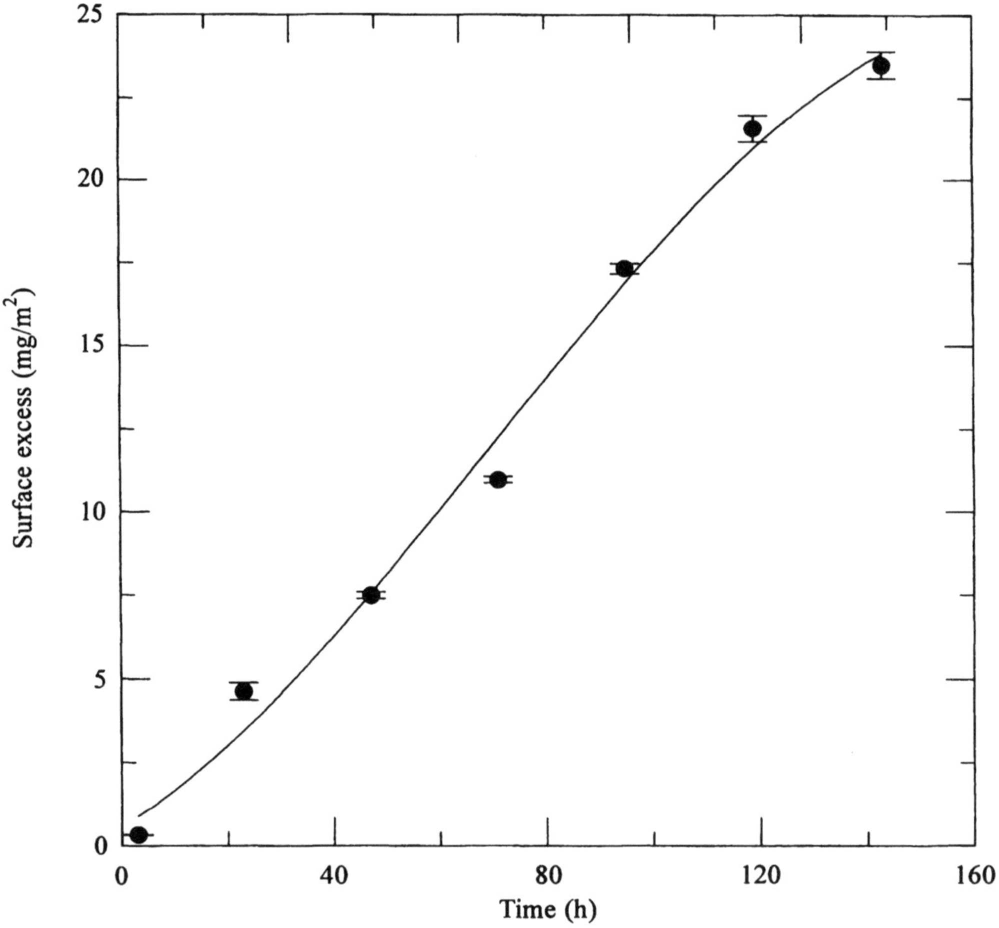

Figure 8. Calculated surface excess as a function of time.

Table 1.
Results from the fitting of the fluorescence data to a step adsorption profile

Time (h)	Thickness (nm)	c_{surf} (wt%)	χ^2	Surface excess (mg/m^2)
3	11.6 ± 0.1	3.5 ± 0.1	0.5	0.32 ± 0.01
23	12.8 ± 0.7	33.9 ± 0.6	0.9	4.61 ± 0.26
47	13.9 ± 0.2	49.8 ± 0.4	1.4	7.48 ± 0.10
71	16.4 ± 0.1	61.7 ± 0.2	3.5	10.95 ± 0.09
95	22.3 ± 0.2	71.7 ± 0.1	5.5	17.31 ± 0.15
119	25.5 ± 0.5	77.9 ± 0.2	18.0	21.56 ± 0.41
143	26.7 ± 0.5	81.1 ± 0.2	18.5	23.50 ± 0.40

to realize that DMSO is a non-solvent for pure polystyrene and that the solubility of PSS-Mn increases with the degree of sulfonation. At 20°C, the surface tension of polystyrene is estimated at 40.7 dynes/cm [23], which is lower than our determined value for the surface tension of DMSO at 20°C, namely 42.75 ± 0.22 dynes/cm. We thus expect that the polystyrene segments will be preferentially attracted to the surface. On the other hand, the sulfonate groups promote solubility and they are repelled by the free surface. The origin of the repulsion may be understood from the concept of image charges. The electric field from a charge, either on the sulfonate or on the manganese counterion, will be reflected from the interface with a strength proportional to the normal reflection coefficient [24], $(\varepsilon_{\text{DMSO}} - \varepsilon_{\text{vap}})/(\varepsilon_{\text{DMSO}} + \varepsilon_{\text{vap}})$, where $\varepsilon_{\text{DMSO}} = 48$ and $\varepsilon_{\text{vap}} \approx 1$ are the dielectric constants for DMSO and the vapor phase, respectively. Hence, an ion located at a distance z from the surface will experience a repulsive force, F [24]:

$$F = \frac{\varepsilon_{\text{DMSO}} - \varepsilon_{\text{vap}}}{\varepsilon_{\text{DMSO}} + \varepsilon_{\text{vap}}} \frac{e^2}{\varepsilon_{\text{DMSO}}(2z)^2} \approx \frac{e^2}{4\varepsilon_{\text{DMSO}} z^2}, \quad (6)$$

where e is the ionic charge in esu.

Thus, an ionomer at the interface must balance (1) the attractive interactions between the polystyrene units and the surface with (2) the repulsive interaction between the sulfonate groups and the surface.

The energy balance between these two competing effects will be reflected in the surface tension. In Fig. 9, the surface tension of the PSS-Mn is shown as a function of concentration at 15% and 7.5% sulfonation. Note that at the concentration used for the study described here, the surface tension does not appear to change from that of the pure solvent. At 20°C, the surface tension was measured about 2.5 h after the sample was introduced into the surface tension cell. This measurement should be compared with the fluorescence data at the earliest time in Table 1. Even though the surface tension data indicate no preferential adsorption of the polymer to the liquid/vapor interface, the fluorescence measurements unambiguously demonstrate absorption.

We note that the image charge repulsion will be reduced as the sulfonation level on the polymer chain is lowered. One would thus predict a lower surface tension for a solution where the polymer has a lower sulfonation level. This is in fact observed in Fig. 9 for the surface tension measurements on the 7.5 mol% sulfonated PSS-Mn in DMSO.

The apparent conflict between the microscopic and macroscopic measurements raises an interesting paradox, which, to our knowledge, has not been previously addressed. First, is the net-zero change in the surface energy a general phenomenon for polyelectrolyte adsorption at low concentrations, even when the polymer adsorbs to the surface? A careful surface tensiometry study of aqueous polyelectrolyte solutions by Okubo [25] suggests that this may the case. For polyelectrolytes with simple counterions, the solution surface tension remained unchanged from that of pure water until a concentration of about 0.1 M, above which a decrease in the surface tension was observed. Like our system, the surface tension did not change at low polymer concentration and we believe that this is due to the balance between the surface energy reduction of the hydrophobic backbone and the repulsion of the ionic charge groups on the chain. As the bulk concentration is increased, the free energy cost of creating

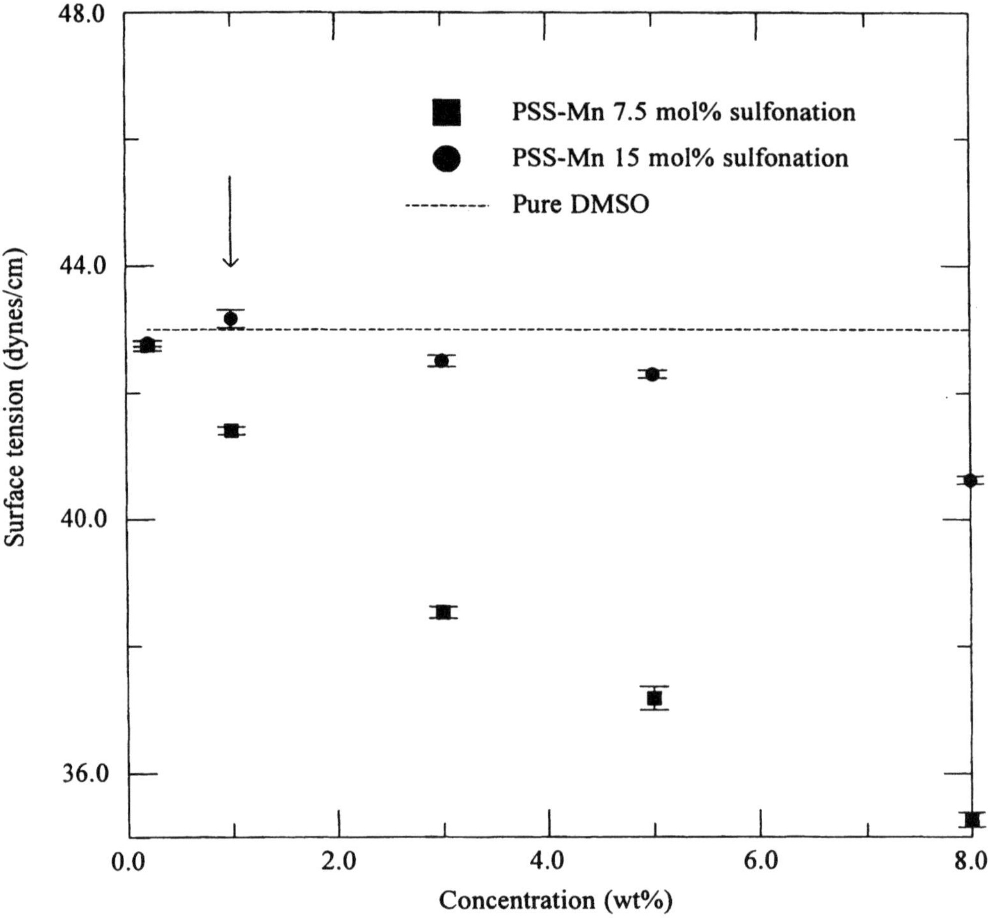

Figure 9. Surface tension of PSS-Mn solutions in DMSO. The data are shown as a function of concentration for 15 mol% and 7.5 mol%. The arrow marks the sample studied by XEWIF in this paper.

the high surface concentration is reduced (see Section 2) and the long-range Coulomb interactions of the charge groups and the surface are more effectively screened. This can account for the eventual decrease in the surface tension at higher concentrations. Unfortunately, only the surface tension was measured in Okubo's study; we plan to use XEWIF to look at the surface activity of some of the sulfonated polyelectrolytes in aqueous solution discussed in [25].

The other point which we wish to address is the relatively long time dependence of the adsorption. We took special pains to ensure that water did not enter the sample chamber during the experiment and thus we do not believe that the time dependence was due to water uptake by the solution. Rather, the time dependence was due to relaxations within the adsorbed film. Long time relaxations were seen in the surface tension measurements of aqueous polyelectrolyte solutions [25] and were presumably due to the energy barrier of overcoming strong charge–charge interactions to move towards an equilibrium conformation. In two previous adsorption studies of PSS-Mn in DMSO [4, 26], the time dependence that we report here was not described; since no mention was made on whether or not the adsorption was followed over time, we cannot make a more quantitative comparison.

6. CONCLUSION

We have shown that XEWIF is a valuable method for determining the adsorption excesses of polymers adsorbed from solution to the liquid/vapor interface. In the PSS-Mn/DMSO system, it is clear that surface tension measurements would mislead one into believing that the ionomer would be relatively surface-inactive, while the XEWIF measurements demonstrate quite the opposite! We suggest that the effect of strong adsorption with a net-zero change in the surface energy may be general to other polyelectrolyte systems.

Acknowledgements

This work was supported by the National Science Foundation (DMR-9113706), the Exxon Education Foundation, and the Materials Research Laboratory at the University of Massachusetts at Amherst.

REFERENCES

1. L. T. Lee, O. Guiselin, B. Farnoux and A. Lapp, *Macromolecules* **24**, 2518–2522 (1991).
2. T. Mansfield, PhD Thesis, Polymer Science and Engineering Department, University of Massachusetts (1993).
3. M. Brunel, *Acta Crystallogr., Sect. A* **42**, 304–309 (1986).
4. J. M. Bloch, M. Sansone, F. Rondelez, D. G. Peiffer, P. Pincus, M. W. Kim and P. M. Eisenberger, *Phys. Rev. Lett.* **54**, 1039–1042 (1985).
5. W. B. Yun and J. M. Bloch, *J. Appl. Phys.* **68**, 1421–1428 (1990).
6. S. W. Barton, L. Bosio, R. Cortes and F. Rondelez, *Europhys. Lett.* **17**, 401–406 (1992).
7. M. S. Kent, L. Bosio and F. Rondelez, *Macromolecules* **25**, 6231–6239 (1992).
8. For details on X-ray and neutron reflectivity, refer to T. P. Russell, *Mater. Sci. Rep.* **5**, 171–271 (1990).
9. R. D. Lundberg and H. S. Makowski, in: *Advanced in Chemistry Series*. A. Elsenberg (Ed.), Vol. 187, pp. 21–36. American Society, Washington, DC (1980).
10. *The Merck Index*, 9th edn. Merc & Co., Rahway, NJ (1976).
11. J. M. H. M. Scheutjens and G. J. Fleer, *J. Phys. Chem.* **84**, 178–190 (1980).
12. J. W. Cahn, *J. Chem. Phys.* **66**, 3667–3672 (1977).
13. P. G. de Gennes, *Macromolecules* **14**, 1637–1644 (1983).
14. R. A. L. Jones, *Phys. Rev. E* **47**, 1437–1440 (1993).
15. A. H. Compton and S. K. Allison, *X-rays in Theory and Experiment*, 2nd edn. Van Nostrand, New York (1948).
16. W. K. Warburton and K. F. Ludwig, *Phys. Rev. B* **33**, 8424–8432 (1986).
17. W. H. McMaster, N. K. del Grande, J. H. Mallett and J. H. Hubbell, *Compilation of X-ray Cross Sections*, ref. UCRL-50174 Sec. II Rev. 1 (1969). Available from National Technology Information Service, Springfield, VA.
18. G. H. Vineyard, *Phys. Rev. B* **26**, 4146–4159 (1982).
19. M. Born and E. Wolf, *Principles of Optics*, 6th edn. Pergamon, New York (1983).
20. A. W. Adamson, *Physical Chemistry of Surfaces*, 4th edn. John Wiley, New York (1982).

21. W. D. Harkins and H. F. Jordan, *J. Am. Chem. Soc.* **52**, 1751–1772 (1930).
22. M. W. Kim, D. G. Peiffer and P. Pincus, *J. Phys. Lett.* **45**, L953–L959 (1984).
23. S. Wu, *Polymer Interface and Adhesion*. Marcel Dekker, New York (1982).
24. L. Onsager and N. T. Samaras, *J. Chem. Phys.* **2**, 528–536 (1934).
25. T. Okubo, *J. Colloid Interface Sci.* **125**, 386–398 (1987).
26. M. W. Kim, D. G. Peiffer, W. Chen, H. Hsiung, Th. Rasing and Y. R. Shen, *Macromolecules* **22**, 2682–2685 (1989).

Fundamentals of Adhesion and Interfaces, pp. 125–142
D. S. Rimai, L. P. DeMejo and K. L. Mittal (Eds)

Surface modification of ultra-high molecular weight polyethylene fibers by γ-radiation-induced grafting

A. B. BRENNAN,* J. J. ARNOLD and M. P. ZAMORA
Department of Materials Science and Engineering, 317 MAE Building, University of Florida, Gainesville, FL 32611, USA

Revised version received 5 August 1994

Abstract—A technique for grafting acrylic polymers on the surface of ultra-high molecular weight polyethylene (UHMWPE) fibers utilizing ^{60}Co gamma radiation at low dose rates and low total dose has been developed. Unlike some of the more prevalent surface modification schemes, this technique achieves surface grafting with complete retention of the exceptional UHMWPE fiber mechanical properties. In particular, poly(butyl acrylate) and poly(cyclohexyl methacrylate) were successfully grafted onto UHMWPE fibers with no loss in tensile properties. The surface and tensile properties of the fibers were evaluated using Fourier transform infrared/photoacoustic spectroscopy (FTIR/PAS), X-ray photoelectron spectroscopy (XPS), and tensile tests. The reinforcement efficiency of untreated, polymer-grafted, and plasma-treated UHMWPE fibers in polystyrene and a poly(styrene-co-butyl acrylate-co-cyclohexyl methacrylate) statistical terpolymer was characterized using mechanical tensile tests. The thermoplastic matrix composites were prepared with 4 wt% discontinuous (10 mm), randomly distributed UHMWPE fibers. An approximate 30% increase in composite strength and modulus was observed for poly(cyclohexyl methacrylate)-grafted fibers in the terpolymer and polystyrene matrices. A comparable improvement was realized with the plasma-treated fibers. On the other hand, poly(butyl acrylate) grafts induced void formation, i.e. energy dissipation through plastic deformation and volume expansion at the fiber/matrix interface in terpolymer composites. The latter resulted in a 75% increase in the elongation to failure. The effect of polymer grafts on fiber/matrix adhesion is discussed in terms of the graft and matrix chain interactions and solubility, graft chain mobility, and fracture surface characteristics as determined by scanning electron microscopy (SEM).

Keywords: Polyethylene fibers; surface treatment; composites; fiber/matrix adhesion; gamma radiation.

1. INTRODUCTION

Ultra-high molecular weight polyethylene (UHMWPE) fibers exhibit outstanding specific strength and modulus, exceeding those of nearly all other fibers available (see Table 1) [1]. In addition, their excellent chemical, fatigue, and abrasion resistance make

*To whom correspondence should be addressed.

Table 1.
Tensile and specific properties of commercially available fibers [1]

	UHMWPE		Kevlar 49	PET Dacron	S-glass	Graphite HM
	Spectra 900	Spectra 1000				
Density (g/cm^3)	0.97	0.97	1.44	1.38	2.55	1.81
Tensile strength (MPa)	2586	2999	2758	1157	4585	2413
Specific strength	272	315	198	85.6	188	137
Modulus (GPa)	117	172	124	11.9	90	379
Specific modulus	12.3	18.1	8.8	0.880	3.6	21.6
Ultimate elongation (%)	3.5	2.7	2.8	12	5.7	0.6

them an attractive candidate for reinforcement of low-temperature, high-performance composites in marine, biomedical, sporting, and ballistic applications. However, UHMWPE fibers are limited by their low working temperature, their propensity to creep, and their poor adhesion characteristics. Poor adhesion to UHMWPE fibers is attributed to the non-polar nature of the $(-CH_2-CH_2-)_n$ backbone and the smooth, low surface energy fiber surface. Without adequate fiber/matrix adhesion, the excellent fiber properties are not fully utilized and optimization of composite mechanical properties is not achieved.

Adhesion is a function of the surface characteristics of the material, i.e. (i) surface topography and roughness, (ii) surface energetics and wettability, and (iii) chemical functionality; and the contribution of each to the level of adhesion is additive [2]. Thus, surface treatments which enhance any of these characteristics for UHMWPE fibers will improve their reinforcement efficiency in composite systems.

A variety of surface treatments have been devised for UHMWPE fibers, utilizing, for instance, radio frequency (RF) plasma, chemical, UV radiation, and electron beam (E-beam) sources [3–12]. Surface treatments by plasma exposure in various environments have received considerable attention with mixed results. In particular, Allied Signal uses gas plasma in the commercial production of plasma-treated Spectra® fibers. Allied Signal showed that plasma exposure improved the flexural properties of epoxy composites, but the fiber tensile strength decreased by approximately 10% with this treatment [3]. Ward and co-workers [2, 4] demonstrated that exposure of UHMWPE fibers to oxygen gas plasma increased the surface energy, but caused pitting of the fiber surface. Additionally, in contrast to the research at Allied Signal, these treated fibers did not improve the tensile properties of unidirectional epoxy composites.

Amination by ammonia and allylamine plasma also leads to similar improvements in the surface characteristics of UHMWPE fibers, i.e. an increase in surface energy and wettability, but significantly decreases the mean fiber strength with exposure power [1, 5–7]. In a related study, Brown *et al.* [8] showed that ammonia plasma treatment resulted in increases in the interlaminar shear strength (ILSS), fracture energy, and flexural properties of Spectra® fabric-reinforced epoxy composites. In summary, while the effect of plasma on composite mechanical properties is still being debated, the preponderance of literature indicates that plasma is detrimental to the UHMWPE fiber tensile properties.

Several research groups have analyzed the effect of oxidation by chemical treatments on the adhesion and wetting characteristics of UHMWPE fibers. For instance, Taboudoucht *et al.* [9] reported an increase in the ILSS but cited only minimal improvements in the flexural strength and stiffness of nitric acid-treated UHMWPE fiber/epoxy matrix composites. Similarly, Ulcay *et al.* [10] showed that chemical treatments with a 50 : 50 hydrochloric acid/nitric acid solution lowered the strength of UHMWPE fibers by approximately 10% and reduced the strength of epoxy matrix composites.

As has been shown, the existing surface modification techniques involve oxidation, functionalization, polymer deposition, and/or grafting of the fiber surface via high-energy exposure and/or aggressive chemical environments. As a result, ultra-high molecular weight, highly crystalline polyethylene chains are susceptible to degradation during these treatments. Therefore, the first aim of our work was to develop a surface modification technique that improves fiber/matrix adhesion but is not deleterious to the mechanical properties of the fiber.

One method of doing this is by radiation-induced surface modification using low dose rates and low total dose. Radiation processes such as E-beam, UV, and gamma radiation have been used extensively for the surface modification of polymers [11–16]. Recently, successful grafting of acrylic acid and acrylamide onto UHMWPE fibers using E-beam and UV radiation (320–440 nm) with no loss in fiber tensile properties was reported [11, 12]. However, the effects of the surface treatment on actual composite properties have not yet been evaluated. E-beam is especially attractive because the extremely high dose rates enable shorter production times for industrial processes. Unfortunately, the high dose rates of E-beam techniques generally promote chain scission relative to crosslinking mechanisms [17, 18]. Thus, a low dose rate gamma radiation process was selected which has the potential to be extended to the more commercially viable E-beam grafting process. A gamma radiation process employing low dose rates and low overall doses has been developed by Goldberg and Kumar the surface grafting of numerous polymers [16]. The significance of this method is that the surface properties may be modified with no detectable change in the bulk properties of the polymer substrate. For example, gamma radiation-induced surface modification has been used on films of low- and high-density polyethylene, polypropylene, and poly(ethylene terephthalate) [13–16]. Further, Kaji *et al.* [19] have grafted acrylic acid onto general-use HDPE filaments to impart flame resistance, thermal stability, and hydrophilicity. However, to our knowledge, there have been no reports concerning the surface grafting of UHMWPE fibers using gamma radiation. Moreover, numerous surface treatment schemes for UHMWPE fibers have been evaluated with regard to fiber/matrix adhesion in thermosets, but few studies have considered this aspect in conventional thermoplastics. Consequently, we have developed a method of grafting onto UHMWPE fibers using gamma radiation at relatively low dose rates and low total dose.

UHMWPE fibers were grafted with poly(butyl acrylate) and poly(cyclohexyl methacrylate) and incorporated in polystyrene ($T_g = 105\,°C$) and a poly(styrene-co-butyl acrylate-co-cyclohexyl methacrylate) statistical terpolymer ($T_g = 44\,°C$). In this way, fiber/matrix adhesion was evaluated as a function of the graft mobility at room temperature and matrix ductility, i.e. brittle polystyrene versus a relatively ductile terpolymer matrix. The surface and tensile properties of the UHMWPE fibers along with the tensile properties and fracture analysis of each composite were characterized using XPS, FTIR, and mechanical tensile testing.

2. MATERIALS AND METHODS

2.1. Fibers

Ultra-high molecular weight polyethylene fibers, Spectra 1000® and plasma-treated Spectra 1000®, were supplied by Allied Signal and Himont Plasma Science, Inc., respectively. Segments of as-received Spectra® (650 denier), 6″ in length, were cut and presoaked in test-tubes of 33 wt% monomer/chloroform solutions for 84 h at room temperature. Prior to use, cyclohexyl methacrylate (CHM) and butyl acrylate (BA) monomers were purified by vacuum distillation. Each filled test-tube was purged with argon and subsequently exposed to ^{60}Co gamma radiation at a dose rate of 140 rads/min for a total dose of 0.1 Mrad. After radiation grafting, the fibers were immersed in chloroform for 24–48 h to displace any adsorbed, but not chemically grafted, polymer from the surface. The fibers were then rinsed with methanol and dried at 50°C for 12 h.

2.2. Polymer matrices

Polystyrene and a poly(styrene-co-butyl acrylate-co-cyclohexyl methacrylate) statistical terpolymer were obtained by bulk polymerization at 60°C for 24 h using 0.3 wt% azobisisobutyronitrile (AIBN) as an initiator. All monomers were purified by vacuum distillation prior to use. The polymers were subsequently dissolved in chloroform, reprecipitated in methanol, filtered, and dried at 150°C for 24 h. The T_g's of polystyrene and the statistical terpolymer were determined by DSC to be 105°C and 44°C, respectively.

2.3. Composites

Discontinuous fiber-reinforced composite laminates were prepared by solution techniques. First, fibers were cut into 10 mm segments and manually dispersed in a 5 wt% polymer/chloroform solution at room temperature. The composite solution was then cast into a glass Petri dish, dried under vacuum at 120°C for 12 h, and compression-molded (20–31 MPa) at 120°C into a film approximately 0.35–0.50 mm thick. Finally, the laminate was allowed to cool slowly under pressure.

A simple nomenclature for describing the composites according to matrix material, fiber loading, and fiber treatment will be used here. For example, a terpolymer matrix

Table 2.
UHMWPE fiber composite nomenclature

Fiber treatment	Matrix material	
	Polystyrene	Statistical terpolymer
Untreated	PS/R4/UNPE	TER/R4/UNPE
Poly(butyl acrylate) grafted	PS/R4/BAPE	TER/R4/BAPE
Poly(cyclohexyl methacrylate) grafted	PS/R4/CHMPE	TER/R4/CHMPE
Plasma treated	PS/R4/PTPE	TER/R4/PTPE

composite randomly (R) reinforced with 4 wt% poly(butyl acrylate) grafted fibers is designated TER/R4/BAPE. Table 2 lists each composite and the corresponding designation.

2.4. Characterization

Surface characteristics of grafted fibers were determined by FTIR/photoacoustic spectroscopy (FTIR/PAS) and X-ray photoelectron spectroscopy (XPS). FTIR/PAS analysis was accomplished using an MTEC model 200 photoacoustic detector (spectral resolution of 8 cm^{-1}) interfaced with a Nicolet 20SX FTIR. Fibers were chopped into approximately 10 mm segments and analyzed in an atmosphere of dry helium gas using 32 sample scans. The FTIR gain and mirror velocity were set at 10 and the photoacoustic gain was set at 32. XPS was performed with a Kratos DS800 spectrometer (spectral resolution of 1.1 eV) using Mg K_α incident radiation (1.2536 keV). Chopped fibers were washed with ethanol, dried at 50°C, carefully secured to steel mounts using double-sided tape, and analyzed.

Each composite was analyzed by DSC prior to mechanical characterization to verify the complete removal of chloroform solvent. Representative samples were taken from the composite films and analyzed from 25 to 150°C at a heating rate of 10°C/min under a nitrogen gas flow of 100 ml/min. A decrease in the T_g of the matrix material indicated residual solvent due to the plasticizing effect of solvent molecules. In fact, it was found that small concentrations of solvent molecules (1–2 wt% as determined by thermogravimetric/differential thermal analysis) decreased the T_g by nearly 20°C. Composite films containing residual solvent were reheated under vacuum according to the procedure described above. Upon removal of solvent, strips 10 mm wide cut using a diamond saw under dry conditions. No lubricant was used to prevent modification of the fiber/matrix interface.

Tensile properties were measured using as Instron 1122 equipped with an 890 Newton load cell at ambient conditions. Five samples were tested for each material according to ASTM specifications. UHMWPE single filaments were tested according to ASTM D2256 using a gage length of 254 mm and a crosshead rate of 254 mm/min. Terpolymer specimens were characterized according to ASTM D638M (Type M-III die) because of their high ductility. A crosshead rate of 1.27 mm/min was chosen so that the composite and homopolymer testing rates were comparable. Polystyrene and all composite samples were evaluated using a rectangular geometry

(10 × 75 mm) according to ASTM D882. The sample thicknesses were approximately 0.35–0.50 mm as noted above. The initial distance between the grips was 50 mm and the crosshead rate was 5.08 mm/min.

Fracture surfaces of failed tensile specimens were analyzed by scanning electron microscopy (SEM). A JEOL JSM35CF microscope using secondary electron imaging at an accelerating voltage of 5 kV was employed. The fracture surfaces were coated with Au/Pd for 30 s at a coating rate of approximately 5 nm/min prior to analysis.

3. RESULTS AND DISCUSSION

3.1. Characterization of fibers

3.1.1. FTIR. High-energy radiation such as ^{60}Co gamma radiation (1.17–1.33 MeV) may be used to induce surface as well as bulk grafting in UHMWPE fibers. If the chemical structure of the graft material is different from that of the substrate, different IR spectra may be expected due to different bond deformations and corresponding variations in characteristic absorption regions.

The photoacoustic IR spectra of ungrafted fibers, PBA-grafted fibers, and PBA homopolymer are illustrated in Fig. 1. Spectra 1 and 2 represent the photoacoustic IR spectra for ungrafted, gamma-irradiated and PBA-grafted UHMWPE fibers, respectively. Ungrafted, gamma-irradiated fibers were used as a reference to account for any oxidation or modification of the polyethylene structure during irradiation. Unlike the gamma-irradiated-only fibers (spectrum 1), the spectrum of the PBA-grafted fibers (spectrum 2) exhibits an additional peak at 1736 cm^{-1} corresponding to C=O bond stretching. The carbonyl peak is attributed to the ester groups in PBA and PCHM grafts. Subtraction of the ungrafted fiber spectrum (spectrum 1) from the grafted fiber spectrum (spectrum 2) should yield a spectrum representative of the graft polymer (spectrum 3). Indeed, spectrum 3 is in good agreement with the IR spectrum for PBA homopolymer (spectrum 4). Similar results were obtained for PCHM-grafted UHMWPE fibers.

3.1.2. XPS. XPS is commonly used to analyze the surface composition of various materials. Depending on the substrate material, the depth analyzed by XPS may include the first 1–25 nm of surface material, whereas the sampling depth of FTIR is typically 1–10 μm [20]. Hence, XPS is more representative of the surface composition of the substrate material. This does not suggest that surface grafts are the only material detected by XPS, nor does it imply that the gamma radiation process only induces surface grafting. Rather, diffusion of monomer into the amorphous portion of fiber during the pre-soak step is critical to successful grafting [12]. In fact, we were unable to detect graft material on fibers which were not pre-soaked.

The high-resolution XPS C(1*s*) spectra of the as-received, irradiated (no monomer), and grafted UHMWPE fibers are exhibited in Fig. 2. As was done for FTIR/PAS, the results for fibers irradiated in chloroform with no monomer present were used as

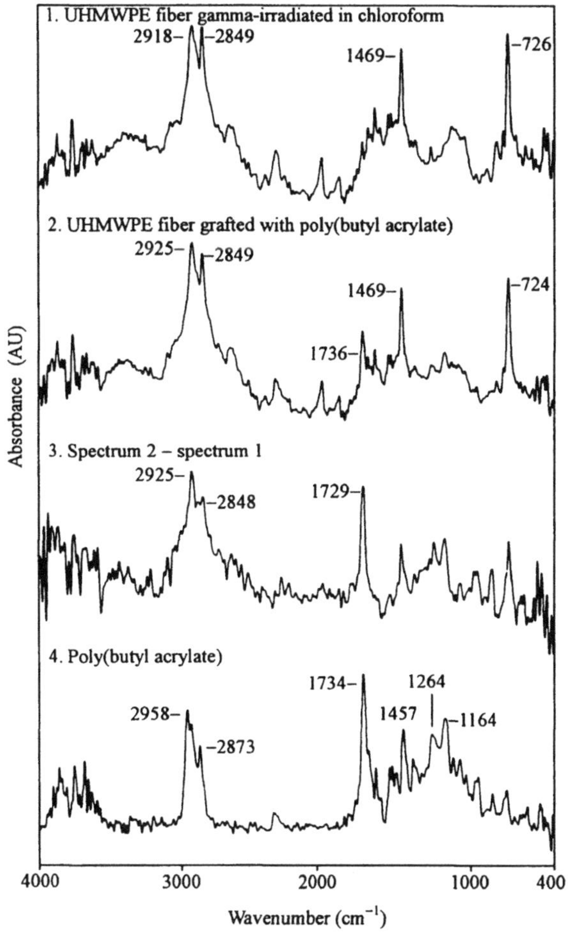

Figure 1. FTIR/photoacoustic spectroscopy of (1) UHMWPE fibers irradiated in chloroform with no monomer present, (2) poly(butyl acrylate)-grafted UHMWPE fiber, (3) subtraction spectrum (spectrum 2–spectrum 1), and (4) poly(butyl acrylate) homopolymer.

a reference to account for any modifications due to irradiation. The main C(1*s*) peak and a Mg satellite peak are observed at approximately 285 and 277 eV, respectively, in all spectra. As is shown, the spectra of the as-received (spectrum 1) and irradiated without monomer (spectrum 2) fibers are essentially the same. This is expected since the fibers and solution were purged to remove oxygen prior to radiation exposure. However, an additional peak is exhibited by the PBA-grafted fibers (spectrum 3) with a shift in C(1*s*) binding energy of approximately 4.5 eV. It has been shown that oxygen induces a C(1*s*) peak shift to higher binding energies by *ca.* 1.5 eV per shared electron pair between carbon and oxygen [21]. Hence, a shift of 4.5 eV is consistent with the presence of ester side-groups in the poly(butyl acrylate) grafts. Similar results were obtained for PCHM-grafted fibers.

XPS analysis of the plasma-treated fibers was also performed but is not included. The oxygen content of plasma-treated fibers was comparable to that of the PBA-

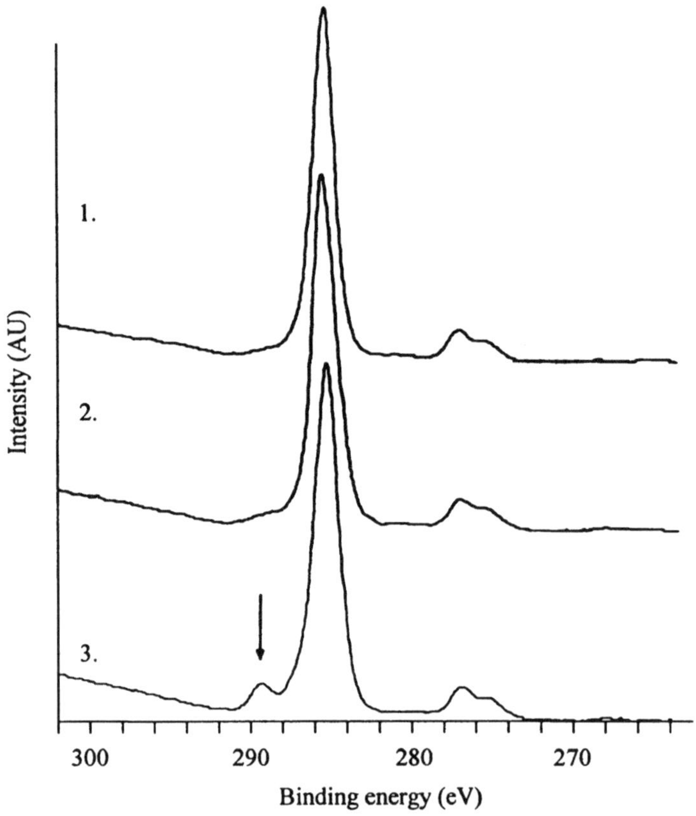

Figure 2. High-resolution XPS C(1*s*) spectra of (1) as-received, (2) gamma-irradiated in chloroform, and (3) poly(butyl acrylate)-grafted UHMWPE fibers. The additional peak in (3) indicates the chemical shift in C(1*s*) binding energy due to ester groups in PBA surface grafts.

and PCHM-grafted fibers and the nitrogen content was negligible, suggesting that an oxygen gas plasma treatment was probably used in their manufacture.

3.1.3. Tensile properties of UHMWPE fibers. Radiation grafting involves hydrogen abstraction and scission of the polyethylene backbone to produce free radicals, upon which graft polymerization is initiated [18]. However, other competing processes exist. For instance, a radical may recombine, terminate by disproportionation or chain transfer, or attach itself to an adjacent chain to form a crosslink junction. Crosslinking is generally viewed favorably since it tends to lead to an increase in mechanical properties. For example, E-beam radiation-induced crosslinking enhances the lackluster creep resistance of melt-spun polyethylene fibers [18]. Chain scission, on the other hand, may lead to a catastrophic decline in properties, especially for ultra-high molecular weight, highly crystalline and oriented polyethylene fibers. Klein *et al.* [18] demonstrated, for instance, that chain scission is more apt to occur at high E-beam radiation doses (9 Mrad) in high molecular weight gel spun fibers. Similarly, plasma treatment causes a reduction in fiber tensile properties which is generally

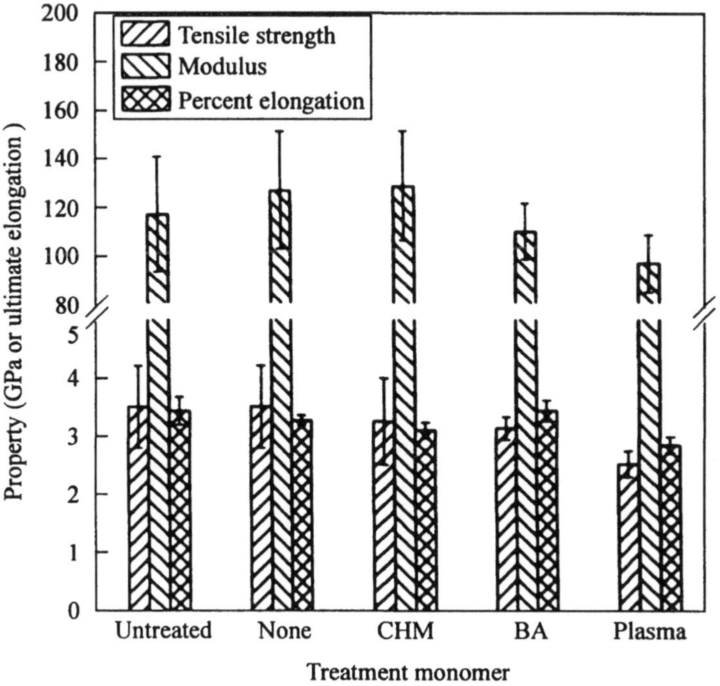

Figure 3. Tensile properties of untreated, gamma-irradiated in chloroform only (denoted 'none'), PCHM-grafted, PBA-grafted, and plasma-treated Spectra® 1000 fibers.

attributed to a predominance of chain scission [7]. Hence, the tensile properties of gamma-irradiated UHMWPE fibers were measured to monitor this adverse effect.

The tensile properties of plasma-treated, gamma-irradiated, and grafted UHMWPE fibers are illustrated in Fig. 3. In agreement with previous literature results, plasma-treated fibers exhibited a lower strength and modulus than those of untreated fibers [3]. It is evident that, within error, the high strength and modulus are retained after grafting with PBA and PCHM. Hence, chain scission does not appear to dominate at the relatively low dose rate and low total doses used for our grafting procedure. It is also worth noting that neither an increase in the fiber diameter nor a weight gain was observed following radiation grafting, suggesting that a thin graft layer was formed with no adverse effect on the fiber properties.

3.2. Polystyrene composite properties

The tensile properties, including the tensile strength, elongation to failure, and elastic modulus of polystyrene composites are listed in Table 3. In all composites, linear stress–elongation behavior was exhibited up to the point of failure. Failure was followed by fiber pull-out, not fiber fracture, suggesting that the fiber length was below the critical fiber length for fiber failure to occur. The critical fiber length is defined as the minimum fiber length required for the fiber stress to be equal to the fiber ultimate strength at the mid-length [22]. The addition of as-received and PBA-grafted fibers resulted in a decrease in the stress at break of the composites

Table 3.
Tensile properties[a] of polystyrene composites

Sample	Tensile strength (MPa)	Percent elongation	Elastic modulus (GPa)
Polystyrene	33.5 ± 1.7	2.2 ± 0.2	1.58 ± 0.10
PS/R4/UNPE	27.7 ± 4.4	2.1 ± 0.4	1.52 ± 0.12
PS/R4/BAPE	29.2 ± 1.9	1.9 ± 0.1	2.11 ± 0.12
PS/R4/CHMPE	36.1 ± 2.0	2.1 ± 0.1	1.99 ± 0.08
PS/R4/PTPE	39.8 ± 3.1	2.5 ± 0.1	1.98 ± 0.12

[a] Results represent the mean and standard deviation for a minimum of five samples.

relative to the polystyrene homopolymer. In contrast, a strengthening effect was observed for PCHM-grafted and plasma-treated fibers as indicated by a respective 30% and 44% increase in tensile strength. Similarly, PS/R4/BAPE, PS/R4/CHMPE, and PS/R4/PTPE all exhibited nearly a 33% increase in modulus. However, the increase in modulus for PS/R4/BAPE composites was viewed as somewhat anomalous behavior since the rubbery PBA grafts were not expected to enhance reinforcement efficiency.

The SEM micrographs of polystyrene composite fracture surfaces illustrate the relatively poor adhesion in the case of PS/R4/UNPE (Fig. 4a). Improved bonding is evident in the micrographs for PCHM-grafted and plasma-treated fibers (Figs 4c and 4d), which supports the higher composite modulus values.

The interfacial properties of UHMWPE fibers in polystyrene are critical due to the brittle nature of the matrix material. If fiber/matrix adhesion is weak, fibers may act as sites of stress concentration, inducing voids at the fiber ends from which crazing of the brittle polymer matrix is initiated [23]. Moreover, fibers oriented parallel to the crack plane can act like a rod-shaped microcrack, which further accelerates propagation of the crack front [23].

In terms of polymer–polymer interactions, molecular mixing is thermodynamically favored when the solubility parameters, δ_1 and δ_2, of two polymers are similar. For example, miscibility is favored when the δ values are within one unit of each other [24]. Typical δ values for polyethylene and polystyrene are 17.0 and 18.5 $MPa^{1/2}$, respectively [25]. Likewise, the accepted δ value for PBA is 18.5 and a calculated value for PCHM is 18.2 $MPa^{1/2}$ [25, 26]. Thus, we would expect more favorable interactions, i.e. possible graft/matrix molecular entanglement and mixing, for the PBA- and PCHM-grafted fibers with polystyrene relative to the incompatible polyethylene surface. Figures 4b and 4c demonstrate that a somewhat tighter interface is achieved for both PS/R4/BAPE and PS/R4/CHMPE. In the case of PS/R4/CHMPE composites, a reinforcing effect was observed as indicated by the increase in strength and modulus. However, we did not expect to see a significant reinforcing effect with PBA-grafted fibers based on the low glass transition temperature of PBA (−56°C). We hypothesized that the rubbery PBA grafts would not promote stress transfer between the fiber and matrix. As expected, PBA-grafted fibers did not reinforce in terms of ten-

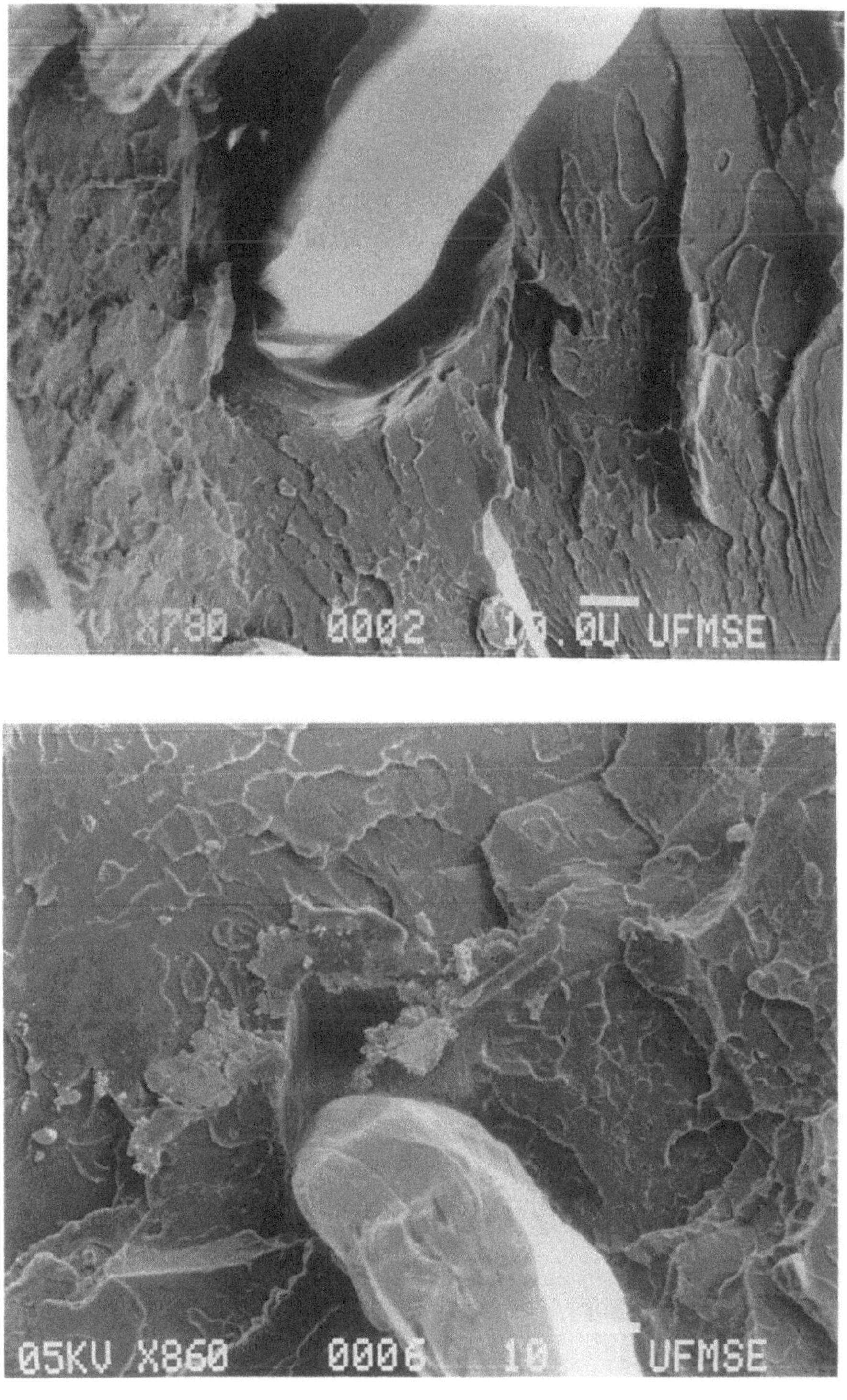

Figure 4(a–b). Tensile fracture surfaces of (a) PS/R4/UNPE (780×) and (b) PS/R4/BAPE (860×).

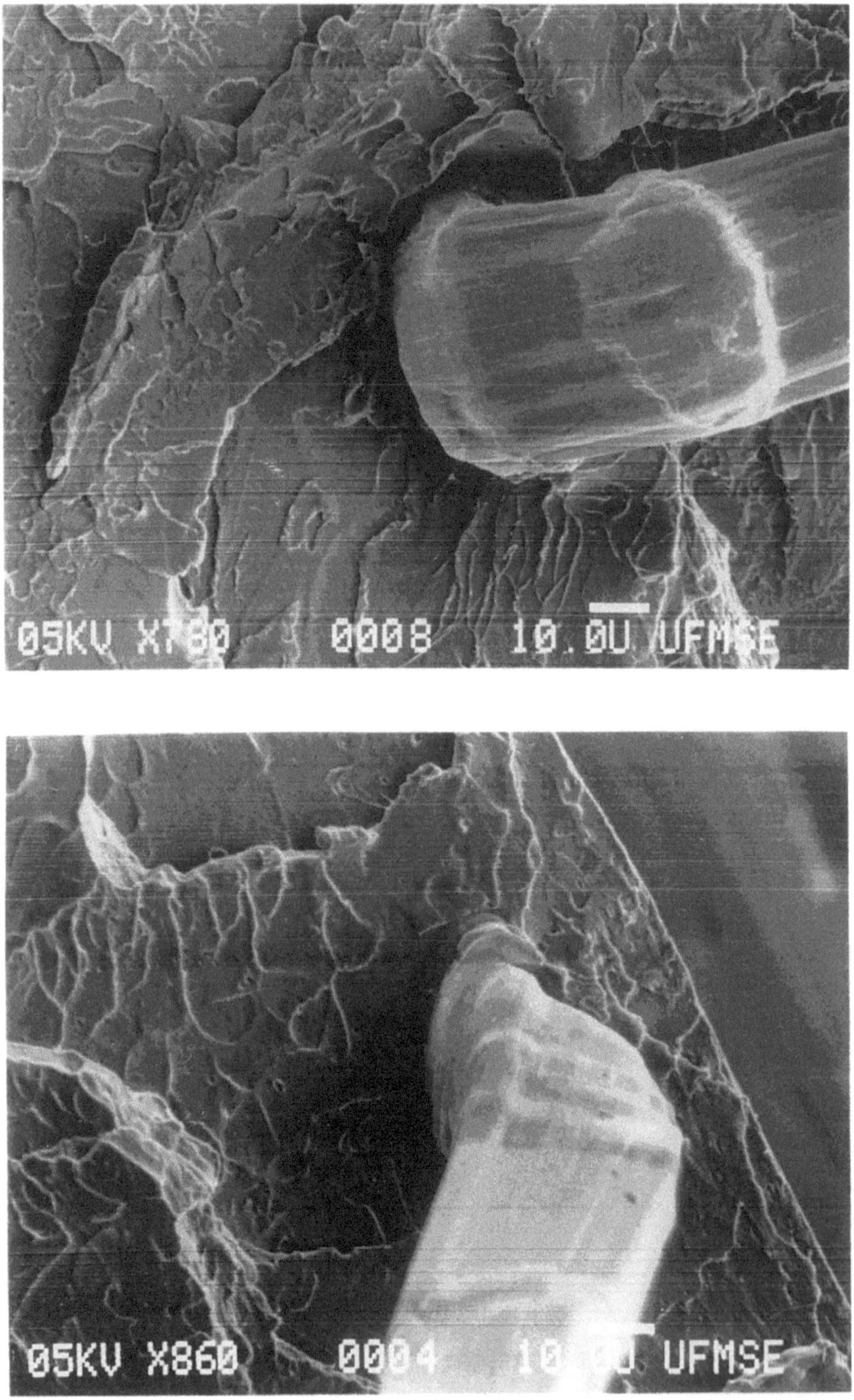

Figure 4(c–d). Tensile fracture surfaces of (c) PS/R4/CHMPE (780×) and (d) PS/R4/PTPE (860×).

sile strength. However, PS/R4/BAPE composites displayed a somewhat surprising increase in stiffness. It was believed that this behavior was somewhat anomalous based on the dynamic mechanical spectrometry (DMS) response and analysis of these composites [27].

3.3. Terpolymer composite properties

The terpolymer composite tensile properties are listed in Table 4. All the composites except TER/R4/BAPE displayed linear stress–elongation curves up to failure, upon which fiber pull-out commenced. In contrast, TER/R4/BAPE composites exhibited a yield point upon which the samples elongated at approximately a constant load until failure. Hence, the tensile strength value reported in Table 4 for TER/R4/BAPE represents the yield stress. A reinforcement effect was not observed for the as-received and PBA-grafted fibers (TER/R4/UNPE and TER/R4/BAPE, respectively), and in fact a decrease in strength was observed relative to the matrix properties. On the other hand, TER/R4/BAPE composites exhibited a 75% increase in the elongation to failure relative to TER/R4/UNPE composites. A higher level of reinforcement was observed for PCHM-grafted and plasma-treated fiber composites, TER/R4/CHMPE and TER/R4/PTPE, respectively. In particular, a 26% increase in modulus and a 20% increase in tensile strength were observed for TER/R4/CHMPE composites.

Fracture surfaces of the terpolymer composites are illustrated in Fig. 5. Note that TER/R4/UNPE and TER/R4/BAPE (Figs 5a and 5b, respectively) exhibit relatively poor adhesion indicated by the large interfacial gap between the fibers and the matrix. In a ductile such as the terpolymer, fiber reinforcement can be especially ineffective when fiber/matrix debonding is possible due to poor adhesion [23]. As a result, fibers may act as sites for stress concentration from which brittle failure is initiated. Figure 5a illustrates that brittle failure occurs due to poor bonding in TER/R4/UNPE, in contrast to the plastic deformation of the matrix exhibited by TER/R4/CHMPE and TER/R4/PTPE in Figs 5c and 5d, respectively. The low strength and modulus of TER/R4/UNPE composites are consistent with these observations. Although

Table 4.
Tensile properties[a] of S/BA/CHM terpolymer composites

Sample	Tensile strength (MPa)	Percent elongation	Elastic modulus (GPa)
S/BA/CHM terpolymer	23.6 ± 0.4[b]	3.1 ± 0.1[b]	1.11 ± 0.35
TER/R4/UNPE	21.5 ± 3.1	2.1 ± 0.5	1.29 ± 0.15
TER/R4/BAPE	21.6 ± 2.4[b]	3.5 ± 0.6	1.32 ± 0.08
TER/R4/CHMPE	25.8 ± 2.4	2.2 ± 0.2	1.62 ± 0.14
TER/R4/PTPE	29.7 ± 2.6	2.5 ± 0.2	1.55 ± 0.05

[a] Results represent the mean and standard deviation for a minimum of five samples.
[b] Values at yield.

Figure 5(a–b). Tensile fracture surfaces of (a) TER/R4/UNPE (720×) and (b) TER/R4/BAPE (720×).

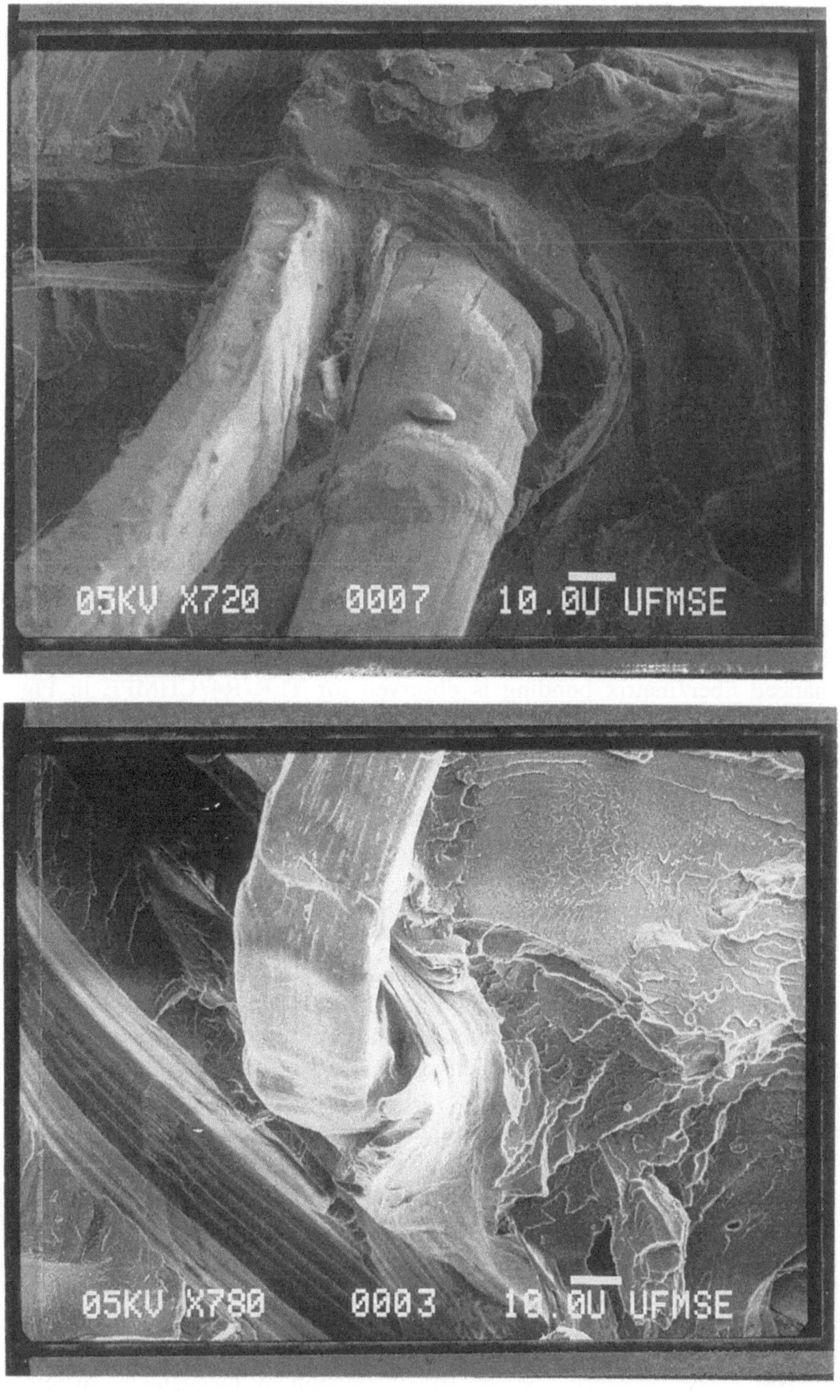

Figure 5(c–d). Tensile fracture surfaces of (c) TER/R4/CHMPE (720×) and (d) TER/R4/PTPE (780×).

PBA-grafted fibers exhibit poor bonding, they elicit a very different matrix response. Closer examination of Fig. 5b reveals significant void formation in the interfacial region. It is important to note that this void formation was volumetric; that is, it occurred throughout the composite along the fiber/matrix interface and was observed as a whitening of the fibers during tensile loading. Thus, the large increase in elongation to failure can be attributed to the volume expansion associated with void formation.

The manifestations of such a dramatic increase in the elongation to failure combined with invariable strength and modulus are a corresponding increase in the work to fracture, i.e. area under the stress–elongation curve. The work to fracture is directly related to the fracture energy, G_{Ic}, which may be related to the fracture toughness, K_{Ic}, and modulus, E, by the expression

$$K_{Ic} = \sqrt{G_{Ic}E}. \tag{1}$$

A useful assessment of a material's toughness may be obtained with these parameters [23]. This suggests that the rubbery PBA grafts (T_g of $-56°C$) are capable of toughening the matrix by inducing an energy dissipative mechanism via void formation in the interfacial region. This same mechanism was observed in PMMA-based, rubber-reinforced particulate composites and was associated with improved impact and fracture toughness [28].

Enhanced fiber/matrix bonding is observed for TER/R4/CHMPE in Fig. 5c as indicated by a much tighter interface and plastic deformation of matrix material around the fiber. This is expected since PCHM ($T_g = 84°C$) is a stiffer polymer with less molecular mobility at room temperature than PBA grafts. Thus, entanglements which form between grafted PCHM chains and matrix molecules during thermal processing are less likely to disentangle and therefore are more capable of transferring stress from the matrix to the fiber. The tensile properties of TER/R4/CHMPE composites indicate that more efficient reinforcement is achieved in this system.

Adhesion is a function of the surface topography, chemical functionality, and surface energetics, and each contributes to fiber/matrix bonding [2]. It should be pointed out that the mechanisms which influence adhesion for plasma-treated fibers may be completely different from those which affect adhesion for grafted fibers. For example, a decrease in the tensile properties of plasma-treated fibers is shown in Fig. 3. Ward and co-workers [2, 4] have postulated that severe etching causes this decrease in fiber strength due to stress concentration at the pits. However, etching may also enhance mechanical bonding with these fibers. Additionally, XPS reveals that the oxygen content of plasma-treated fibers is comparable to that of the grafted fibers. Since the atomic oxygen in plasma-treated fibers is not contained within polymeric grafts, the functionality of the surface is expected to be vastly different from that of PBA- and PCHM-grafted fibers. Considering these differences, the reinforcement efficiency of PCHM-grafted fibers is comparable to that of the plasma-treated fibers but PCHM grafts influence fiber–matrix interaction by mechanisms different from those operating for plasma-treated fibers.

4. CONCLUSIONS

Surface grafting of poly(butyl acrylate) and poly(cyclohexyl methacrylate) onto UHMWPE fibers has been achieved by our technique of ^{60}Co gamma irradiation at a relatively low dose rate and low total dose. Unlike some of the more prevalent surface modification schemes, this technique achieves surface grafting with complete retention of the exceptional UHMWPE fiber mechanical properties.

The effect of the relative molecular mobility of the polymer grafts on the level of fiber/matrix adhesion in thermoplastic matrix composites was evaluated by grafting two materials with very different glass transition temperatures. Poly(cyclohexyl methacrylate) ($T_g = 84\,°C$)-grafted fibers are an effective reinforcement in the terpolymer and polystyrene matrices. In particular, an approximately 30% increase in both the strength and the modulus was observed. On the other hand, poly(butyl acrylate) ($T_g = -56\,°C$) grafts were found to induce void formation, i.e. energy dissipation through plastic deformation and volume expansion, at the fiber/matrix interface in terpolymer composites. The latter results in a 75% increase in the elongation to failure.

REFERENCES

1. Z. F. Li, A. N. Netravali and W. Sachse, *J. Appl. Polym. Sci.* **44**, 333–346 (1992).
2. M. Nardin and I. M. Ward, *Mater. Sci. Technol.* **3**, 814–826 (1987).
3. S. L. Kaplan, P. W. Rose, H. X. Nguyen and H. W. Chang, *Proc. 33rd Int. SAMPE Symp.*, pp. 551–559 (1988).
4. B. Tissington, G. Pollard and I. M. Ward, *J. Mater. Sci.* **26**, 82–92 (1991).
5. Z. F. Li and A. N. Netravali, *J. Appl. Polym. Sci.* **44**, 319–331 (1992).
6. S. Holmes and P. Schwartz, *Composites Sci. Technol.* **38**, 1–21 (1990).
7. Z. F. Li, A. N. Netravali and W. Sachse, *J. Mater. Sci.* **27**, 4625–4632 (1992).
8. J. R. Brown, P. J. C. Chappell and Z. Mathys, *J. Mater. Sci.* **27**, 3167–3172 (1992).
9. A. Taboudoucht, R. Opalko and H. Ishida, *Polym. Composites* **13**, 81–85 (1992).
10. Y. Ulcay, I. Block and B. Pourdeyhimi, *PP (Am. Soc. Mech. Eng.) 32 (Composite Mater. Technol.)* **131**, 23–26 (1990).
11. Z. Feng and B. Ranby, *Angew. Makromol. Chem.* **195**, 17–33 (1992).
12. K. Kaji, Y. Abe, M. Murai, N. Nishioka and K. Kosai, *J. Appl. Polym. Sci.* **47**, 1427–1438 (1993).
13. A. Singh and J. Silverman (Eds), *Radiation Processing of Polymers*, Vol. 3. Hanser Publishers, New York (1992).
14. D. Cohn, A. S. Hoffman and B. D. Ratner, *J. Appl. Polym. Sci.* **33**, 1–20 (1987).
15. D. H. Cohn, A. S. Hoffman and B. D. Ratner, *J. Appl. Polym. Sci.* **29**, 2645–2663 (1984).
16. E. P. Goldberg and S. Kumar, unpublished results (1985).
17. P. G. Klein, J. Gonzalez-Orozco and I. M. Ward, *Polymer* **32**, 1732–1736 (1991).
18. P. G. Klein, D. W. Woods and I. M. Ward, *J. Polym. Sci., Part B: Polym. Phys.* **25**, 1359–1379 (1987).
19. K. Kaji, T. Okada and I. Sakurada, *Radiat. Phys. Chem.* **18**, 503–510 (1981).
20. B. D. Ratner, *J. Biomed. Mater. Res.* **27**, 837–850 (1993).
21. D. Briggs and M. P. Seah, *Practical Surface Analysis, Volume 1, Auger and X-ray Photoelectron Spectroscopy*, 2nd edn, pp. 444–447. John Wiley, New York (1990).
22. P. K. Mallick, in: *Fiber-Reinforced Composites*, L. L. Faulkner and S. B. Mankes (Eds), pp. 73–167. Marcel Dekker, New York (1988).
23. K. Friedrich, *Composites Sci. Technol.* **22**, 43–74 (1985).
24. L. H. Sperling, *Introduction to Physical Polymer Science*, pp. 65–71. John Wiley, New York (1992).

25. A. F. M. Barton, *CRC Handbook of Polymer–Liquid Interaction Parameters and Solubility Parameters*. CRC Press, Boca Raton (1990).
26. C. W. Frank and M. A. Gashgari, *Macromolecules* **12**, 163–165 (1979).
27. J. J. Arnold, M. P. Zamora and A. B. Brennan, unpublished results (1993).
28. P. A. Lovell, J. McDonald, D. E. J. Saunders, M. N. Sherratt and R. J. Young, in: *Rubber-Toughened Plastics*, C. K. Riew (Ed.), Adv. Chem. Ser. No. 222, pp. 61–77. American Chemical Society, Washington, DC (1989).

Fundamentals of Adhesion and Interfaces, pp. 143–160
D. S. Rimai, L. P. DeMejo and K. L. Mittal (Eds)

Probing adhesion spectroscopically: rheo-photoacoustic FT-IR spectroscopy and work of adhesion

B. D. PENNINGTON and M. W. URBAN*

Department of Polymers and Coatings, North Dakota State University, Fargo, ND 58105, USA

Revised version received 13 March 1995

Abstract—Physico-chemical changes at the interface between two polymer surfaces have a significant effect on adhesion and the molecular level processes associated with it. Although rheo-photoacoustic (RPA) Fourier transform infrared (FT-IR) spectroscopy was developed to establish the origin of polymer–polymer interfacial interactions, no correlations were made between the spectroscopic data and the work of adhesion. In this study, we attempt to provide a relationship between the vibrational energy changes in the C—H stretching region in polyethylene, induced by the interfacial stresses resulting from elongations of the polyethylene–polydimethylsiloxane double-layer system, and the work of adhesion. Stresses induced in the interfacial regions allow a correlation between the film thickness and the vibrational energy changes in the symmetric C—H bands.

Keywords: Rheo-photoacoustic FT-IR; molecular level adhesion; work of adhesion; thermo-plastic/thermosetting systems.

1. INTRODUCTION

Although measurement and the molecular-level understanding of adhesion have important implications, typically adhesion is monitored on a macroscopic level. However, macroscopic methods commonly used to measure practical adhesion, such as tensile and lap shear pull-off, or stress–strain measurements, provide information concerning only the overall forces that hold two substances together. Owing to the nature of such measurements, they do not provide the origin of the physico-chemical processes responsible for adhesion [1, 2]. In contrast, spectroscopic methods, if properly utilized, are capable of monitoring molecular-level mechanisms and the forces responsible for adhesion, and thus further enhance our knowledge of interactions promoting or disrupting adhesion [3, 4]. Although there are numerous spectroscopic techniques that can provide molecular-level information in the interfacial regions, in most cases, adhered substances need to be separated, thus destroying the interfacial interactions prior to the analysis. Among the various spectroscopic ap-

*To whom correspondence should be addressed.

proaches, Fourier transform infrared (FT-IR) spectroscopy, especially when equipped with reflection–absorbance (RA) or attenuated total reflectance (ATR) attachments, can provide useful information about the interfaces. However, RA requires highly reflective substrates [5–7], while ATR requires a coating to be in an intimate contact with the ATR crystal [8–10]. In contrast, photoacoustic (PA) FT-IR requires minimal or no sample preparation [11]; is capable of surface depth profiling [12, 13], and allows detection of non-equilibrium processes [14, 15]. For these reasons PA FT-IR spectroscopy, especially when combined with stress–strain measurements, can be used to monitor molecular-level changes while shear stresses are imposed in the interfacial regions [16, 17]. As this approach provides a correlation between the spectral changes and interfacial properties at the coating–substrate interface, it has potential for studying the work of adhesion [18]. In this study, we take advantage of the previous experimental findings that the vibrational energy changes occur when stresses are imposed in the interfacial regions, and using basic concepts of potential energy changes, we try to establish the relationship between the work of adhesion and vibrational energy changes. Before such a relationship can be established, it is necessary to develop a theoretical foundation for the approach which ultimately leads to determination of the work of adhesion. In the first part, we will outline the theoretical foundation of the approach, whereas the remaining sections will address the experimental results and discuss their significance in the context of theoretical considerations.

2. THEORETICAL FOUNDATION: WAVENUMBER SHIFT AND WORK OF ADHESION

Let us envision two materials attached to each other which are being sheared along their contact interface. This can be accomplished by designing an experiment in which a substrate is stretched and a top film contract in the opposite direction, thus inducing shear stresses in the interfacial regions. This is depicted schematically in Fig. 1. If the interfacial interactions are such that during application of the external forces the potential energy of the bonds involved in adhesion changes, the vibrational frequencies of these bonds will also change. If this is the case, the relationship between the wavenumber shift of a specific bond involved in that interfacial interaction and the forces imposed at the interface may be correlated to the work of adhesion. With this in mind, we will attempt to establish a relationship that correlates the wavenumber shift and the work of adhesion between two polymer surfaces.

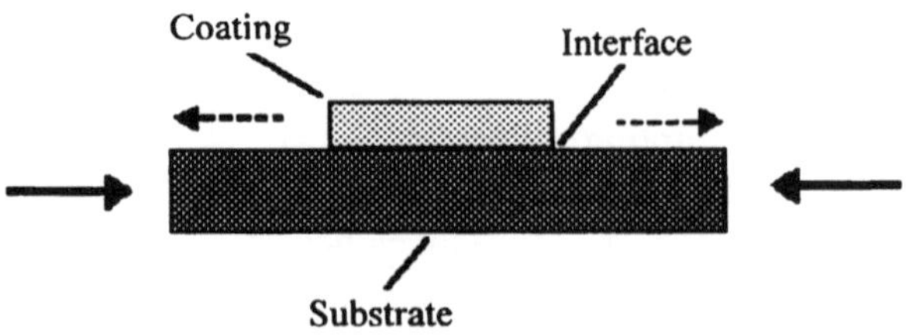

Figure 1. Diagram of the combined coating–substrate system with interfacial region for which the adhesion was studied.

A bilayer of two substances can be represented as a single interaction between a high molecular weight substrate molecule designated as 1, being perturbed from equilibrium by an interaction with a top film, designated as 2. Interaction between the top film and the substrate will result in the introduction of internal stresses into the bilayer system. If the two substances are in intimate molecular contact, the intermolecular potential of the unperturbed state of the substrate can be expressed as [19]

$$V_{11} = \frac{1}{2} f(r_1)^2 + \frac{1}{2} g(r_1)^3 + \cdots, \tag{1}$$

where r_1 is the change in distance of a substrate bond from equilibrium $(r - r_e)$, f is the quadratic force constant of the unperturbed substrate, and g is the cubic force constant. In this approach, it is assumed that segments closest to the interfacial region will be mostly affected, and, therefore, the vibrational energy will be perturbed to a larger extent. This assumption is quite reasonable in the context of numerous studies which illustrated that, indeed, the energies are mostly affected by the surrounding environment [20, 21]. As the substrate molecule (1) is perturbed from equilibrium resulting from an interaction with the top film (2), the vibrational potential of interaction upon perturbation will change and can be expressed as [19]

$$V_{\text{int}} = F(r_1) + G(r_1)^2 + \cdots, \tag{2}$$

where F is evaluated as $\langle(\partial V_{\text{int}}/\partial r)_e\rangle$, an ensemble average of the quadratic force over the perturbation coordinates, and G is the ensemble average for the cubic force over the perturbation coordinates. The total potential energy for such a system can be expressed as the sum of all interactions:

$$V_{\text{total}} = V_{11} + (V_{\text{int}})_{12} + V_{22}, \tag{3}$$

where V_{11} is the interaction potential between the substrate molecules, $(V_{\text{int}})_{12}$ is the interfacial interaction potential between the substrate (1) and top film (2) molecules, and V_{22} is the interaction potential between the film molecules themselves.

Since the second derivative of the potential energy with respect to a bond distance $(\partial^2 V/\partial r^2)$ is the force constant, let us determine the relationship between the force constant and the vibrational energy of a given bond. The substrate bond force constant can be calculated from a simple harmonic oscillator approximation which correlates the wavenumber with the bond strength and the reduced mass of the system. For such a case, the vibrational frequency of a bond is given as

$$\nu = \frac{1}{2}\pi c \left[\frac{k}{m_A m_B} (m_A + m_B) \right]^{1/2} = \frac{1}{2}\pi \left[\frac{k}{\mu} \right]^{1/2}, \tag{4}$$

where ν is the frequency in cm^{-1}, k is the force constant of the bond, m_A and m_B are the masses of the atoms in the substrate diatomic-like bond, μ is the reduced mass, and c is the speed of light. Qualitatively, the vibrational wavenumber of a bond will increase as the force constant of the bond increases. The change of the force constant as a function of the bond length can be determined from Badger's rule [22] using the expression $k\ (r_e - d_{ij})^3 = 1.86 \times 10^5$ (where k is in dynes/cm, r_e is the equilibrium distance in Å, and d_{ij} is a constant specific for the system under investigation).

The total force at the interface experienced by the interfacial bilayer can be determined by differentiating the total potential energy with respect to the change in bond distance to give

$$\frac{\partial V_{\text{total}}}{\partial r_1} = \frac{\partial V_{11}}{\partial r_1} + \frac{\partial (V_{\text{int}})_{12}}{\partial r_1}. \tag{5}$$

The total force is the sum of the force in the unperturbed substrate molecule and the force resulting from 1–2 interactions. For example, the intermolecular potential for an outer diatomic segment of a substrate will change as a result of interactions with a top film layer. Therefore, a vibrational frequency perturbation is anticipated. The bilayer interaction force and the force constant can be related to the frequency perturbation of the bond vibrational mode using [19]

$$\Delta\nu = \frac{1}{8}\pi c^2 \mu\nu \left[-\frac{3g}{f}\frac{\partial (V_{\text{int}})_{12}}{\partial r_1} + \frac{\partial^2 (V_{\text{int}})_{12}}{\partial r_1^2} \right], \tag{6}$$

where $\partial(V_{\text{int}})_{12}/\partial r_1$ is the nuclear configurational change effect and $\partial^2(V_{\text{int}})_{12}/\partial r_1^2$ represents the harmonic interaction effect. The unperturbed frequency and the frequency shift are designated as ν and $\Delta\nu$, respectively, while c is the speed of light and μ is the reduced mass. It should be realized that equation (6) was derived based on the following assumptions: the substrate bond is approximated by a diatomic-like molecule with a single vibrational degree of freedom; the vibrational movement is much faster than the translational or rotational motions, which, in our case, are negligible contributions; and the neighboring vibrations to the substrate bond will have a negligible effect on the bond's vibrational state. It should be noted that the above equation for $\Delta\nu$ indicates that the frequency shift is a function of the derivatives of the potential energies, but not necessarily the energies themselves.

With this in mind, let us identify those contributions to the overall potential energy that are expected to play a role when two molecular interfacial entities are close to each other. Because dispersive, repulsive, dipolar orientational, and inductive interactions have a significant effect on the net potential energy, let us estimate each of these energies and establish their contributions to the work of adhesion.

The dispersion forces are created by the Coulombic interactions of electrons and nuclei in atoms. This motion produces an instantaneous dipole that induces a secondary dipole in the neighboring atoms. Because the dispersion energy is inversely proportional to the sixth power of the interatomic distance, the dispersion component can be estimated using the London relationship [23]

$$V_{\text{dispersion}} = \frac{\alpha_1\alpha_2}{R_{12}^6}\frac{I_1 I_2}{I_1 + I_2}, \tag{7}$$

where α_1 and α_2 are the bond polarizabilities of components 1 and 2, I_1 and I_2 are the bond ionization energies, and R_{12} is the 'effective' interbond distance or the distance between the center of the substrate bond and the center of the coating bond.

The interaction of two molecules, one with a permanent dipole moment and the other with a zero dipole moment, will result in a secondary (induced) dipole resulting from the electrical field existing around the first molecule. The magnitude of the induced dipole can be shown as $V = \alpha E$, where E is the electric field strength. The attraction energy between the dipole-induced dipole was shown by Debye [24] to be

$$V_{\text{induction}} = \frac{\mu_1^2\alpha_2 + \mu_2^2\alpha_1}{R_{12}^6}. \tag{8}$$

Another interaction is a dipole–dipole interaction in which molecules with a permanent dipole moment interact with each other. The energy of this interaction for the dipolar orientation of species 1 and 2 can be expressed as [19]

$$V_{\text{dipolar orientation}} = \frac{\mu_1\mu_2}{R_{12}^3}\varphi, \tag{9}$$

where μ_1 and μ_2 are bond dipole moments and φ is the orientation factor.

While the dispersion, dipolar orientational, and inductive components can be classified as attractive contributions between two molecules, the opposing effect of repulsion is expected between the electrostatic charges in two interacting molecules. The repulsion component can be determined from

$$V_{\text{repulsion}} = a\exp\left(-bR_{12}\right), \tag{10}$$

where a and b are empirically derived constants. The repulsive component is seen to result from the force due to the overlap of the electronic orbital structures between the two molecules.

The component of the interaction force along a bond length r, when the substrate bond is at its equilibrium distance, can be described by

$$V' = \left\langle \frac{\partial V_{\text{int}}}{\partial r_e} \right\rangle. \tag{11}$$

The force for each type of interaction can be determined by taking the derivative of each potential energy term with respect to r_1, giving

$$V'_{\text{dis}} = \left[\frac{\alpha'_1}{\alpha_1} - 6\frac{R'_{12}}{R_{12}} + \frac{I'_1}{I_1} - \frac{I'_1}{I_1 + I_2}\right] V_{\text{dis}}, \tag{12a}$$

$$V'_{\text{ind}} = \left[2\frac{\mu'_1}{\mu_1} + \frac{\alpha'_1}{\alpha_1} - 6\frac{R'_{12}}{R_{12}}\right] V_{\text{ind}}, \tag{12b}$$

$$V'_{\text{or}} = \left[\frac{\mu'_1}{\mu_1} - 3\frac{R'_{12}}{R_{12}} + \frac{\varphi'}{\varphi}\right] V_{\text{or}}, \tag{12c}$$

$$V'_{\text{rep}} = \left[\frac{a'}{a} - (b'R_{12}) - (R'_{12}b)\right] V_{\text{rep}}, \tag{12d}$$

where the prime notation indicates the first derivative with respect to r_1.

The overall force of interactions is the sum of attractive and repulsive forces:

$$F = F_{\text{att}} + F_{\text{rep}} = (V'_{\text{dis}} + V'_{\text{or}} + V'_{\text{ind}}) + V'_{\text{rep}}. \tag{13}$$

The prevalence of an attractive or a repulsive force will depend on the separation distance between the interacting molecules, as shown in Fig. 2. At relatively large separation distances, attractive forces (F_{att}) will dominate the interactions. The contribution of the attractive force will decrease, however, when two species are a very short distance apart, and the repulsive forces (F_{rep}) will become dominant. At the point where the F_{att} and F_{rep} terms are equal, the equilibrium bond distance will be achieved, thus giving a total force on the system equal to zero.

As a result of perturbation, the vibrational frequency will change. At the same time, the bond distance and the force constant will also change. To account for these changes, one can adopt an approach similar to that proposed for solute–solvent interactions and assume that the substrate frequency shift produced by the attractive and repulsive forces of the surrounding medium are proportional to the changes in bond length [25, 26]:

$$\frac{\Delta\nu}{\nu_0} = -a\frac{r_1}{r_e}. \tag{14}$$

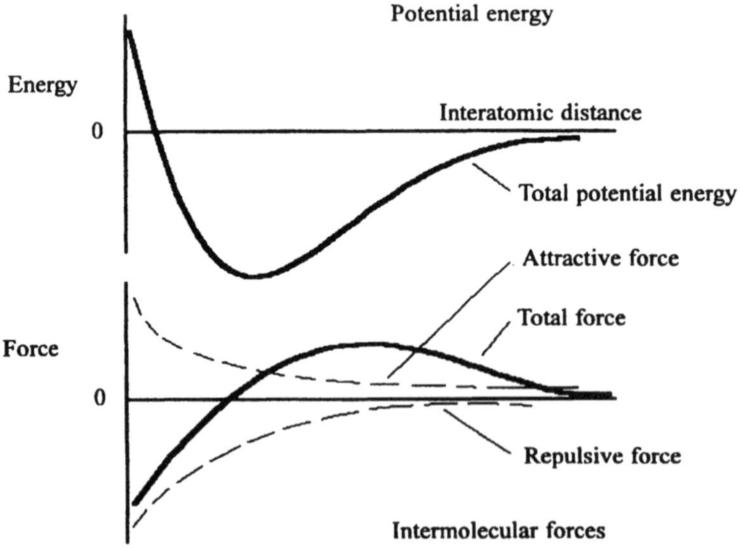

Figure 2. Diagram of the intermolecular forces and the corresponding potential energy between two molecules.

The equilibrium distance, r_e, is regarded as the distance at which the potential energy well reaches a minimum. The quadratic force constant of the bond, f, is related to the curvature of the well in equation (1). As the bond is displaced from equilibrium, the potential energy well can be estimated by combining the equilibrium intermolecular potential, V_{11} [equation (1)], and the interaction vibrational potential, V_{int} [equation (2)]. Under such circumstances, the ensemble average of the force, F, can be determined by differentiating the intermolecular potential energy of the unperturbed substrate with respect to the perturbation conditions. Differentiating the harmonic oscillator model equation [equation (1)], and neglecting the cubic term, the first derivative $(\partial V_{11}/\partial r_1)$ is evaluated to give

$$F = \frac{\partial V_{11}}{\partial r_1} = -f(r_1) = \left\langle \left(\frac{\partial V_{\text{int}}}{\partial r} \right)_{\text{e}} \right\rangle. \tag{15}$$

The force constant, f, is a property of the unperturbed molecule, whereas F is an ensemble average over the process of the perturbation.

The proportionality constant, a, in equation (14) depends on the properties of the interacting molecules. The g/f term can be determined by utilizing Badger's rule [27], whereas the G/F term can be evaluated using the empirical correlation from Oxtoby [28] and Herzfeld [29], giving

$$a = r_e \left[-\frac{3}{2}\frac{g}{f} + \frac{G}{F} \right], \tag{16}$$

where G is the ensemble average for the cubic force over the perturbation coordinates and g is the cubic force constant.

From the above relationships, W_A will be determined by integrating the force of interaction, F, over the limits of the equilibrium bond distance to infinity, giving the work of adhesion:

$$\begin{aligned} W_A &= -\int F\,\mathrm{d}r = -\int F_{\text{att}} + F_{\text{rep}}\,\mathrm{d}r \\ &= -\int V'_{\text{dis}} + V'_{\text{or}} + V'_{\text{ind}} + V'_{\text{rep}}\,\mathrm{d}r = \int f(r_1)\,\mathrm{d}r, \end{aligned} \tag{17}$$

where F represents the ensemble average of the quadratic force over the perturbation coordinates and f is the quadratic force constant of the unperturbed substrate.

3. EXPERIMENTAL

3.1. Sample preparation

Vinyl dimethyl-terminated poly(dimethyl siloxane) (PDMS) oligomer (molecular weight 17 250), neutral platinum–divinyl tetramethyl disiloxane complex, and tetrakis(dimethyl siloxy)-silane crosslinker were purchased from Hüls America. All materials were used as received. The polyethylene (PE) substrate (First Brands Co.) was prepared by compression molding 0.013 mm thick sheets to form a 0.3 mm thick substrate.

The vinyl-PDMS coating was prepared in the following manner. A 3.3 w/w% solution of 1 ppm platinum–divinyl tetramethyl disiloxane complex in xylene, 1.0 w/w% of tetrakis(dimethyl siloxy)-silane crosslinker in xylene, and 95.7 w/w% of PDMS oligomer in xylene were thoroughly mixed and applied to the PE substrate with a draw-down bar to obtain coatings with thicknesses of 120, 150, 220, 260, 300, and 340 μm. In an effort to prevent evaporation of reactants during the crosslinking reaction, the double-layer samples were placed in a sealed container.

3.2. Spectral measurements

Figure 3 shows a photoacoustic cell adapted for rheo-photoacoustic measurements. The sample compartment utilizes a stationary clamping block to anchor the sample and a movable clamping block which is attached to the lead screw. The lead screw and clamping block elongate the sample while holding it in a position for IR data acquisition. The lead screw, with 40 threads per inch, provides elongation of 0.635 mm

A Microphone compartment
B Lead screw
C Moveable clamping block
D Stationary clamping block
E Sample / sample area
F Helium inlet

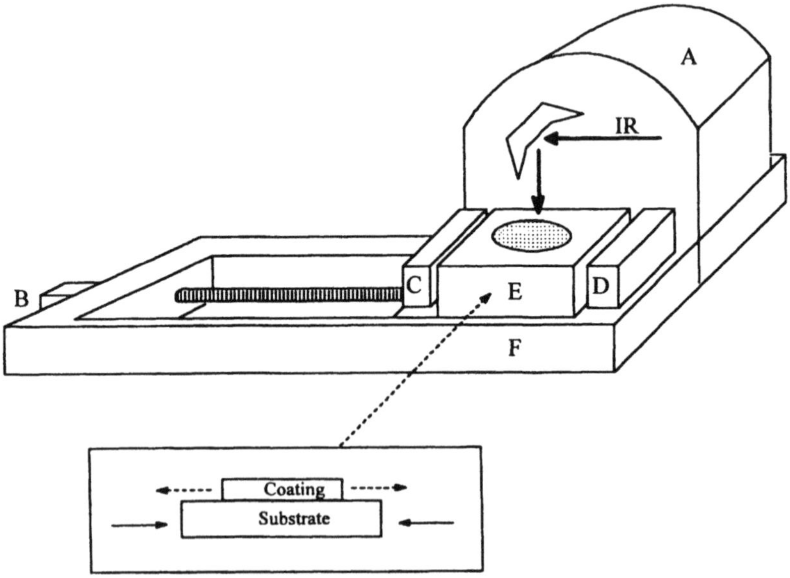

Figure 3. Schematic diagram of the photoacoustic FT-IR cell (Digilab) adapted for rheo-photoacoustic measurements. (A) Microphone compartment; (B) lead screw for sample elongation; (C, D) moveable and stationary clamping blocks; (E) sealed sample area; (F) gas inlet for purging the cell.

per turn and an approximate resolution of 40 μm. The cell design is such that the sample is sealed during data collection by clamping it between o-rings with the cell top. The cell top, which allows IR light to pass through the sample, contains a KBr window. The cell top, clamping blocks, and lead screw are attached by cap screws to the main cell frame and block. The cell block contains holes tapped into it which allow helium or other gases to purge and pressurize the cell, as well as to provide a path for photoacoustic waves to reach the microphone detector.

The cell design is universal, allowing measurements of a variety of samples such as fibers, composites, and films. The procedure of loading the sample, followed by data acquisition, consisted in removing the cell top, clamping the sample in the clamping blocks, turning the lead screw the desired amount to elongate the sample, purging the cell with helium for several minutes, pressurizing, and collecting 400 scans at a 4 cm^{-1} resolution on a Digilab FTS-10M FT-IR spectrometer. The modulation frequency of the spectrometer is a function of the frequency of the light source in wavenumbers and the optical velocity of the interferometer. For the Michelson interferometer, the relationship between the modulation frequency, f_1, wavenumber, ν, and mirror velocity, v, is expressed by the relationship $f_1 = 2\pi \nu v$ [30]. The co-added sample scans were ratioed to a carbon reference and then transferred to a computer for further spectral manipulations utilizing Spectra Calc software (Galactic Industries).

The rheo-PA FT-IR spectra were collected at 0.83% elongation increments. All measurements were carried out in triplicate to verify their reproducibility. Prior to the experiments on the double-layer samples, rheo-PA FT-IR measurements were conducted on polyethylene substrate and the PDMS coating alone to serve as references. The spectra were normalized to the 1265 cm^{-1} band due to the Si—C bending mode of $Si—CH_3$ groups, assuming that constant strain is maintained during data collection.

4. RESULTS

Figure 4 shows the frequency shift for the C—H symmetric stretching band in PE substrate plotted as a function of the elongation for PDMS layers with various thicknesses. It appears that the magnitude of the frequency shift of the symmetric C—H stretch is affected by the thickness of the PDMS film and by the degree of elongation of the PE substrate.

The PDMS–PE double layer at 0% elongation shows a frequency shift less than that for the substrate alone. At the early stages of elongation, as the substrate is stretched, the PDMS top layer will adhere to it and will undergo the same displacements. As the top layer follows the substrate elongations, its interfacial stresses decrease, which ultimately leads to an increase of the C—H bond vibrational energy. For thicker coatings, the decrease of this interfacial stress becomes less pronounced and a smaller shift of the C—H bond vibrational energy is observed. The magnitude of the C—H frequency shift reaches a maximum at 4.2% elongation. When interfacial attraction begins to diminish, the coating and the substrate are separated and the C—H frequency shift decreases to a value equal to that of the substrate.

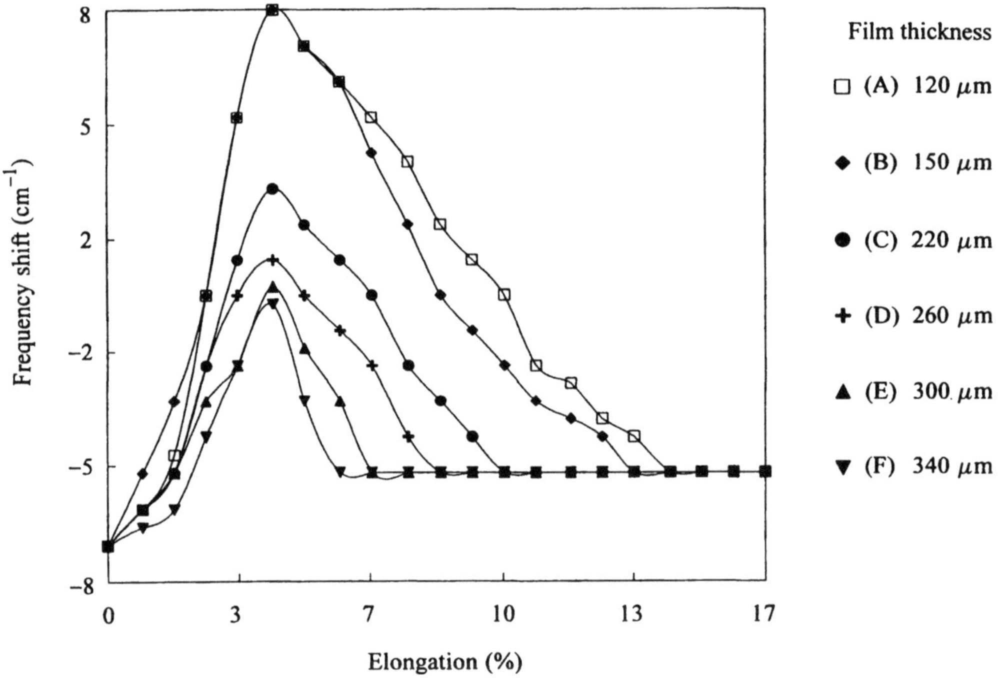

Figure 4. Frequency shift plotted as a function of the elongation for various coating thicknesses: A (□), 120 μm; B (◆), 150 μm; C (●), 220 μm; D (+), 260 μm; E (▲), 300 μm; F (▼), 340 μm.

Figure 5 shows the frequency shift plotted as a function of the film thickness for various elongations. It appears that the dependence of the frequency shift of the symmetric C—H stretch is mostly affected by the thickness of the PDMS films at 4.2% elongation. Elongations above and below 4.2% show less pronounced frequency shifts as a function of the coating thickness.

5. DISCUSSION

The central theme of this study was to elucidate the effect of different intermolecular forces on the work of adhesion. The spectroscopic results presented in Figs 4 and 5 show small changes in the substrate bond length produced by elongations of the adhered system inducing changes in C—H stretching up to 15 cm^{-1}. These, in turn, will result in large potential energy changes of the system. At the same time, the potential energy also changes with the bond distance.

The properties of the system can be determined by relating the ratio of the cubic force constant to the quadratic force constant (g/f) [equation (1)] by use of Badger's rule to the ratio of the corresponding forces (G/F) [equation (2)] using the empirical correlation expressed by equation (16), to give the proportionality constant, a. The experimentally measured frequency shifts shown in Fig. 4 can be used in equation (14) with the constant a to determine the change in C—H bond distance from equilibrium, r_1 (or $r - r_e$), which becomes the independent variable in the potential energy

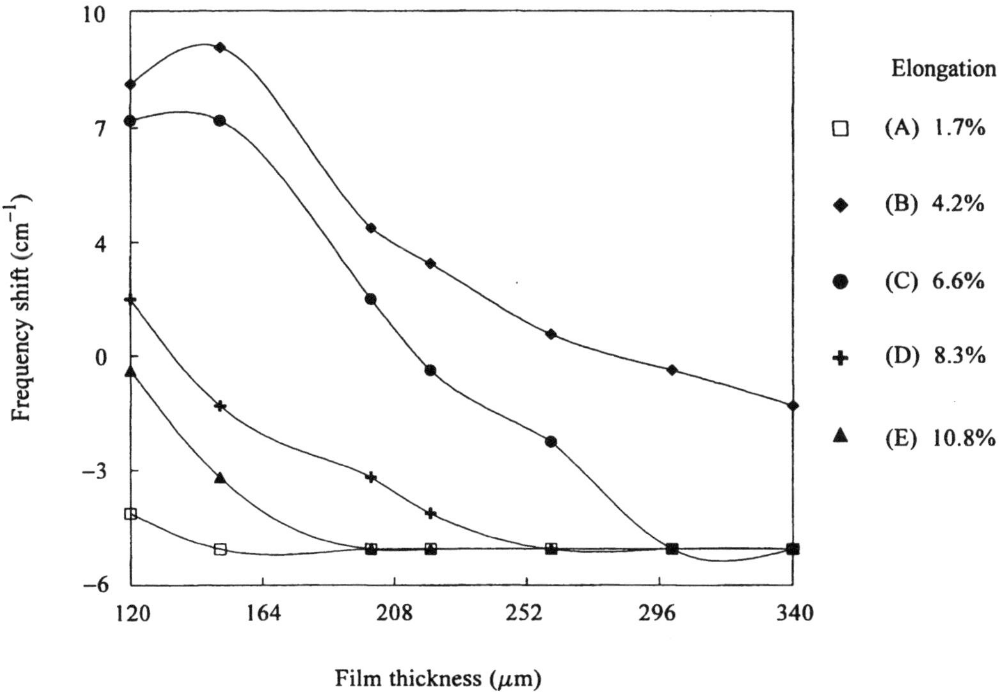

Figure 5. Frequency shift plotted as a function of the coating thickness for various elongations: A (□), 1.7%; B (◆), 4.2%; C (●), 6.6%; D (+), 8.3%; E (▲), 10.8%.

surface plots. The attractive or repulsive forces due to individual contributions can be calculated using the first derivative relations given by equations (12a)–(12d). With the corresponding force (or first derivative of the potential energy), the quadratic force constant for each contributing component to the total potential energy function can be determined from equation (15). The change of the quadratic force constant can be calculated from Badger's rule, which relates the changes of the force constant to the bond distance changes. Utilizing the harmonic oscillator relation [equation (1)], the potential energy at a specific bond length change with the corresponding force constant at that point can be derived, generating the three-dimensional potential energy surfaces for each of the attractive or repulsive forces.

In view of the above considerations and according to equations (12a)–(12d), there are dispersive, inductive, dipolar orientational, and repulsive intermolecular forces. The first three intermolecular forces are also collectively referred to as the van der Waals forces. The intermolecular force contributions to the work of adhesion can be attractive or repulsive. All intermolecular forces are proportional to the distance between molecules and depend on the geometry of the two interacting surfaces. The attractive terms can be determined by the summation of the contributions of the dispersive, inductive, and dipole orientational potential energies, giving a total attractive energy. The dispersive potential energy and dispersive attractive force contributions determined from relationships given by equations (7) and (12a) can be used to obtain the potential energy surface plot shown in Fig. 6. The inductive potential energy and

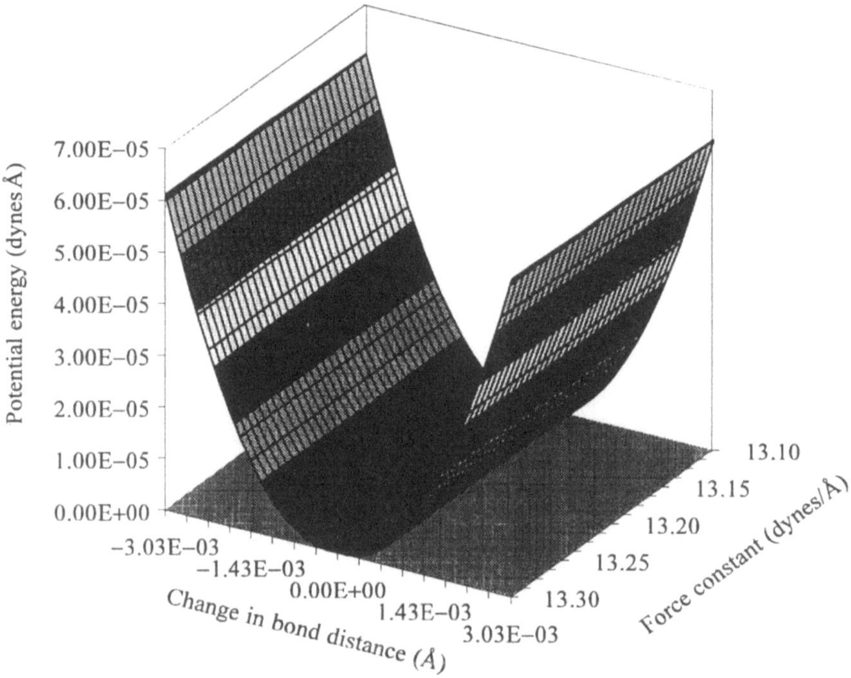

Figure 6. Dispersion potential energy surface calculated for an equilibrium bond distance of 1.11 Å, perturbed by ±0.00303 Å, and an equilibrium force constant of 13.2 dynes/Å.

force calculated from equations (8) and (12b), respectively, result in the potential energy surface plot illustrated in Fig. 7. Since the system under study does not involve dipole–dipole interactions, equation (9) indicates that the energy of this interaction is essentially zero.

Before we continue discussion concerning intermolecular forces at the interfaces, it is important to realize the significance of the potential energy curves shown in Figs 6–9. Their importance comes from the fact that they serve as a measure of how each interaction is affected by changes in the bond distance and the force constant. However, van der Waals forces are not the only forces that are responsible for interfacial interactions. As the substrate and top layer molecules approach each other, the interfacial distance (r) decreases, and at some point, the repulsive forces will counteract the attractive forces resulting from electronic structure overlap of the two surfaces. Although the repulsive behavior is far more complex than illustrated here, it can be approximated by several repulsion models [31]. The repulsive potential energy and repulsive force are obtained from equations (10) and (12d) and the resulting potential energy surface plot is shown in Fig. 8. Combining the results illustrated in Figs 6–8, the plot representing the total potential energy surface can be obtained and is shown in Fig. 9. The scaling of the plot axis will vary each specific interaction.

Having established each potential energy component contributing to the interfacial interactions, let us now calculate the actual work of adhesion. Due to the absence of high polarity at the interfaces, the attractive interactions are expected to be of primarily

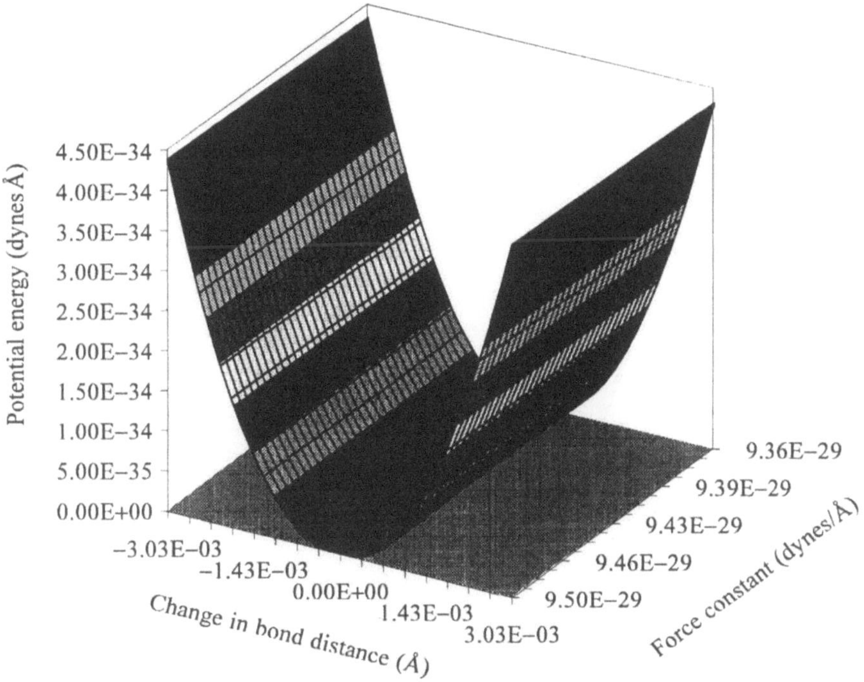

Figure 7. Induction potential energy surface calculated for an equilibrium bond distance of 1.11 Å, perturbed by ±0.00303 Å, and an equilibrium force constant of 9.43×10^{-29} dynes/Å.

dispersive nature. The presence of polar groups would introduce a permanent dipole moment, giving a more significant contribution to the total forces from the inductive and dipolar orientational effects. Using the results presented in Figs 6–8 and combining with equation (17), the estimated value of the work of adhesion for the dispersive component (Fig. 6) is 0.0539 J/m^2, which accounts for nearly 100% of the total attractive work of adhesion. The repulsive interaction component (Fig. 8) contributing to the total work of adhesion has a value of 0.0570 J/m^2, and although the magnitude is similar to the dispersive component, it will act in an opposing manner. It should be realized that the work of adhesion is the reversible work necessary to separate a unit area of the interface between two solids into two solid–gas interfaces and in the case of adhesion between PDMS and PE, its magnitude is significantly lower due to the nature of the adhesion forces between the two polymer films. Since PE and PDMS are not expected to adhere strongly, the work of adhesion values are expected to be small in relation to other systems, for example polymer–metal interfaces.

Table 1 summarizes the results for the work of adhesion from all contributing components. If one envisions the repulsive and attractive components in a state of equilibrium, the intermolecular distance will determine to what extent each component dominates. For an equilibrium situation, the attractive and repulsive forces are equal. As external work is imposed on the system, the bond distance decreases and the repulsive component will dominate. In contrast, as the bond distance increases from the equilibrium bond distance, the attractive forces will prevail. This offsetting

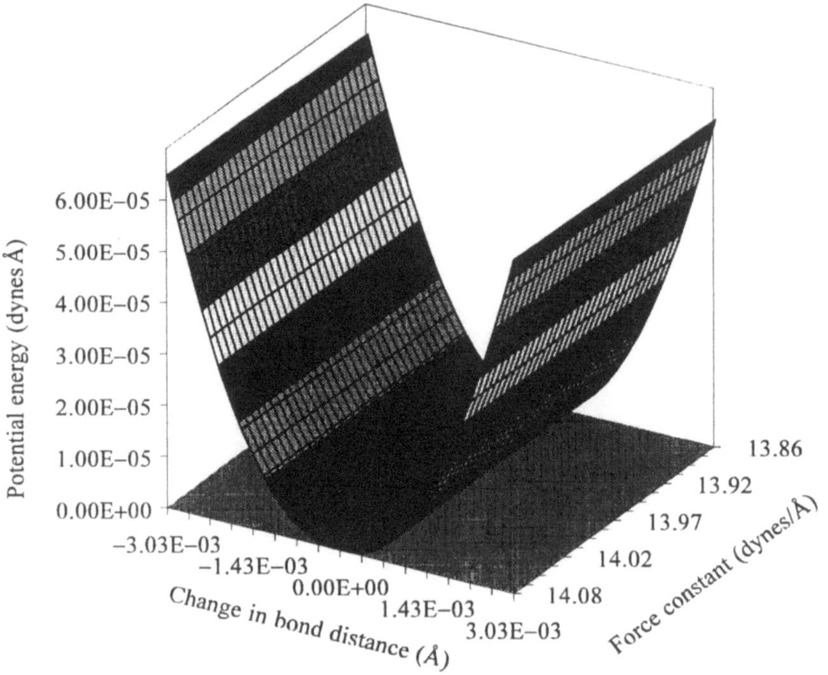

Figure 8. Repulsion potential energy surface calculated for an equilibrium bond distance of 1.11 Å, perturbed by ±0.00303 Å, and an equilibrium force constant of 14.0 dynes/Å.

Table 1.
Work of adhesion results from the theoretical model

Potential energy (ergs/bond interaction)
$V_{\text{dispersion}} = 2.20 \times 10^{-17}$
$V_{\text{repulsion}} = 1.37 \times 10^{-10}$
$V_{\text{orientation}} = 0 \ (\mu_1 = 0)$
$V_{\text{induction}} = 8.42 \times 10^{-16}$
Force required to separate (dynes/bond interaction)
$V'_{\text{dispersion}} = 5.67 \times 10^{-3}$
$V'_{\text{repulsion}} = 6.00 \times 10^{-3}$
$V'_{\text{orientation}} = 0$
$V'_{\text{induction}} = 4.05 \times 10^{-32}$
$V'_{\text{total}} = 3.30 \times 10^{-4}$
Work of adhesion (J/m^2)
$W_{\text{A (dispersion)}} = -5.39 \times 10^{-2}$
$W_{\text{A (repulsion)}} = 5.70 \times 10^{-2}$
$W_{\text{A (orientation)}} = 0$
$W_{\text{A (induction)}} = -3.85 \times 10^{-31}$
$W_{\text{A (total)}} = 0.0031$

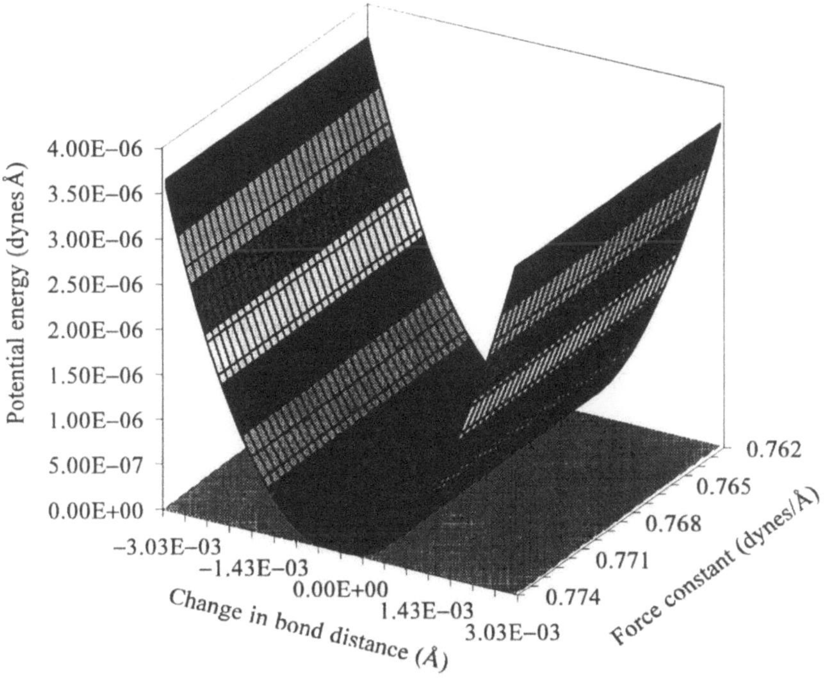

Figure 9. Total potential energy surface calculated for an equilibrium bond distance of 1.11 Å, perturbed by ±0.00303 Å, and an equilibrium force constant of 0.768 dynes/Å.

relationship of the attractive and repulsive forces to produce the total force of a system is shown in Fig. 2, along with the corresponding potential energy function. The attractive and repulsive interactions will offset at the equilibrium bond distance and no perturbation of the C–H stretching bond is anticipated, thus giving no frequency shift for the thickest film as illustrated in Fig. 4.

At this point, let us go back to the experimental results shown in Fig. 4 and correlate these results with our theoretical considerations. Due to shear forces induced between PDMS film on the PE substrate, the C–H bond distance of the PE substrate changes. Upon elongation, as illustrated in Fig. 1, an initial increase in frequency is observed (Fig. 4). Since the PDMS–PE system attains a degree of interfacial stress during the crosslinking process, an initial increase of the C–H stretching frequency is detected. As the interfacial stress is released by the sample elongation, the bond distance decreases, giving rise to the frequency shift increase. As the frequency shifts level off, the bond distance will approach the equilibrium distance. Further elongation of the substrate produces a decrease from the equilibrium frequency due to elongation of the bonds in the substrate.

The change in the bond distance can be used to calculate the work of adhesion as a function of the film thickness. Assuming an equilibrium C–H bond length equal to 1.11 Å, the shift of the C–H stretching frequency can be correlated to the change of the bond length [equation (14)], and the effect of film thickness on the work of adhesion can be studied by determining the change in bond distance from its

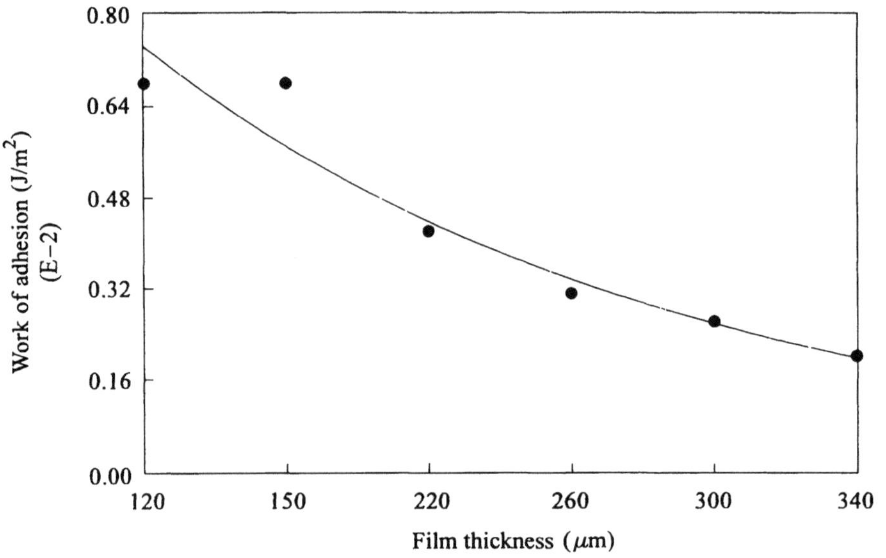

Figure 10. Work of adhesion at the maximum frequency shift plotted as a function on the film thickness.

equilibrium. Because the point of maximum frequency shift in Fig. 4 is a transition point between the release of the interfacial stress and the perturbation due to the elongational forces, these frequency shifts can be used to calculate the relationship between the work of adhesion and film thickness. The results are shown in Fig. 10.

Although these results may be a bit surprising, one needs to consider viscoelastic relaxation processes in and near the interfacial regions. The increase in film thickness will introduce a larger number of molecular layers that are not in close proximity to the interfacial region. This will tend to decrease the work of adhesion for thicker films because the strength of interaction decays as the molecules are further away from the interfacial region which contributes a smaller effect to the work of adhesion. The cohesive work becomes more pronounced as the distance from the interface is increased. Relaxation by a viscoelastic mechanism will also have a significant effect on the interacting groups resulting from the increased film thickness. The viscoelastic response allows the shear forces from elongation of the substrate to dissipate, thus resulting in smaller interfacial perturbations. This relaxation is expected to increase with thicker top films due to larger viscoelastic contributions, thus affecting the work of adhesion. Although one can bring up the issue that the practical adhesion may increase with film thickness, it should be kept in mind that cohesive forces may also contribute to this process. Considering our experimental set-up, no forces applied to the coating except for interfacial stresses were induced by elongation of the substrate.

Finally, it should be noted that the depth of penetration, in the present experimental set-up, is not only a function of the modulation frequency, but also depends very strongly on thermal and optical properties. Therefore, in order to be able to estimate the actual penetration depth, a knowledge of these material constants is necessary. This is, however, beyond the scope of this study, but it should be realized that both PDMS and PE are optically transparent and thermally thick. Therefore, the depth of light penetration will be high, but the amount of heat at the surface will be low. In our case, we are not concerned with the thickness of the PDMS layer because the passage of the thermal energy does not occur through the top layer. Instead, we take advantage of 'interfacial leakage' of the thermal signal.

6. CONCLUSIONS

The utilization of rheo-photoacoustic stress–strain measurements allows molecular-level forces at bilayer interfaces to be monitored. By measuring the vibrational energy changes upon perturbation of the interfacial region, correlations can be made between the spectroscopic data and the work of adhesion. In this study, we showed that the total potential energy changes in the C–H stretching region in polyethylene result from attractive and repulsive intermolecular forces occurring in the interfacial region between PDMS and PE. The shear forces introduced by elongation will produce small changes in the substrate bond length, affecting the potential energy of the system to a greater extent. The changes in the bond length are related to the change in the force constant, through Badger's rule, which, in turn, allows us to obtain three-dimensional potential energy diagrams. Using these plots, the work of adhesion can be calculated. The transition point between the release of interfacial stress and the onset of elongational perturbation is seen as the point of maximum frequency shift, which allows correlation between the film thickness and the work of adhesion. The trend of decreasing work of adhesion as a function of the film thickness is a result of viscoelastic relaxation processes and the relative proximity of molecular layers to the interfacial regions. This measurement of W_A is dependent on the experimental set-up being used and will reflect the region of the sample being monitored; in this case, the interfacial region.

REFERENCES

1. S. G. Croll, in: *Adhesion Aspects of Polymeric Coatings*, K. L. Mittal (Ed.), pp. 107–129. Plenum Press, New York (1983).
2. R. O. Carter, R. A. Dickie, J. W. Holubka and N. E. Lindsay, *Ind. Eng. Chem. Prod. Res. Dev.* **29**, 48 (1989).
3. K. L. Mittal, *Polym. Eng. Sci.* **17**, 467 (1977).
4. M. W. Urban, *J. Adhesion Sci. Technol.* **7**, 1–47 (1993).
5. W. C. Wake, *Adhesion and the Formulation of Adhesives*, 2nd edn. Applied Science, New York (1982).
6. J. D. Swalen and J. F. Rabolt, in: *Fourier Transform Infrared Spectroscopy: Applications to Chemical Systems*, J. R. Ferraro and L. J. Basile (Eds), p. 283. Academic Press, New York (1979).
7. J. Farenfort, *Spectrochim. Acta* **17**, 698 (1961).
8. A. M. Tiefenthaler and M. W. Urban, *Appl. Spectrosc.* **42**, 163 (1988).
9. J. B. Huang and M. W. Urban, *Appl. Spectrosc.* **46**, 1666 (1992).
10. J. B. Huang and M. W. Urban, *Appl. Spectrosc.* **46**, 1014 (1992).
11. A. Rosencwaig, *Photoacoustics and Photoacoustic Spectroscopy*. Wiley, New York (1980).
12. M. W. Urban, *Chemtech* **21**, 38 (1991).
13. M. W. Urban and C. D. Craver (Eds), *Structure–Property Relations in Polymers*, ACS Adv. Chem. Ser. No. 236. American Chemical Society, Washington, DC (1993) (and refs cited therein).

14. J. B. Huang and M. W. Urban, *J. Chem. Phys.* **98**, 5259 (1993).
15. J. B. Huang and M. W. Urban, *J. Chem. Phys.* **100**, 4509 (1994).
16. W. F. McDonald, H. Goettler and M. W. Urban, *Appl. Spectrosc.* **43**, 1387 (1989).
17. W. F. McDonald and M. W. Urban, *J. Adhesion Sci. Technol.* **4**, 751 (1990).
18. M. W. Urban and H. Goettler, US Patent 5,036,708 (1991).
19. A. M. Benson and H. G. Drickamer, *J. Chem. Phys.* **27**, 1164 (1957).
20. M. H. Brooker, D. J. Jobe and V. C. Reinsborough, *J. Chem. Soc., Faraday Trans. 2* **80**, 73 (1984).
21. M. W. Urban, in: *Polymers in Aqueous Media*, J. E. Glass (Ed.), ACS Adv. Chem. Ser. No. 223, Ch. 15. American Chemical Society, Washington DC (1989).
22. R. M. Badger, *J. Chem. Phys.* **2**, 128 (1934).
23. F. London, *Trans. Faraday Soc.* **33**, 8 (1937).
24. P. J. W. Debye, *Phys. Z.* **21**, 178 (1920).
25. M. R. Zakin and D. R. Herschbach, *J. Chem. Phys.* **83**, 6540 (1985).
26. M. R. Zakin and D. R. Herschbach, *J. Chem. Phys.* **85**, 2376 (1986).
27. D. R. Herschbach and V. W. Laurie, *J. Chem. Phys.* **35**, 458 (1961).
28. D. W. Oxtoby, *J. Chem. Phys.* **70**, 2605 (1979).
29. R. N. Schwartz, Z. I. Slawsky and K. F. Herzfeld, *J. Chem. Phys.* **20**, 1591 (1952).
30. M. W. Urban and J. L. Koenig, *Appl. Spectrosc.* **40**, 994 (1986).
31. J. O. Hirschfelder, C. F. Curtiss and R. B. Bird, *Molecular Theory of Gases and Liquids*. Wiley, New York (1954).

Fundamentals of Adhesion and Interfaces, pp. 161–177
D. S. Rimai, L. P. DeMejo and K. L. Mittal (Eds)

Formation of polyimide–Cu complexes: improvement of direct Cu-on-PI and PI-on-Cu adhesion

K.-W. LEE,* G. F. WALKER and A. VIEHBECK

IBM Research Division, T. J. Watson Research Center, Yorktown Heights, NY 10598, USA

Revised version received 25 January 1995

Abstract—Polyimides containing triazole or imidazole functionalities have been synthesized. Poly(3,3′,4,4′-benzophenone tetracarboxylic dianhydride-3,5-diamino-1,2,4-triazole) (BTDA-TADA) contains triazole groups as repeat units and poly(4,4′-oxydiphthalic anhydride-1,3-aminophenoxybenzene-8-azaadenine) (ODPA-APB-8-azaadenine) consists of triazole (8-azaadenine) groups as the end caps. While the BTDA-TADA polyimide starts to decompose at 350°C, the ODPA-APB-8-azaadenine polyimide is thermally stable at 400°C. The peel strength of copper to BTDA-TADA polyimide without surface modification is 200–400 J/m^2. For the adhesion of polyimide to copper, ODPA-APB-8-azaadenine polyisoimide (1.0 μm) and poly(pyromellitic dianhydride-oxydianiline) (20 μm) were coated onto the copper substrate and then the two layers were cured together to polyimides at 400°C. Peel strengths of 500–800 J/m^2 were obtained. The failure of the copper/polyimide interface by peeling either the copper or the polymer layer occurred in the near-interface region of the polymer. Both PI/Cu and Cu/PI adhesion was enhanced due to the formation of Cu–polyimide complexes. In the case of PI/Cu adhesion, the polymer chain flexibility of ODPA-APB-8-azaadenine polyimide also plays a significant role in improving adhesion.

Keywords: Polyimide; copper; adhesion; passivation; lamination.

1. INTRODUCTION

Polyimides are used as dielectric layers in a variety of microelectronic applications, as they have good processability, a low dielectric constant, high thermal stability, low moisture absorption, and good mechanical properties [1–3]. As shown in Scheme 1, copper patterns are deposited on polyimide surfaces and in vias to form electrically conducting lines in electronic packages and chips. However, polyimide materials having the best chemical and mechanical properties often give poor adhesion. Specifically, the adhesion of copper directly deposited onto high-temperature polyimides used in industry is always weak, regardless of the surface modification techniques. To improve the adhesion of copper to polyimide (metal-on-polymer or metal/polymer), a polymer surface is modified and then a thin layer of Cr, Al, Ti, Ni, Co, W, Ta, Pd

*To whom correspondence should be addressed.

or alloys of these metals is employed as an adhesion-promoting layer [4, 5] or a seed layer [6, 7]. The application of these adhesion-promoting layers requires deposition equipment and a number of process steps. By improving copper adhesion to polyimide without an adhesion-promoting metal layer, the cost of the metallization process can be reduced.

In multilevel thin film packaging, the polyimide precursor is spin-coated onto the substrate consisting of metal and polyimide (polymer-on-metal or polymer/metal) and then cured at high temperature. If a polyimide precursor is directly coated onto copper, either the adhesion of polyimide to copper is weak or the polyamic acid reacts with copper [8, 9], leaving copper oxide particles in the cured polyimide layer, which results in an increase of the dielectric constant and the degradation of material properties. Kim *et al.* [5, 10] reported that poly(pyromellitic dianhydride-oxydianiline) (PMDA-ODA) amic acid in a solvent reacts with copper during a high-temperature ($\geqslant 300\,^{\circ}C$) curing process to form copper oxide particles which diffused into the bulk of PMDA-ODA polyimide. Transmission electron microscopy (TEM) showed very fine copper oxide particles in the PMDA-ODA layer. However, PMDA-ODA poly(amic ethyl ester) (ester form of polyamic acid) did not react with copper. Adhesion of PMDA-ODA poly(amic ester) on copper provided $\leqslant 100$ J/m^2 peel strength, which is a relatively weak adhesion in electronic applications. Shih *et al.* [11] also reported a similar study which also included measurement of dielectric constants. The

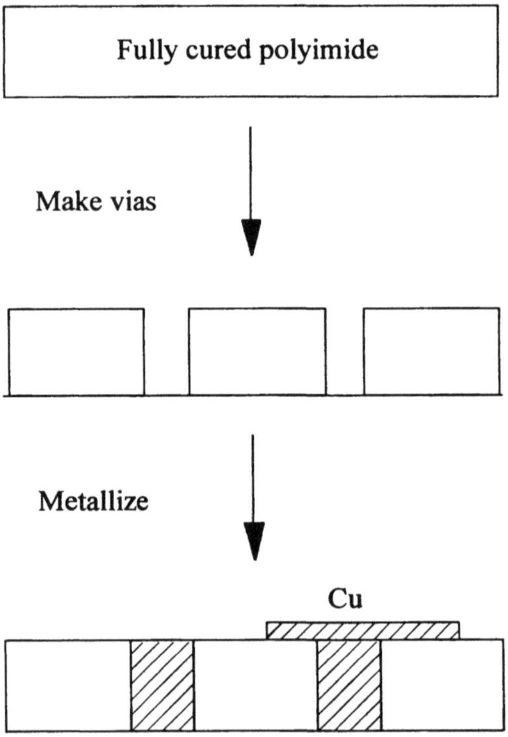

Scheme 1. Thin film packaging process.

PMDA-ODA film with copper oxide particles had almost twice the dielectric constants (≅7) compared with the pure material.

To prevent the reaction of poly(amic acid) with copper as well as to improve the adhesion of polyimide to copper, a thin passivation/adhesion-promoting layer of metals such as Cr, Ni, Ta, Co, etc. is coated onto copper. To simplify the manufacturing process, we have studied a new process for both organic passivation and adhesion promotion, particularly using polymers. As shown in Scheme 2, a thin layer of a capping polymer is coated onto a copper substrate followed by a thick layer of a di-

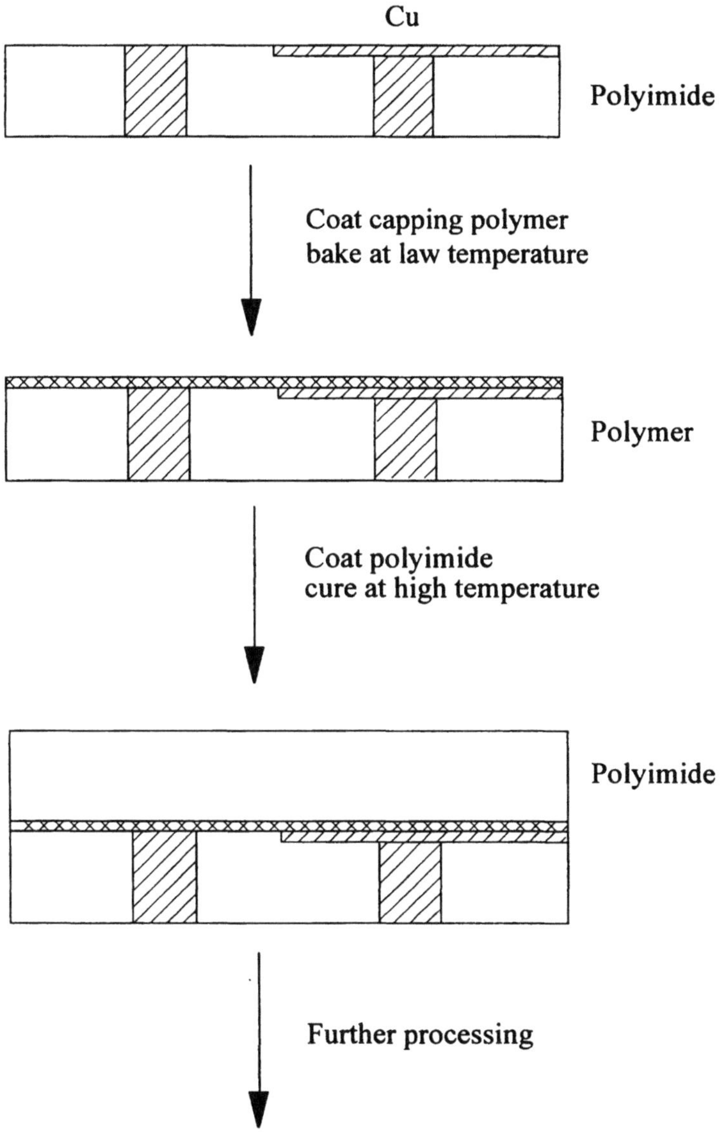

Scheme 2. Polymer capping process.

electric polyimide. The thin polymer layer will prevent permeation of polyamic acid solution used for the thick layer in the coating and baking processes and will form a stable copper complex to enhance the adhesion of a dielectric polyimide to copper by acting as an adhesion promoter. Thus, this thin layer acts not only as a passivation layer between PI and Cu, but also as an adhesion promoter for the PI/Cu interface.

An organic compound which interacts with copper to form a stable complex is a possible candidate for the passivation/adhesion promotion layer. It is known that compounds containing a sulfide, triazole, or imidazole functionality can form a complex with copper, and these compounds have been used to protect the copper surface from corrosion [12–19]. Xue *et al.* [16, 17] proposed that the N—H bond of triazole or imidazole reacts with copper to form an ionic complex as shown in Scheme 3. If a polymer consists of triazoles or imidazoles, it is expected that copper will form complexes with the polymer. It has been postulated that polybenzimidazole and mercaptobenzimidazole coated on a copper substrate form complexes with copper [19]. Although these polymers prevent copper from oxidation, a certain amount of oxidation was observed upon heating the sample at 200 °C for over 10 h. A uniform coating without a significant number of pinholes was important. Eng and Ishida [20] reported that poly(*N*-vinylimidazole) and poly-(4-vinylimidazole) inhibited the corrosion of copper at elevated temperature up to 330 °C. The capability of corrosion inhibition degraded as the polymer degraded at ⩾400 °C. Ishida and co-workers [21, 22] recently reported that poly(γ-methacryloxypropyl trimethoxysilane-co-*N*-vinylimidazole) was hydrolyzed to form poly(γ-methacryloxypropyl trihydroxysilane-co-*N*-vinylimidazole) and the latter protected copper from corrosion. They proposed that trihydroxysilane groups in the polymer chain also improved the adhesion to copper.

Recently we have found that the adhesion of a polymer to copper (cleaned with acid to remove excess copper oxides) is better with a compound containing a triazole or an imidazole functionality than with a silane coupling agent such as hydrolyzed 3-aminopropyltriethoxysilane. If the triazole or imidazole functionality of a polymer forms a complex with copper, adhesion between the polymer and the copper will be enhanced. A processable polyimide which contains the triazole functional groups, especially with the free N—H bonds that will form a complex with copper as shown in Scheme 3, has been developed for use as a passivation/adhesion promotion layer.

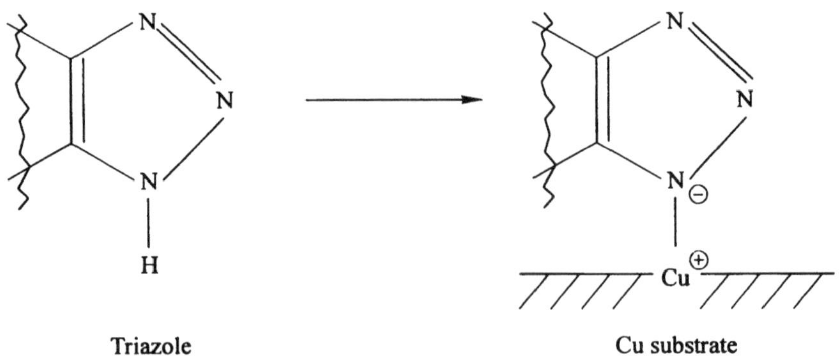

Scheme 3. Reaction of a triazole functionality with copper.

(a)

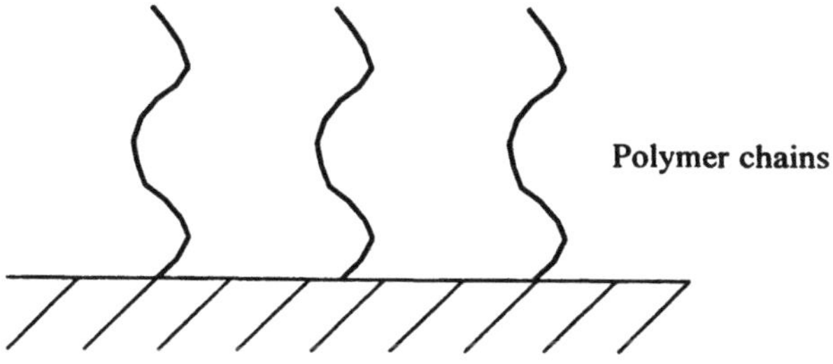

(b)

Figure 1. Two different types of polymer–copper interaction.

Here we report on the synthesis of new polyimides containing triazole functional groups and the adhesion between these polyimides and copper. The triazole functional groups which interact with copper can be either repeat units or end-groups in the polymer. In order to enhance the thermal stability of a polymer, the triazole or imidazole functionality can be introduced onto the end of a thermally stable polymer (this work focused on polyimide) in the expectation that at least one end of the polymer will anchor on the copper surface as displayed in Fig. 1a. If the triazole or imidazole functionality is a repeat unit of a polymer, numerous contact points between the polymer chain and the copper substrate are anticipated, as shown in Fig. 1b to provide enhanced polymer/metal adhesion.

2. EXPERIMENTAL

2.1. Materials

3,5-Diamino-1,2,4-triazole (TADA), anhydrous 1-methyl-2-pyrrolidinone (NMP), 8-azaadenine, 3,3′,4, 4′-benzophenone tetracarboxylic dianhydride (BTDA), and *N,N*′-dicyclohexylcarbodiimide (DCC) were obtained from Aldrich. 1,3-Aminophenoxybenzene (APB) and 4,4′-oxydiphthalic anhydride (ODPA) were supplied by National Starch and Chemical Corp. and Occidental Chemical Corp., respectively. Poly(biphenyl dianhydride-*p*-phenyl diamine) (BPDA-PDA)amic acid in NMP, Pyralin

PI 2611, was purchased from DuPont. A high-purity ODPA-APB-8-azaadenine was provided by National Starch and Chemical Corp.

2.2. *Methods*

Syntheses of polyimides were carried out under a dry nitrogen atmosphere. Infrared spectra were recorded on a Nicolet 510P FTIR spectrometer equipped with an MCT-B detector. A Harrick Segal attachment was used for external reflectance infrared (ERIR) spectroscopy. X-ray photoelectron spectra (XPS) were taken on a Perkin-Elmer Physical Electronics Model 5500 spectrometer with Mg K_{α} excitation. The electron take-off angle was 45° unless specified. The peel strengths were measured by 90° peel using an Instron with a 1.5 mm/min peel rate. The reported values are the average of at least three measurements. Transmission electron micrographs were obtained using a Philips 400T TEM microscope and the specimens for the polyimide/copper interface were prepared by cutting the polyimide/Cu films with a Reichert Ultracut ultramicrotome.

2.3. *Synthesis of BTDA-TADA polyimide*

Into a 300 ml flask which had been dried and flushed with nitrogen for 20 min, 3,5-diamino-1,2,4-triazole (TADA) (50 g; 0.155 mol) was added and then the flask was flushed with nitrogen again. BTDA (15.4 g; 0.155 mol) was dissolved in anhydrous NMP and this solution was cannulated into flask containing TADA. The mixture was stirred until all the BTDA had reacted. This reaction was followed by IR spectroscopy. The mixture was kept stirring for 24 h more to make sure that the condensation reaction was complete. The product was BTDA-TADA poly(amic acid). IR (KBr) 1719 (s), 1667 (s), 1640 (s), 1603 (m), 1543 (m), 1511 (m), 1428 (m), 1406 (m), 1300 (s), 1263 (s), and 1238 (s) cm^{-1}. The BTDA-TADA poly(amic acid) solution was spin-coated onto a substrate (quartz, Si wafer, or KBr disc) and cured at 300 °C for 1 h. Imidization was complete. IR (KBr) 1792 (m), 1738 (vs), 1674 (m), 1561 (m), 1514 (m), 1458 (m), 1426 (w), 1391 (m), 1348 (s), 1291 (m), and 1244 (s) cm^{-1}.

2.4. *Synthesis of ODPA-APB-8-azaadenine polyimide*

One equivalent of APB and 0.1 equivalent of 8-azaadenine were placed in a 300 ml flask equipped with a magnetic stirrer. 50 ml of NMP was cannulated into this flask and the mixture was stirred until they dissolved. A solution of ODPA in anhydrous NMP (50 ml) was cannulated into the flask and stirred overnight. The product was poly(ODPA-APB-8-azaadenine)amic acid, which was converted into the corresponding polyisoimide by reaction with DCC at 5 °C for 12 h. The polyisoimide product was purified by recrystallization. A higher-purity material was later provided by Dr Pat Jobe of National Starch. IR (KBr) 1792 (m), 1738 (vs), 1674 (m), 1561 (m), 1514 (m), 1458 (m), 1426 (w), 1391 (m), 1348 (s), 1291 (m), 1244 (s), 1181 (w), 1156 (w), 1096 (m), and 1042 (w) cm^{-1}.

2.5. Adhesion measurement

The Cu/PI adhesion sample was prepared as follows. A solution of BTDA-TADA poly(amic acid) was spin-coated onto a quartz substrate which previously had been treated with a 0.1% aqueous solution of γ-aminopropyltriethoxysilane coupling agent, and then curred at 300°C for 60 min. A 2.0-μm-thick copper layer was evaporated onto the polyimide film without any surface treatment and then an 8.0-μm-thick copper layer was electroplated. The copper layers (2–3 mm wide) were repeatedly peeled from the polyimide film. To test the PI/Cu adhesion, a solution of ODPA-APB-8-azaadenine in NMP and cyclopentanone (50 : 50) was spin-coated onto the copper substrate and baked at 200°C for 30 min. The thickness of this layer was 0.5–1.0 μm. Around 15-μm-thick (after curing) PMDA-ODA (or BPDA-PDA if necessary for comparison) layer was overcoated. The whole layer was cured at 150°C for 30 min, 230°C for 30 min, 300°C for 30 min, and 400°C for 60 min. The polyimide layers (5 mm wide) were peeled from the copper substrate.

3. RESULTS AND DISCUSSION

Three different types of system have been studied for adhesion in this work: the first was copper-on-polyimide adhesion, in which copper was evaporated onto a fully imidized polyimide; the second polyimide-on-copper adhesion, in which a solution of a polyimide precursor was spin-coated onto a copper substrate and then cured to polyimide; and the third adhesion of a polyimide film and Au-Cu film laminate.

3.1. Copper-on-polyimide adhesion

Poly(3,3′,4,4′-benzophenone tetracarboxylic dianhydride-3,5-diamino-1,2,4-triazole) (BTDA-TADA) amic acid was prepared. A solution of 3,3′,4,4′-benzophenone tetracarboxylic dianhydride (BTDA) in anhydrous 1-methyl-2-pyrrolidinone (NMP) was cannulated into a flask which contained a solution of 1 equivalent of 3,5-diamino-1,2,4-triazole (TADA). The mixture was stirred for 24 h. The reaction gave BTDA-TADA poly(amic acid), which was thermally (at 300°C) cured to BTDA-TADA polyimide. The reaction scheme is shown in Scheme 4. Figure 2 is the infrared (IR) spectrum of BTDA-TADA polyimide and displays the bands at 1792 (m), 1738 (vs), 1674 (m), 1561 (m), 1514 (m), 1458 (m), 1426 (w), 1391 (m), 1348 (s), 1291 (m), 1244 (s), 1181 (w), 1156 (w), 1096 (m), and 1042 (w) cm^{-1}. The peaks at 1792, 1738, and 1348 cm^{-1} correspond to typical imide carbonyl and C—N stretchings.

To evaluate the adhesion of copper onto polyimide, copper (2 μm thick) was electron-beam evaporated onto the cured polyimide under vacuum at 200°C followed by electroplating copper (8 μm) from Sel-Rex plating solution. Copper strips (3 mm wide) were made by a dicer and then peeled at a 90° peel angle. The peel strengths were in the range of 200–400 J/m^2. Cohesive failure of the polymer was observed visually and confirmed by X-ray photoelectron spectroscopy (XPS). It is noteworthy that the peel strength of copper onto poly(pyromellitic dianhydride-oxydianiline) (PMDA-ODA) was less than 10 J/m^2 using the same method of sample preparation

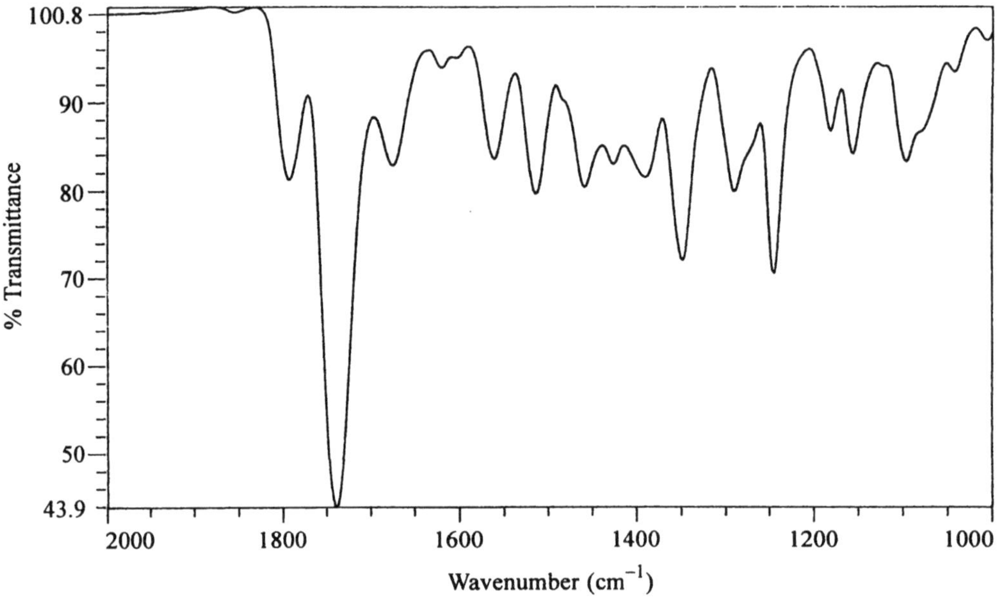

Figure 2. Transmittance IR spectrum of the cured BTDA-TADA polyimide.

and peeling. Thus, the peel strength is enhanced by more than 20 times. It is proposed that the improved adhesion is due to the formation of a copper–polyimide complex.

To determine the locus of failure in the 90° peel, the surface of the copper film after peeling the copper film was analyzed by XPS. As shown in Fig. 3a, the XPS spectrum of the copper side displays the O 1*s*, N 1*s*, and C 1*s* peaks which correspond to the substrate polyimide, but it does not show any copper peaks. The C 1*s* XPS spectrum (Fig. 3b) looks like a typical spectrum of BTDA-TADA polyimide. The XPS spectrum with 90° electron take-off angle (from the sample surface) does not display a peak corresponding to copper. These results clearly indicate that the failure occurred in the polyimide region, and that the polyimide thickness left on the copper peel layer is greater than 7 nm. While the peel test was being carried out, valleys were formed on the polyimide surface by the peel propagation, which has been observed in the case of relatively strong metal/polyimide adhesion [22].

3.2. *Polyimide-on-copper adhesion*

There are three different chemical forms of polyimide precursosrs: poly(amic acid), poly(amic ester), and polyisoimide. Poly(amic acid), upon curing, reacts with copper to leave copper oxide particles in the cured polyimide layer. Poly(biphenyl dianhydride-*p*-phenylene diamine) (BPDA-PDA) amic acid was spin-coated onto a copper substrate and baked at 85 °C for 30 min on a covered hot plate with a continuous flow of nitrogen gas, followed by curing under a nitrogen atmosphere at 400 °C for 60 min. Figure 4a shows the cross-sectional transmission electron micrograph (TEM). The top white area corresponds to the cured polyimide and the bottom black area to the copper substrate. The dots in the polyimide layer correspond to the

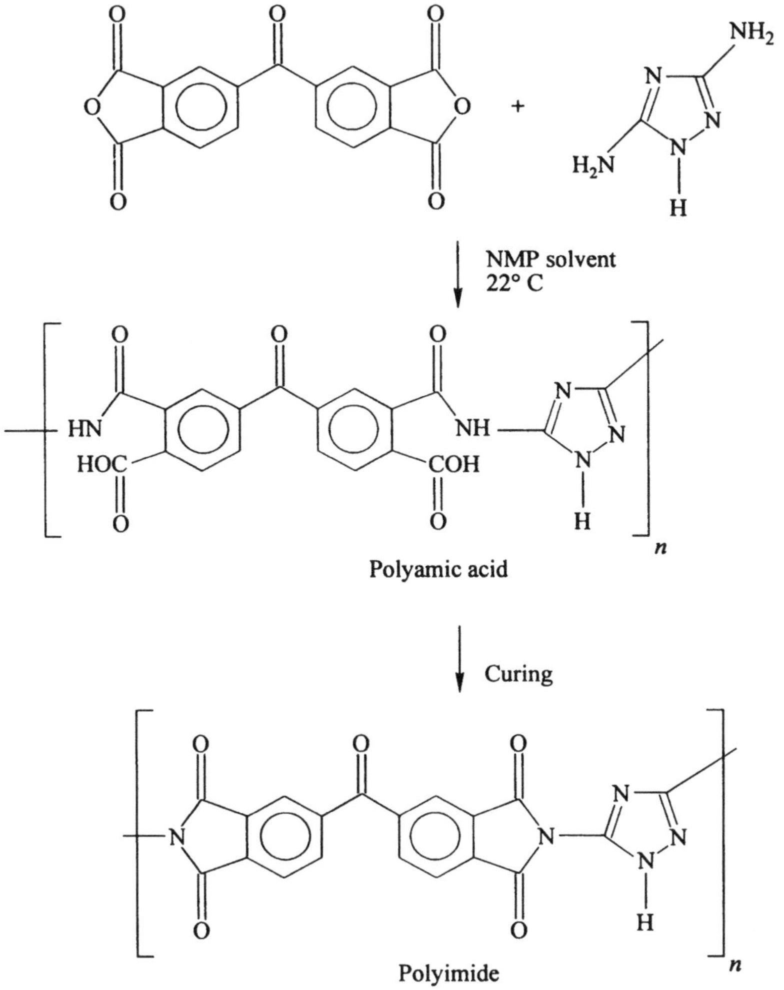

Scheme 4. Preparation of BTDA-TADA polyimide.

copper oxide particles. The carboxylic acid groups are known to react with copper to form copper oxides which diffuse into the bulk of polyimide with the help of a solvent. On the other hand, poly(amic ester) and polyisoimide do not give such copper oxide particles because of the absence of carboxylic acid groups. To prevent such a reaction and to enhance the adhesion to copper, poly(4,4′-oxydiphthalic anhydride-1,3-aminophenoxybenzene) isoimide with 8-azaadenine end groups (ODPA-APB-8-azaadenine) was prepared. 8-Azaadenine consists of a triazole functional group and it is thermally stable. Another reason why polyisoimide was prepared for this purpose is because it is soluble in NMP.

4,4′-Oxydiphthalic anhydride (ODPA), 1,3-aminophenoxybenzene (APB), and 0.1 equivalent of 8-azaadenine were reacted at 22°C in NMP for 24 h to produce a polyamic acid which was converted to polyisoimide by reaction with *N,N′*-dicyclohe-

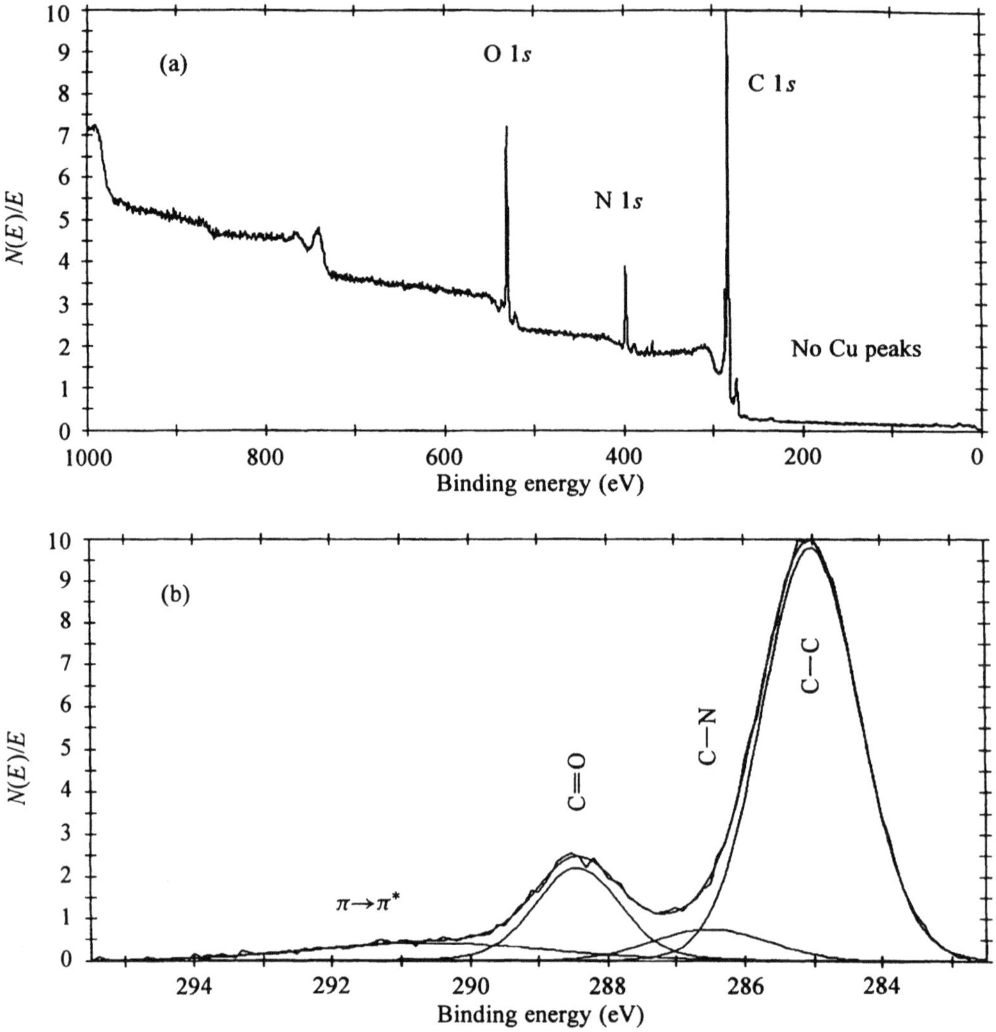

Figure 3. XPS spectra of the copper side after peeling the Cu/BTDA-TADA polyimide system. (a) Survey spectrum; (b) high-resolution C 1*s* region. The C 1*s* XPS spectrum looks like a typical BTDA-TADA polyimide spectrum.

xylcarbodiimide (DCC) at 5 °C for 8 h. The product, which was purified by recrystallization, was a mixture of polyisoimide and polyimide. The IR spectrum, shown in Fig. 5a, displays the bands corresponding to both polyisoimide and polyimide. The peaks at 1806, 1690, 1175, 1146, 941, and 895 cm^{-1} are due to polyisoimide, while those at 1782 (shoulder), 1723, 1372, and 744 cm^{-1} are due to polyimide. Curing this polyisoimide at 400 °C gives polyimide, the IR spectrum of which is shown in Fig. 5b. The whole reaction sequence is exhibited in Scheme 5.

Various properties were evaluated using the 15-μm-thick cured polyimide films. The thermal stability of the film was good as designed. The weight loss at 400 °C measured by TGA was 0.3%/h for the first 8 h period. The glass transition temperature measured

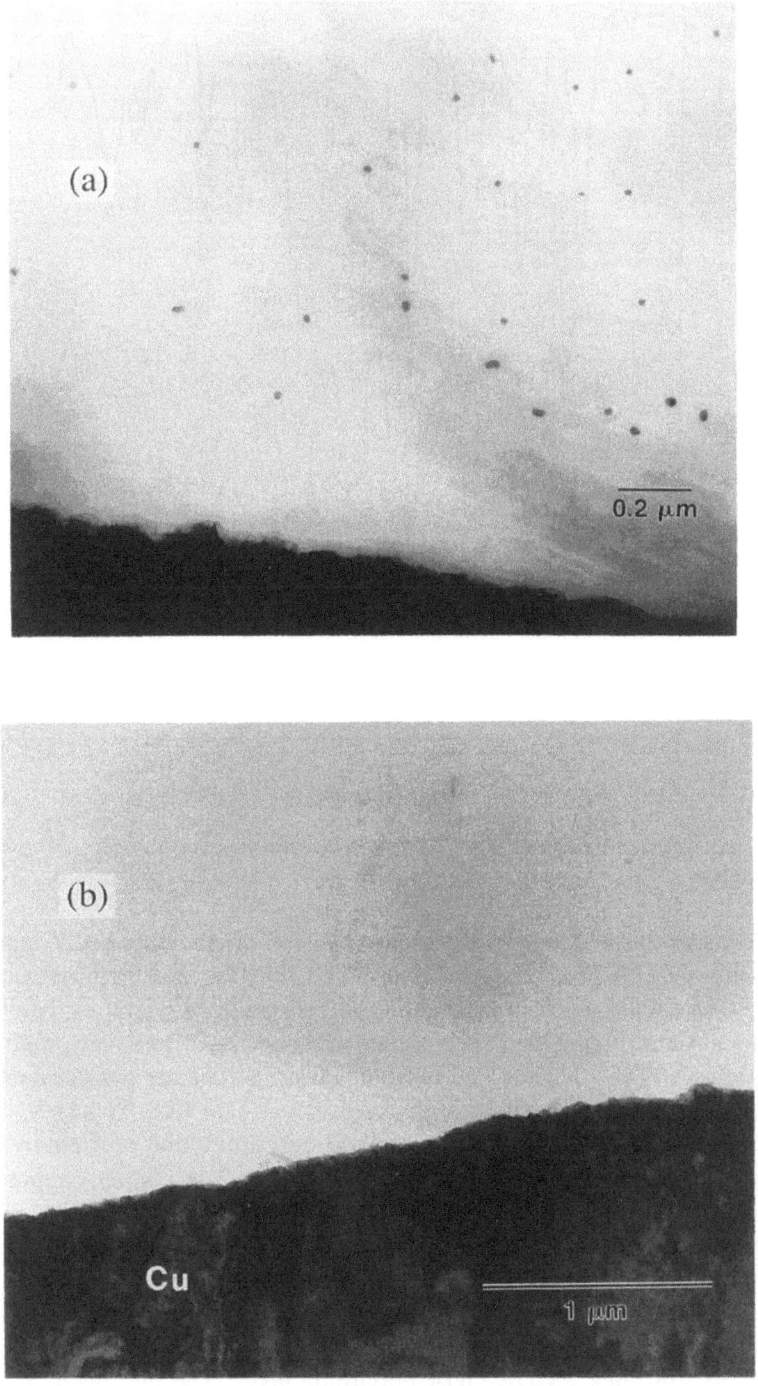

Figure 4. Cross-sectional TEM of (a) the BPDA-PDA polyamic acid/Cu interface and (b) the ODPA-APB-8-azaadenine polyisoimide/Cu interface. Polyimide precursors were spin-coated onto the copper substrate, and then cured at 400°C for 1 h.

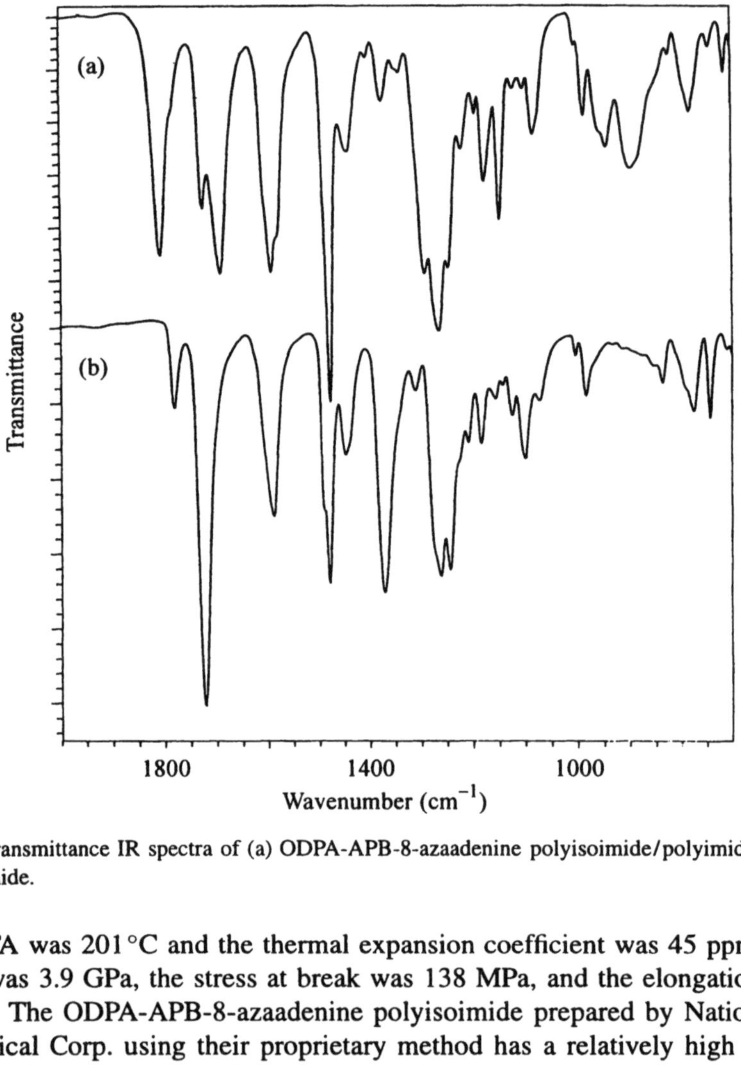

Figure 5. Transmittance IR spectra of (a) ODPA-APB-8-azaadenine polyisoimide/polyimide and (b) the cured polyimide.

with DMTA was 201 °C and the thermal expansion coefficient was 45 ppm/°C. The modulus was 3.9 GPa, the stress at break was 138 MPa, and the elongation at break was 28%. The ODPA-APB-8-azaadenine polyisoimide prepared by National Starch and Chemical Corp. using their proprietary method has a relatively high purity and good properties. These thermal and mechanical properties are not good enough for it to be used as a dielectric layer in high-performance thin film packages, but they are good enough for it to be employed as an adhesion-promoting or a passivation layer.

As previously discussed and illustrated in Scheme 2, a copper capping layer can be employed to simplify the multilevel thin film packaging process. The copper capping layer acts as an adhesion promoter as well as a passivation layer between copper and the polyimide precursor. These two properties were evaluated. A solution of ODPA-APB-8-azaadenine polyisoimide in NMP/cyclopentanone was applied onto the copper substrate and baked at 230 °C for 30 min (1.0 μm thick), followed by coating with a 20 μm-thick PMDA-ODA polyimide layer. The two layers together were cured at 400 °C for 60 min. Polyimide-to-Cu adhesion was evaluated by peeling the polymer layer. Peel strengths in the range of 500–800 J/m^2 were obtained. Cross-sectional TEM of the polymer/Cu interface showed no copper oxide precipitates in

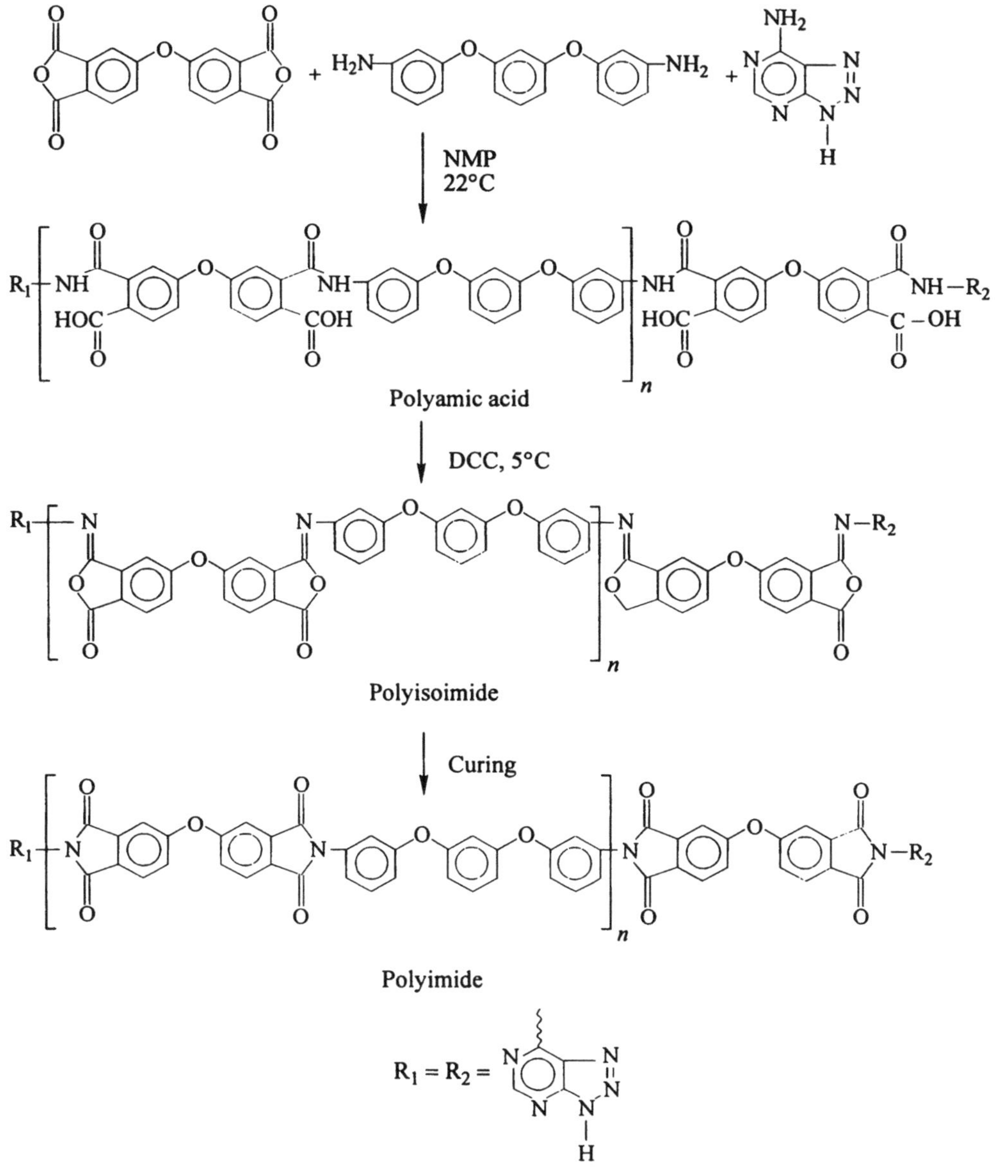

Scheme 5. Reaction sequences for the synthesis of ODPA-APB-8-azaadenine polyimide.

the polyimide as expected (Fig. 4b). After peeling off the polyimide (ODPA-APB-8-azaadenine and PMDA-ODA) layer from the copper substrate, the copper side was analyzed by XPS. Figure 6 shows the survey and C 1*s* spectra. The C 1*s* spectrum looks like a typical ODPA-APB-8-azaadenine polyimide spectrum. The external reflectance infrared (ERIR) spectrum of the same sample shows mostly ODPA-APB-8-azaadenine. These results indicate that the failure occurred within ODPA-APB-8-azaadenine polyimide or at the ODPA-APB-8-azaadenine/PMDA-ODA interface. The corresponding

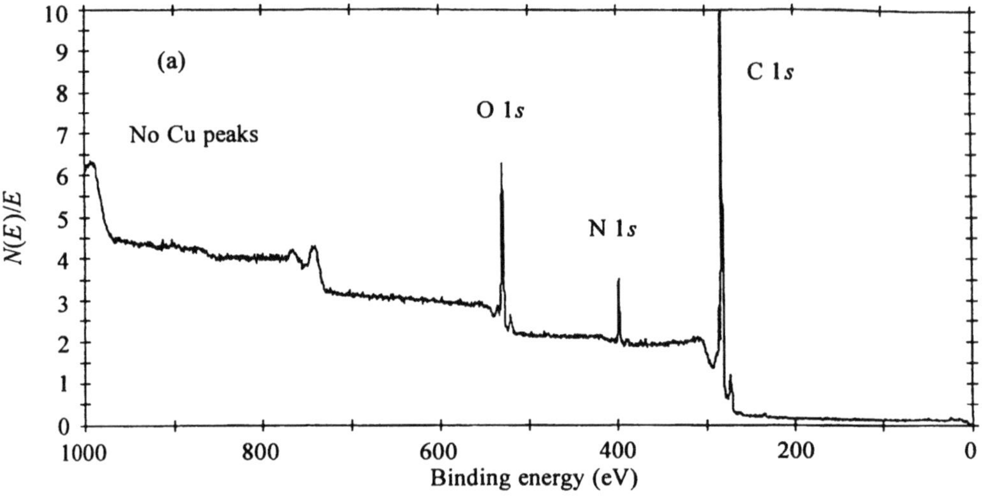

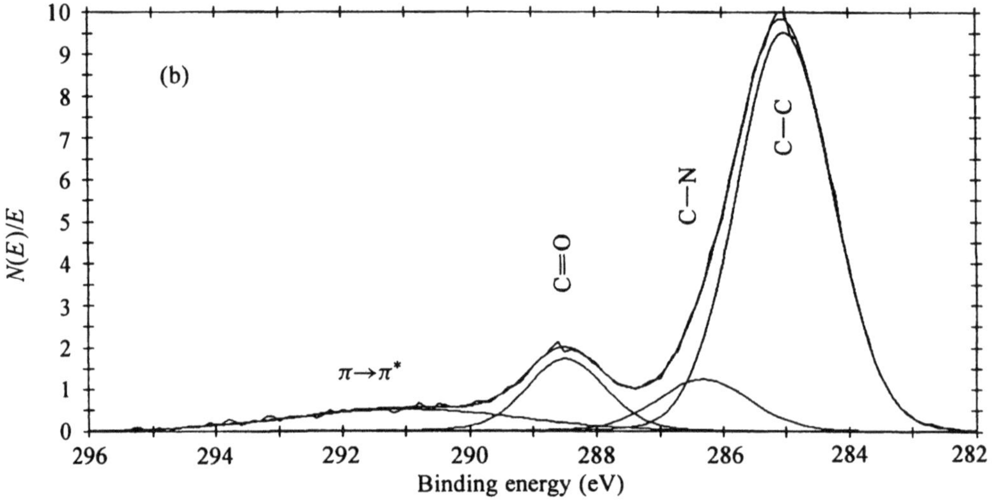

Figure 6. XPS spectra of the copper substrate side after peeling the ODPA-APB-8-azaadenine polyimide/Cu system. (a) Survey spectrum; (b) high-resolution C 1*s* region. The C 1*s* XPS spectrum looks like a typical ODPA-APB-8-azaadenine polyimide spectrum.

polyisoimide supplied by National Starch and Chemical Corp. gave essentially the same adhesion and TEM results. When the samples were treated under 81% relative humidity at 85°C for 500 h, there was no significant change in the peel strengths.

To determine whether 8-azaadenine chemically adsorbs on the copper surface, a copper substrate cleaned with an acid was dipped into 0.3% 8-azaadenine solution in NMP for 30 min and then rinsed with NMP to remove the unreacted compound from the substrate. The sample was rinsed with methylene chloride to remove the NMP solvent and dried under vacuum at room temperature overnight. The XPS result showed a N 1*s* band at 396 eV, which probably corresponds to 8-azaadenine, indicating that 8-azaadenine forms a complex with copper as shown in Fig. 7.

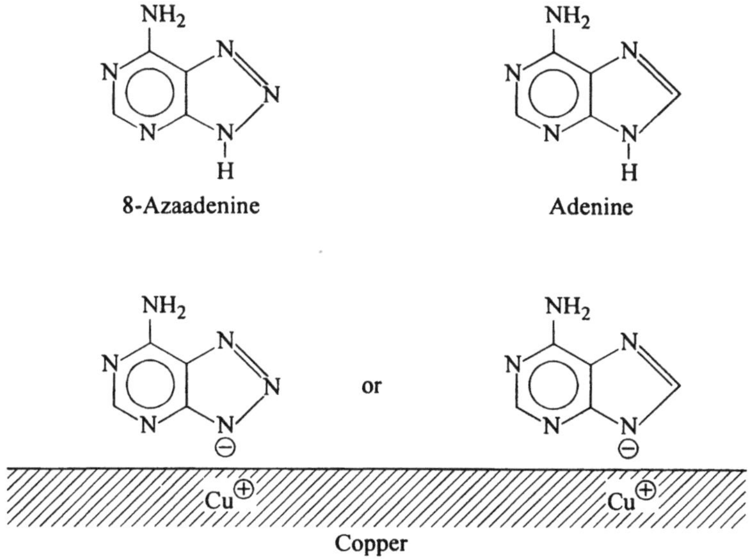

Figure 7. Chemical interaction of 8-azaadenine and adenine on the copper substrate.

3.3. Polyimide/Au-Cu lamination adhesion

Lamination is often employed in manufacturing printed circuit boards. Thus, the adhesion between polymer dielectrics and metals by lamination is important. PMDA-ODA and BPDA-PDA polyimides have excellent thermal and mechanical properties, but they have poor adhesion properties. A thin layer of ODPA-APB-8-azaadenine can be employed as an adhesion-promoting layer between a major dielectric polyimide and metal(s).

A thin layer of ODPA-APB-8-azaadenine (2.0 μm thick after curing) was coated onto a Si wafer and cured at 230 °C. A thick layer of PMDA-ODA (20 μm after curing) was applied and then the whole layer was cured at 400 °C for 60 min to obtain the optimum properties of polyimides. The film was peeled off from the Si wafer. For a metal substrate, Cr (0.05 μm), Cu (2.0 μm), and Au (0.2 μm) were evaporated onto a quartz substrate in that order. The polyimide film was laminated to the Au-Cu substrate under a pressure of 100 psi at 375 °C for 60 min. ODPA-APB-8-azaadenine was in contact with Au/Cu. The peel strength, which was measured by peeling the polyimide film, was 500–600 J/m^2. When the adhesion samples were treated under 81% relative humidity at 85 °C for 500 h, there was no significant drop in peel strength.

4. CONCLUSION

Polyimides containing a triazole or an imidazole functionality were synthesized. The adhesion of copper to BTDA-TADA polyimide and of ODPA-APB-8-azaadenine polyimide to copper was significantly enhanced. It is postulated that the enhanced adhesion is due to the formation of copper–polyimide complexes. The N–H group in triazole or imidazole is the active site for the complex (Cu–N) formation. A thin layer of ODPA-APB-8-azaadenine can be employed as a copper capping layer in multilevel thin film packaging technology, since it acts as a passivation layer between the polyimide precursor and Cu, as well as an adhesion-promoting layer between a dielectric polyimide and Cu. The failure during the peel test occurred in the near-interface region of the polyimide.

Acknowledgements

We would like to acknowledge Drs T. O'Toole, C. Feger, R. Chen, S. Purushothaman, and Mr E. Adamopoulos of IBM for measuring the polymer properties and their help in sample preparations, and Drs Patrick Jobe and Robert Rossi at National Starch and Chemical Corp. for their helpful discussions and kind supply of the APB and ODPA-APB polyisoimide materials. ODPA was kindly supplied by Occidental Chemical Corp.

REFERENCES

1. C. Feger and C. Feger, in: *Multichip Module Technologies and Alternatives*, D. A. Doane and P. D. Franzon (Eds), pp. 311–348. Van Nostrand Reinhold, New York (1993).
2. K. L. Mittal (Ed.), *Polyimides: Synthesis, Characterization, and Applications*, Vols 1 and 2. Plenum Press, New York (1984).
3. D. Wilson, H. D. Stenzenberger and P. M. Hergenrother (Eds), *Polyimides*. Chapman & Hall, New York (1990).
4. R. R. Tummala and E. J. Rymaszewski (Eds), *Microelectronics Packaging Handbook*. Van Nostrand Reinhold, New York (1989).
5. Y.-H. Kim, J. Kim, G. F. Walker, C. Feger and S. P. Kowalczyk, *J. Adhesion Sci. Technol.* **2**, 95–105 (1988).
6. A. Viehbeck, C. A. Kovac, S. L. Buchwalter, M. J. Goldberg and S. L. Tisdale, in: *Metallization of Polymers*, E. Sacher, J.-J. Pireaux and S. P. Kowalczyk (Eds), ACS Symposium Series No. 440, Ch. 29. American Chemical Society, Washington, DC (1990).
7. R. F. Saraf, *Macromolecules* **26**, 3623–3630 (1993).
8. T. Sugama, L. E. Kukacka and N. Carciello, *J. Mater. Sci.* **19**, 4045 (1984).
9. M. C. Burrell, P. J. Codella, J. A. Fontana, J. J. Chera and M. D. McConnell, *J. Vac. Sci. Technol. A* **7**, 55–58 (1989).
10. S. P. Kowalczyk, Y.-H. Kim, G. F. Walker and J. Kim, *Appl. Phys. Lett.* **52**, 375–376 (1988).
11. D. Y. Shih, J. Paraszcak, N. Klymko, R. Flitsch, S. Nunes, L. Lewis, C. Yang, J. Cataldo, R. McGouey, W. Graham, R. Serino and E. Galligan, *J. Vac. Sci. Technol. A* **7**, 1402 (1989).
12. J. R. Schaeffer, US Patent 2,618,608 (1952).
13. I. Dugdale and J. B. Cotton, *Corros. Sci.* **3**, 69 (1963).
14. H. Ishida and R. Johnson, *Corros. Sci.* **26**, 657 (1986).
15. S. L. Cohen, V. A. Brusic, F. B. Kaufman, G. S. Frankel, S. Motakef and B. Rush, *J. Vac. Sci. Technol. A* **8**, 2417 (1990).
16. G. Xue, Q. Dau and S. Jiang, *J. Am. Chem. Soc.* **110**, 2393 (1988).
17. G. Xue, J. Dong and J. Zhang, *Macromolecules* **24**, 4195 (1991).
18. K. T. Carron, G. Xue and M. L. Lewis, *Langmuir* **7**, 2 (1991).
19. K. T. Carron, M. L. Lewis, J. Dong, J. Ding, G. Xue and Y. Chen, *J. Mater. Sci.* **28**, 4099–4103 (1993).
20. F. P. Eng and H. Ishida, *J. Mater. Sci.* **21**, 1561 (1986).

21. H. Ishida and K. Kelley, *J. Adhesion* **36**, 177–191 (1991).
22. J. Jang and H. Ishida, *J. Appl. Polym. Sci.* **49**, 1957–1962 (1993).
23. K. S. Kim and J. I. Kim, *J. Eng. Mater. Technol. ASME Trans.* **110**, 266 (1988).

Fundamentals of Adhesion and Interfaces, pp. 179–203
D. S. Rimai, L. P. DeMejo and K. L. Mittal (Eds)

Fracto-emission and electrical transients due to interfacial failure

J. T. DICKINSON, L. C. JENSEN, SUNKYO LEE, L. SCUDIERO and S. C. LANGFORD

Physics Department, Washington State University, Pullman, WA 99164-2814, USA

Revised version received 2 March 1994

Abstract—Interfacial failure is often accompanied by intense photon and electron emissions. When interfacial failure involves the separation of a conducting material from an insulator, transient electrical signals are generated in the conducting phase; these signals are ultimately due to contact charging across the interface, followed by charge separation during failure. We review the application of these signals as probes of the failure process in two adhesive systems: a steel rod embedded in an epoxy matrix, and a pressure sensitive adhesive on a copper substrate. The intensity of the photon and electron emissions (in vacuum) from the steel rod-epoxy system allows us to determine where debonding begins and the direction and approximate velocity of the resulting debond crack. Electron emission is particularly sensitive to the production of exposed fracture surface, and thus indicates when the debond crack reaches the epoxy surface. The peeling of pressure sensitive tape from copper yields a fluctuating current signal that reflects the rate of adhesive removal from the copper, and is thus particularly sensitive to micromechanical events along the peel front. Spectral analysis of these fluctuations show an interesting transition at peel speeds of about 6 mm/s. Numerical analysis of the current signal also shows fractal behavior, suggesting the presence of a chaotic process.

Keywords: Interfacial failure; fracture; peel; fracto-emission; epoxy; pressure sensitive adhesive.

1. INTRODUCTION

Interfacial failure is a serious problem in our technological society, reaching into our daily lives in our transportation vehicles, our home appliances, and even our clothing. Exposure to mechanical stresses, radiation, and chemicals (especially water) is frequently unavoidable and often leads to interfacial crack growth, and ultimately to catastrophic failure. As we push our technology to higher and lower temperatures, higher levels of stress, and wider ranges of environments, demands on bonded systems become continually more intense. In this paper, we would like to explore some fundamental aspects of interfacial failure which may stimulate and encourage extension of mechanical analyses to this complex problem. After a brief introduction, this will include discussion of time and spatially resolved measurements of debonding and pullout of steel rod from an epoxy matrix, and peeling of a pressure sensitive adhesive from a copper substrate.

The chemical forces which bind dissimilar materials together are electronic in nature, and thus the making and breaking of adhesive bonds will involve electrical effects. These effects depend on the electron distributions associated with the atoms along the interface and the type of bonding (e.g. ionic, covalent, metallic or van der Waals), all of which involve dipole interactions or net charge transfer. When these bonds are subsequently strained, the electrons are redistributed. Critical regions of strain exist where debonding and rebonding may occur at equal rates, essentially in equilibrium. Fluctuations due to thermal motion and stress (e.g. reflected stress waves in the bulk) can also influence the rates of bond breaking/making. These fluctuations may be described in terms of phonon emission and absorption from the bonds. The literature on contact charging, reviewed by Lowell and Rose-Innes [1], indicates that 'patchiness' in both charge density and sign is quite common and poorly understood.

Since our observations are generally experimental in nature, we summarize our laboratory techniques in the following section.

2. EXPERIMENTAL METHODS

Interfacial failure is often accompanied by intense emissions of charged and neutral particles, as well as photons. Collectively, we refer to these emissions as fracto-emission (FE). Charged particles can be detected in vacuum with a Channeltron electron multiplier (CEM), which produces fast (10 ns) pulses with high detection efficiencies for both electrons and positive ions. Excited neutrals with sufficient internal energy can also ionize upon collision with the CEM and be detected. By varying the CEM bias voltage and the voltages on nearby grids or barriers, the charge of the detected particles can be determined unambiguously.

Photon emission (phE) is usually detected with a photomultiplier tube (PMT), which also produces fast (30 ns) pulses with reasonable efficiencies (up to 25%), especially in the blue and ultraviolet regions. Photomultiplier tubes can be cooled (in vacuum, with liquid nitrogen) to reduce background noise levels. One of our photon detectors is housed within milimeters of the working area of an Applied Test Systems Universal Testing Machine, which in turn is light shielded as well as electronically shielded for time resolved phE and electrical transient measurements. Spectra of the phE can be obtained with a reasonable resolution spectrometer in a time resolved fashion using a gate-able, image-intensified diode array and an optical multichannel analyzer (EG&G OMA-III).

As mentioned above, interfacial failure often results in highly charged surfaces. One kind of charge or the other will predominate on each surface, due to the contact potential difference between the dissimilar materials making up the interface. However, local variations in surface defect density can result in charged patches on both surfaces. Even one excess charge in a million surface sites yields very strong electric fields at the surface. A coil antenna placed near the sample is used to couple to the changing magnetic fields associated with electromagnetic emissions often associated with electrical discharges within the crack and on the fracture surfaces. (We call the detected emission RE — for Radiofrequency Emission.) Care must be taken to distinguish between near field vs. far field radiation, the former accompanying *any* changing electric field (e.g. due to the rapid motion of charged surfaces), whereas the latter in this context is associated with electrical breakdown events.

3. GENERAL FEATURES OF FE/ELECTRICAL TRANSIENT STUDIES

In investigating the links between particle emissions/electrical signals and failure mechanisms, it is helpful to review some of the observations to date on these phenomena. As mentioned above, FE includes the emission of electrons, positive and negative ions, neutral species in both ground states and in excited states, and visible photons. These emissions are often especially intense when failure occurs at or near a material interface. In general, the emissions are very sensitive to the locus of fracture, and velocity of crack growth, and electronic structure of the region near the crack tip. For systems involving **interfacial failure**, the following model is appropriate:

(a) During crack propagation, charge separation occurs as the freshly created fracture surfaces are drawn apart. The resulting surfaces often display localized patches of especially high charge densities.

(b) The rapid migration of charge carriers (usually electrons) collisionally excites the surfaces. This migration is most intense during fracture due to the high electric fields. Many charge carriers are subsequently trapped in thermally unstable states, often producing long-lived photon emission (phE) and electron emission (EE) which can be observed seconds or hours after the failure event [2].

(c) In some systems, gas molecules emitted during fracture or entering the crack from the environment (e.g. N_2) can precipitate small gaseous discharges during fracture, yielding charged particles (generally electrons and positive ions), phE (line spectra), and far field, long wavelength electromagnetic radiation.

(d) Patches of high charge density along the fracture surfaces accelerate some of the emitted electrons to high energies. A large fraction of these electrons is drawn back to the surface resulting in a self-bombardment process, producing further emission of negative and positive ions (NIE and PIE) and excited neutrals via electron stimulated desorption (ESD).

Gaseous discharges and surface breakdown events can produce intense bursts of phE, EE, and changing magnetic fields, which accompany the fracture of many materials. For instance, the spectra of many triboluminescent materials show strong nitrogen lines due to the ionization of atmospheric or occluded nitrogen. Recent results in our laboratory indicate that charge flow among locally charged regions in the absence of nitrogen can also directly excite surface states to yield intense luminescence (a form electroluminescence). These events can be spectrally distinguished from gaseous discharges, which yield sharp spectral lines characteristic of the gases present in the crack.

phE from composite systems with transparent or semi-transparent materials can be exploited to determine the location of failure at embedded interfaces. Multiple PMT measurements have been used to determine the direction and speed of debond cracks, for instance. Furthermore, comparison of phE signals with simultaneous EE measurements allows us to distinguish between interfacial failure along embedded interfaces and failure which exposes a failed interface to the outside world. As electrons cannot penetrate the matrix to any significant extent, EE is observed only from failure events at external surfaces.

Interfacial failure and subsequent frictional pullout of deformation play key roles in the toughening of polymer-based composites, along with microcracking, crack branching, and crack deflection. A classic example involves crack propagation in fiber-reinforced, brittle-matrix composites, where loading of the crack leads to debonding of crack-bridging fibers, followed by frictional pullout. The resulting dissipation of strain energy (by heat generated along the sliding interfaces) increases the total fracture energy (area under the load/deflection curve) considerably. In the early stages of crack growth, energy dissipation under these conditions can actually increase as the crack grows, resulting in slow, stable crack growth (as opposed to the rapid, catastrophic failure characteristic of brittle materials); this property is highly desired in structural materials, where the ability to detect crack growth prior to ultimate failure is an important safety feature. The amount of energy dissipation depends on the inherent strength of the matrix and reinforcing fibers, as well as the residual stresses in the composite, surface roughness and asperities along the interface, and the shear strength of the interfacial bond [3]. The effective friction coefficient along the debonded interface, and the length of intact, embedded fiber on each side of the crack are very important in determining the amount of friction pullout, and thus the amount of energy which can be dissipated.

Although a number of workers have greatly contributed to the understanding of a number of these effects and have developed methods to exploit them in the processing of actual materials, a better understanding of the contribution of each of these processes, particularly in a time resolved, dynamic fashion in a variety of environments, would be highly beneficial. We are combining our fracto-emission techniques with the detection of *electrical transients* which accompany and follow the failure of interfaces. These techniques offer time and spatially resolved probes of a number of the events responsible for toughening in brittle matrix composites and can prove useful in determining failure mechanisms.

4. FAILURE AT EMBEDDED INTERFACES

4.1. Photon and electron emission [4]

Measurements of phE accompanying failure at internal interfaces are readily made in composites with transparent and semitransparent matrices, which allow for the transmission of light to the detector. Because photomultiplier tubes function in air as well as in vacuum, phE experiments may be performed in both environments. Simultaneously EE (electrons, which require vacuum) can identify failure events which expose new surfaces to the outside world, e.g. a crack reaching an exterior surface. Thus we can distinguish between internal and 'external' failure. We have used these techniques to examine debonding events in glass fiber-epoxy composites [5]. These same methods are also proving helpful in studying the frictional pullout process.

As an example of these types of measurements, Fig. 1 shows phE and EE accompanying the failure of a simple system consisting of a small metal 'pin' whose head was embedded in epoxy, as shown. During loading, sustained phE is observed for approximately 1 s before the drop in force accompanying pullout. Because EE requires a pathway to the 'outside world' and to the detector, EE is not observed until the pin is pulled out of the matrix, i.e. when pullout exposes surfaces capable of releasing electrons. The pre-failure light emission is from *inside* the epoxy. Observations after

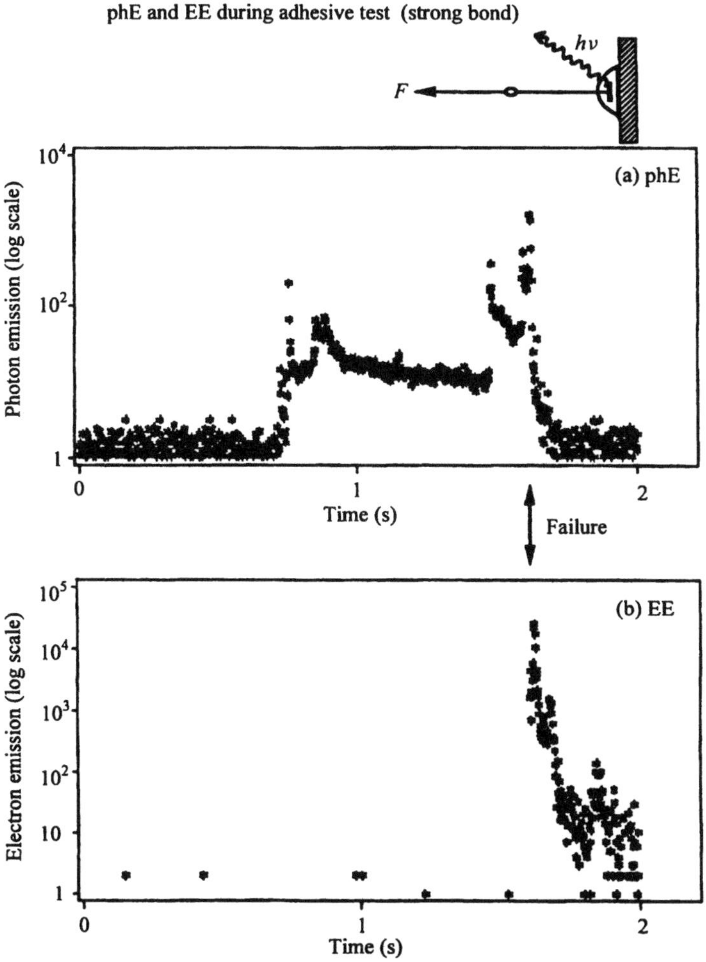

Figure 1. phE and EE during loading and failure of a steel 'pin' embedded in transparent epoxy. The photon emission preceding pullout is due to interfacial failure at the internal metal/epoxy interface.

the onset of phE but prior to pullout show that the end or tip of the pin has indeed detached (observed optically by means of interference fringes viewed from the right hand side of the diagram). Tests with larger, flat surfaces show that the first peak in the phE is due to a debonding crack on the head of the pin. It should be noted that in these experiments the metal pin was removed from the region of the detectors within a few ms after pullout. Thus, the 'tail' observed in both the phE and EE is coming from the epoxy. The hole left behind is glowing and shooting out a spray of electrons. The combined phE and EE measurements serve as sensitive probes of the onset of internal failure and of the arrival of the debond crack at the epoxy surface.

Another geometry which results in internal interfacial failure is a cylindrical rod partially embedded into a matrix and pulled out. When the length of the embedded portion of the rod is shorter than the rod diameter, the phE and EE data are similar to those obtained for the embedded pin; initial failure occurs at the embedded tip, followed by debonding along the sides of the rod. Atkinson *et al.* [6] have discussed the fracture mechanics of this problem and have shown that initial debonding in the absence of residual stresses is favored at the tip of the embedded end of the rod, provided that the rod is not too long. However, when the embedded length of rod is significantly longer than its diameter, they expect initial debonding along the sides of the rod at the point where the rod enters the matrix (the 'free' end of the rod); the debond crack subsequently grows along the sides of the rod towards the embedded end.

More detailed information on the sequence of failure events can be obtained by simultaneously monitoring the phE signal from two photomultiplier tubes; by properly

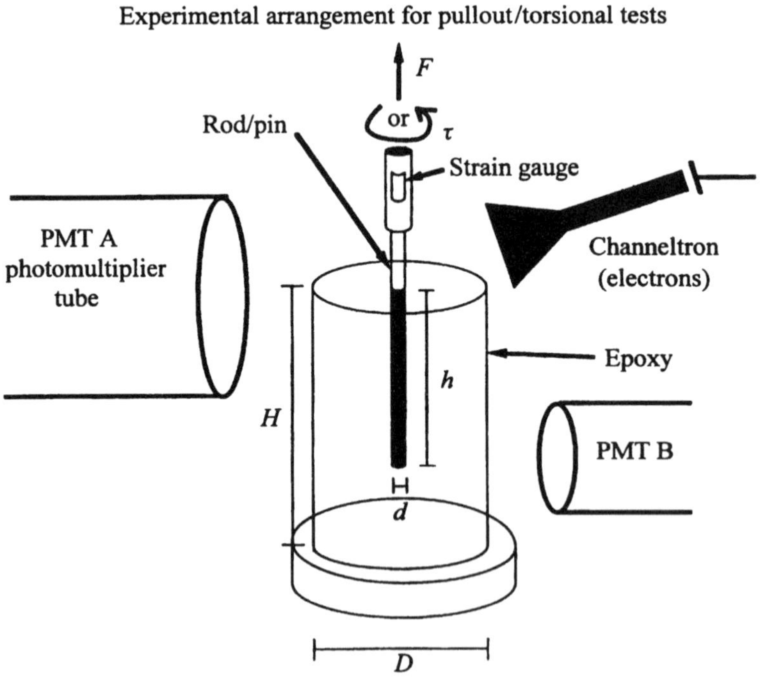

Figure 2. Schematic of the rod debonding and pullout experiment.

masking the sample, the signals from the two PMTs show the light intensities from the two ends of the rod in a time resolved fashion. Figure 2 shows the arrangement of the detectors for a pullout experiment involving a 2.5-mm diameter stainless steel metal rod embedded 50 mm into the epoxy. Loads of about 5 kN were required to produce debonding under these conditions, resulting in very fast debonding cracks followed by rapid rod pullout, where the rod lunges forward as much as 5 mm. Light shields attached to the outer epoxy cylinder limited the field of view of each PMT: PMT A was arranged to favor the detection of photons produced near the top (epoxy free surface) and PMT B (which has a much smaller photocathode) was positioned to favor the detection of photons produced near the embedded tip (bottom) of the rod. In addition to the two PMTs, a CEM was positioned near the epoxy-surface region as shown in Fig. 2.

The resulting signals are shown in Fig. 3. The arrows indicate the onset of significant emission for each of the three detectors. The onset of the signal from PMT A precedes that of PMT B by $\sim$ 68 μs, indicating crack motion from near the free end (top) of the rod towards the embedded end. This time delay corresponds to an average crack velocity of $\sim$ 700 m/s. The leading edge of the large burst in the signal from PMT A (off scale at several points) coincides with the onset of emission from PMT B and with the drop in load, which marks the onset of rod pullout. However, the failed interface is not exposed to the outside world until about 150 μs after the leading edge of the signal from PMT A, as indicated by the onset of EE. Since interfacial failure along the sides of the rod at the point where the rod enters the epoxy matrix should expose the interface to the outside world, we conclude that the initial signal from PMT A is due to the nucleation and growth of a debond crack in the interior of the sample, and that debond crack growth proceeded in both directions. From the timing of the EE signal relative to the phE signals, it appears that crack growth toward the embedded end of the rod was much more rapid than crack growth toward the free end. No evidence of failure at the embedded tip of the rod prior to the fracture along the sides of the rod is observed.

We, therefore, conclude that the normal sequence of events for the tensile loading of the long stainless steel rod embedded in epoxy is as follows:

- a debond crack forms along the metal/epoxy interface near the top but *below* the free surface; (For shallow embedding depths, failure initiated at the embedded tip.)
- the interfacial crack propagates in both directions — rapidly towards the embedded tip of the metal rod and slowly toward the free surface of the epoxy;
- the debonding crack reaches the free surface of the epoxy (yielding EE), and the rod begins to slip along the metal/epoxy interface (yielding intense phE and EE). This initial pullout motion typically lasts $\sim$ 50 μs.

Our conclusion that debond crack nucleation occurs below the epoxy surface conflicts with the fracture mechanics analyses of Atkinson *et al.* [6] and Karbhari and Wilkins [7], who predict that debonding failure along the sides of the rod begins at the point where the rod enters the matrix. (Thus one would expect the onset of EE and phE from PMT A to coincide.) The most likely explanation is that the experimental samples do not conform well to the ideal rod-matrix systems assumed in the

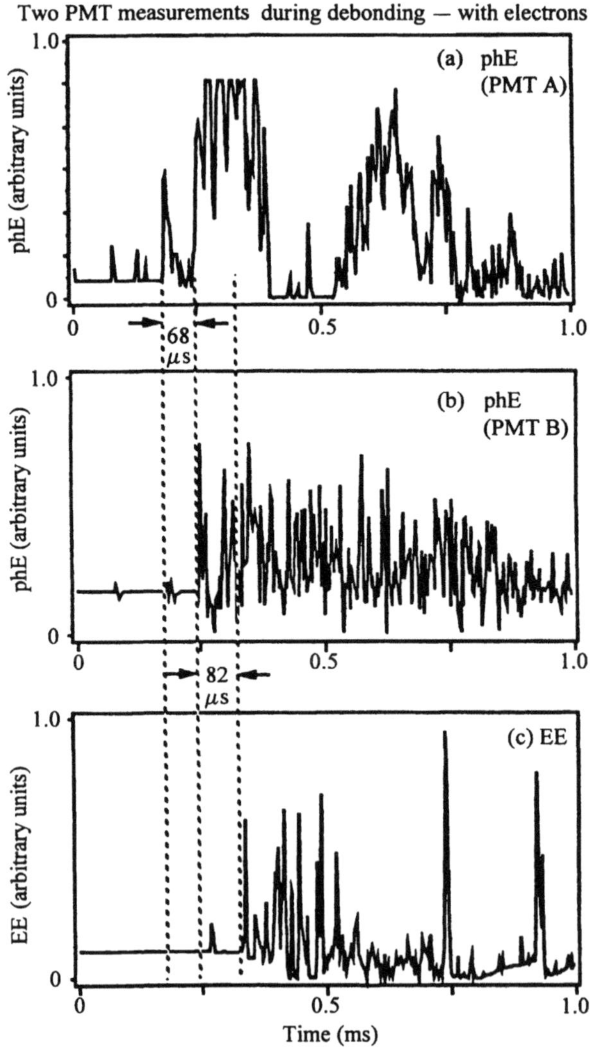

Figure 3. phE and EE accompanying interfacial failure of stainless steel rod in epoxy on a fast time scale. (a) phE detected by PMT A located near the free surface; (b) phE detected by PMT B viewing the lower portion (tip region) of the metal rod; and (c) EE.

fracture mechanics analysis. For instance, the heat generated during epoxy curing may significantly enhance the residual (tensile) stresses in the bulk of the sample due to thermal expansion mismatches upon subsequent cooling, favoring crack nucleation along the interface in the interior.

4.2. Electrical transients accompanying debonding [8]

Debonding and pullout in the stainless steel rod/epoxy matrix system are accompanied by strong electrical transients due to charge separation during the separation of two

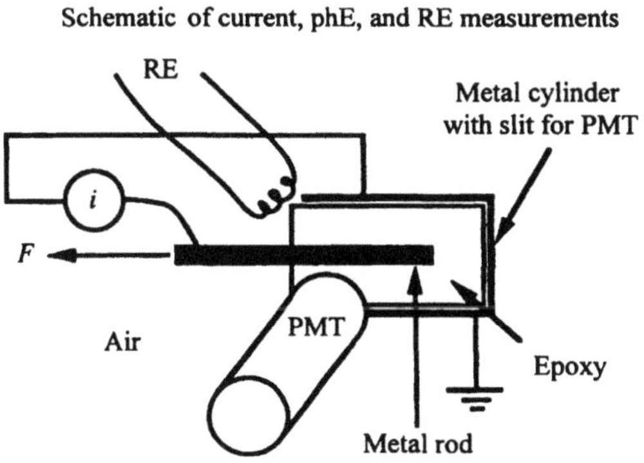

Figure 4. Schematic of the sample geometry and electrical circuit for deriving the observed currents due to debonding and pullout.

dissimilar materials. Charge measurements following fracture show that the stainless steel has a net *positive* charge and the epoxy a net *negative* charge. The arrangement of a typical experiment is shown in Fig. 4; the dimensions of the sample are the same as in the phE and EE measurements, except that a somewhat larger (3.1 mm diameter) rod was used. The rod strain was monitored with a resistive strain gauge (with a time response of 10 μs) attached as shown. A coil antenna was used to detect transient magnetic field bursts (denoted by RE — Radiofrequency Emission) associated with charge separation activity. The tensile testing machine is enclosed in a light-tight, grounded aluminum box to minimize background levels in the current, phE, and RE measurements. The metal rod was insulated from the grips and a wire was attached for the transient current measurement. The epoxy sample was enclosed in a grounded, cylindrical metal electrode which provided the 'return' path for current from the rod. The cylindrical electrode was slotted along the side toward the photomultiplier tube to allow light detection.

Figure 5 shows a typical strain, phE, and current measurements during interfacial failure. The strain measurements suggest that the velocity of the debond crack was initially quite slow, and then accelerated rapidly. Similarly, the current is initially quite small, and then rises rapidly. The current grows to especially high values after the completion of debonding, as the rod lunges forward, out of the matrix. Only a small portion of the rod is removed from the epoxy in this initial pullout event. Post-failure examination typically shows 1–3 mm of rod drawn from the epoxy out of a total embedded length of 50 mm. Although phE is observed throughout the growth of the debond crack, it becomes extremely intense at the completion of debonding, and subsequently shows two large spikes.

The current and phE signals indicate that the epoxy surfaces are electrically charged during interfacial failure. We attribute the large, aligned phE and RE signals accompanying the completion of interfacial failure to electrical discharges between epoxy and rod surfaces as the surfaces are displaced with respect to one another. This displacement brings asperities on the two surfaces in close proximity, allowing the discharge of accumulated charge. phE and RE bursts after failure accompany additional displacements of the rod with respect to the epoxy as it oscillates with its natural

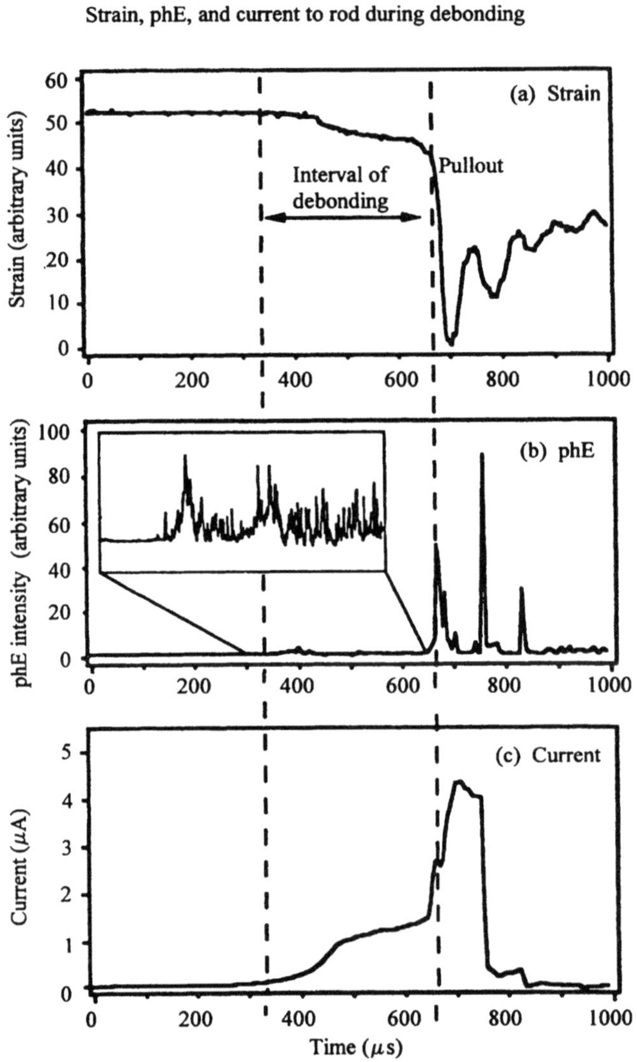

Figure 5. (a) Strain, (b) phE, and (c) current from stainless steel rod during debonding and initial pullout of rod from epoxy. Both rod and epoxy have been rigidly attached to the grips, preventing free motion of the rod upon relaxation of the tensile stress. Insert in (b) shows phE in the initial debond region.

frequency which depends on the clamping arrangements. The total charge involved in phE and RE events is a small fraction of the total charge on the two surfaces, as the large bursts following fracture do not affect the current signal significantly.

The current during the debonding event was modeled using the sample geometry shown in Fig. 6, where the crack propagates from left to right. Assuming that interfacial failure occurs under conditions of constant charge, the opening of the gap between the rod and epoxy changes the potential difference across this gap. In the region where no debonding has occurred, we assume the metal/epoxy interface has a separation of δ_0 (very small) which increases to a constant width $\delta = \delta_0 + \Delta\delta$,

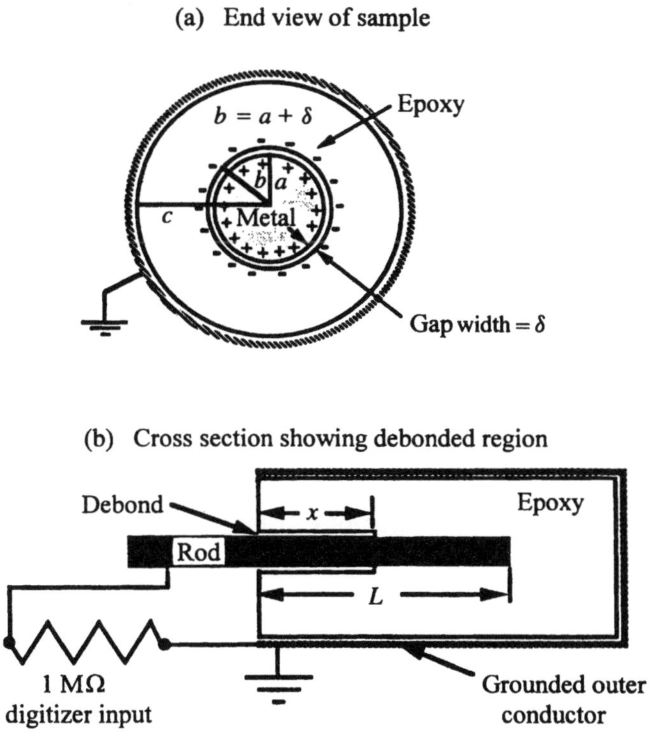

Figure 6. Sample arrangement for transient electrical signal acquisition from interfacial failure between a metal rod and an epoxy matrix. (a) Shows end view of the sample, and (b) depicts cross section showing debonded region.

where $a \gg \Delta\delta \gg \delta_0$. Integrating the electric field from the surface of the rod to the outer conductor, we obtain the potential difference between the rod and outer cylinder; because of the connection through the load resistor, R, shown in Fig. 6(b), this value is zero before and after charge flow. For cylindrical geometry, ignoring edge effects, the radial components of the electric fields in the separation region and in the epoxy at radial distance r are given in equations (1) and (2):

$$\mathrm{E}(r) = \frac{\lambda^+}{2\pi\varepsilon_0 r} \qquad a < r < a + \delta, \tag{1}$$

$$\mathrm{E}(r) = \frac{(\lambda^+ + \lambda^-)}{2\pi\kappa\varepsilon_0 r} \qquad a + \delta < r < c, \tag{2}$$

where λ^+ and λ^- are the charges per unit length on the metal rod and on the epoxy surfaces, respectively. λ^- is the constant, free charge on the epoxy and κ is the relative dielectric constant of the epoxy. We assume that the gap is either vacuum or a low pressure gas with $\kappa = 1$. Before the crack is opened, the potential difference, V, is given by

$$V = \frac{\lambda^+}{2\pi\varepsilon_0} \ln\left(1 + \frac{\delta}{a}\right) + \frac{(\lambda^+ + \lambda^-)}{2\pi\kappa\varepsilon_0} \ln\left(\frac{c}{a + \delta}\right). \tag{3}$$

Under these conditions (before crack growth), $V = 0$, $\lambda^+ = \lambda_0^+$ and $\delta = \delta_0$, so that:

$$\lambda_0^+ = \frac{-\lambda^- \kappa^{-1} \ln\left(\frac{c}{a+\delta_0}\right)}{\ln\left(1 + \frac{\delta_0}{a}\right) + \kappa^{-1} \ln\left(\frac{c}{a+\delta_0}\right)} = -K_0 \cdot \lambda^-, \tag{4}$$

where K_0 is the collection of constants shown. If an annular region debonds and opens to a width $\delta = \delta_0 + \Delta\delta$, then charge must flow to keep $V = 0$, changing the local charge/length in the debonded region: i.e. $\lambda_0^+ \to \lambda_f^+$, where λ_f^+ is given by:

$$\lambda_f^+ = \frac{-\lambda^- \kappa^{-1} \ln\left(\frac{c}{a+\delta_0+\Delta\delta}\right)}{\ln\left(1 + \frac{\delta_0+\Delta\delta}{a}\right) + \kappa^{-1} \ln\left(\frac{c}{a+\delta_0+\Delta\delta}\right)} = -K_f \cdot \lambda^-, \tag{5}$$

where K_f represents the new collection of constants.

Thus, for an increment of length in the crack, Δx, the net charge flowing (through resistance R) is:

$$\Delta q = \left(\lambda_f^+ - \lambda_0^+\right) \cdot \Delta x = \left(K_0 - K_f\right) \cdot \lambda^- \cdot \Delta x. \tag{6}$$

Given a crack velocity $v_c = \Delta x / \Delta t$, the current generated is:

$$i = \left(K_0 - K_f\right) \cdot \lambda^- \cdot v_c. \tag{7}$$

We assume the time constant associated with resistance R and capacitance C of the sample and coaxial cable leads is much less than the duration of fracture; for $R = 1$ MΩ and $C \approx 5$ pF this is well satisfied. The charge density along metal/polymer interfaces has been measured and is typically $\sim 10^{-4}$ C/m^2 [14], which corresponds to $\lambda^- \sim 10^{-6}$ C/m for a 3-mm diameter rod. From the observed phE duration and the strain signal, we obtain average crack velocities of 200 m/s; we estimate $\Delta\delta$ to be ~ 1 μm. This spacing is consistent with Ohara's optical interferometer measurements on polymer/glass structures deformed in shear [9]. Using known values of a (3.1 mm), c, and κ (~ 4), equation (7) yields an estimated current of 0.2 μA and a total charge transfer of 60 pC. This is in good agreement (fortuitous, considering the approximations) with measured current values (0.1 μA $\leqslant i \leqslant$ 1 μA during debonding) and Δq ($\sim$ 50 pC). Furthermore, the model predicts a net decrease in the charge on the rod ($\Delta q < 0$), consistent with the observed direction of charge flow.

Immediately after debonding, the constrained rod undergoes an initial, fast pullout due to the sudden stress release along the interface, followed by a damped longitudinal oscillation of the rod. Additional charge transfer occurs during this fast pullout, which shortens the length of the embedded portion of the rod. To maintain the charge density inside the capacitor defined by the rod and the shield at a constant value (λ_f^+ of equation (5)) while maintaining zero potential difference between the rod and outer shield, additional charge must flow during pullout. Given the initial conditions above, electrons will again flow *to* the rod. Ignoring fringe fields, the current generated as the rod is drawn a distance ΔL in time Δt will be:

$$i = \frac{\Delta q}{\Delta t} = \lambda_f^+ \cdot \frac{\Delta L}{\Delta t} . \quad (8)$$

For $\Delta L \sim 1.5$ mm (from post-failure observations), $\Delta t = 170\ \mu s$, and $\lambda_f^+ = 1\ \mu C/m$, this yields a current of 8 μA. The observed peak currents are typically 1–5 μA (corresponding to electron flow to the rod, as predicted), in good agreement with equation (8).

We attribute most of the increase in current and phE during debonding to increases in crack speed. phE intensities accompanying the fracture of neat epoxies depend strongly on crack velocity [10]. A similar dependence of EE intensities on crack velocity has been observed during the failure of filled polymers [11–13] and polycarbonate [14]. In the final stages of crack growth, the stress intensity at the crack tip should be quite high and the crack velocity correspondingly high. Significantly, the very early stages of debonding appear to be accompanied by a burst of phE. The most natural interpretation of the initial phE bursts is that the initial debonding event involves localized, rapid crack growth at a stress concentrating defect. As the crack grows, it relieves the local stress concentration and slows down to velocities compatible with the average interfacial stress.

Thus, we see that interfacial failure in the metal rod/epoxy matrix system is accompanied by distinctive voltage and current signals which are correlated with interfacial failure events inferred from small changes in rod strain. These transients are accompanied by phE signals which are sensitive indicators of the onset of interfacial crack growth. In the final stages of interfacial failure, intense phE and RE bursts reflect electrical discharges which are associated with the relative motion of rod and matrix accompanying pullout. Subsequent bursts which accompany further displacement are correlated with oscillations in rod strain. If loading is continued (at a constant rate of elongation) after debonding, the motion of the rod along the interface can exhibit a variety of behaviors, including stick-slip behavior. This motion can also be studied by means of transient current and fracto-emission measurements [8].

5. PEELING OF A PRESSURE SENSITIVE ADHESIVE

The extremely nonlinear, viscous behavior of pressure sensitive adhesives results in considerable energy dissipation during peeling over a wide range of peel speeds and substrates. In particular, the formation and elongation of adhesive fibrils within the peel zone is generally the dominant dissipative process during peeling [15], provided that these fibrils do not fail prematurely (cohesively or adhesively). The interplay between fibril release and failure modes and the viscoelastic properties of the adhesive are of fundamental importance in the design and application of pressure sensitive adhesives. Because of the interfacial failure involved in the peel process, removing a pressure sensitive adhesive from a metal substrate will generate electrical signals in the substrate in much the same manner as removing a metal rod from an epoxy matrix. These signals, detected as currents or voltages, reflect the course of adhesive detachment along the 'peel front.'

5.1. Current and phE measurements [16]

Figure 7 shows a diagram of the experimental geometry for current measurements in air and in a vacuum system. This work employed 3M Scotch Brand Magic Tape [No. 810], which consists of a cellulose acetate backing coated with a long-side-chain alkyl acrylate adhesive. Current measurements were made at peel speeds from 0.004 to 10 mm/s. Typical current signal and load signals during the peeling of a 12.7 mm wide strip of adhesive at a speed of 0.5 mm/s are shown in Fig. 8, where the direction of the current is taken to be positive when positive charge flows from the copper to ground. Thus the sign of the observed current represents a flow of positive charge to ground and indicates that negative charge is removed from the copper during peeling. Measurements of the charge remaining on the adhesive after peeling indicate that a negative charge remains on the tape after peeling, consistent with the sign of the observed current.

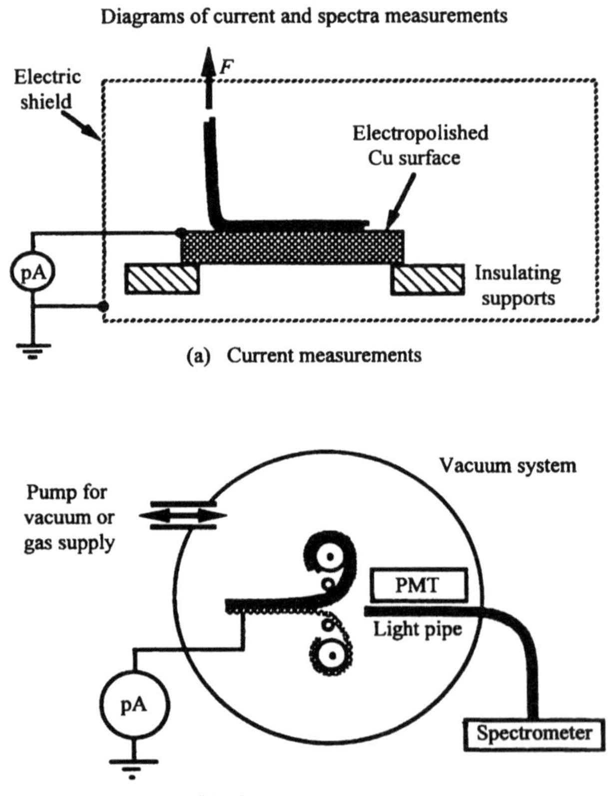

Figure 7. Diagram of apparatus used in (a) current measurements in the 90° peel geometry and (b) spectra measurements in the T-peel geometry.

The tape in Fig. 8 was not under tension at the onset of loading, and loading was continued until the end of the tape was removed from the copper. Although the load begins to rise at Point A, no detectable current is observed prior to actual adhesive detachment, which begins at Point B. Both the load and current signals increase sharply between Points B and C, i.e. during fibril formation. 'Steady state' peeling is achieved after Point C, where the load signal decreases to a steady level. During 'steady state' peeling, the average current gradually rises until detachment. The rise is a consequence of the experimental geometry, where charged adhesive removed early in the peel process continues to contribute to the observed current throughout the experiment. Many negative current spikes are evident throughout the peeling process. When the end of the tape is reached at Point D, the detached tape snaps away from the copper block, yielding an off-scale positive current signal.

Although the average current during peeling does not reach a constant level during the short peels employed in this work, current measurements after the peeling of a fixed length of tape (here, half-way through a 1 cm peel) are comparable. (This is supported by analysis of the model of current generation in [16].) We denote the average current between negative current spikes measured halfway through a 1 cm peel by I_{avg}. Measurements of I_{avg} vs. peel speed yield a linear relation over more than three orders of magnitudes of peel speed, from 0.003 mm/s to 8 mm/s. This is consistent with a constant charge density along the peeled adhesive, independent of peel speed.

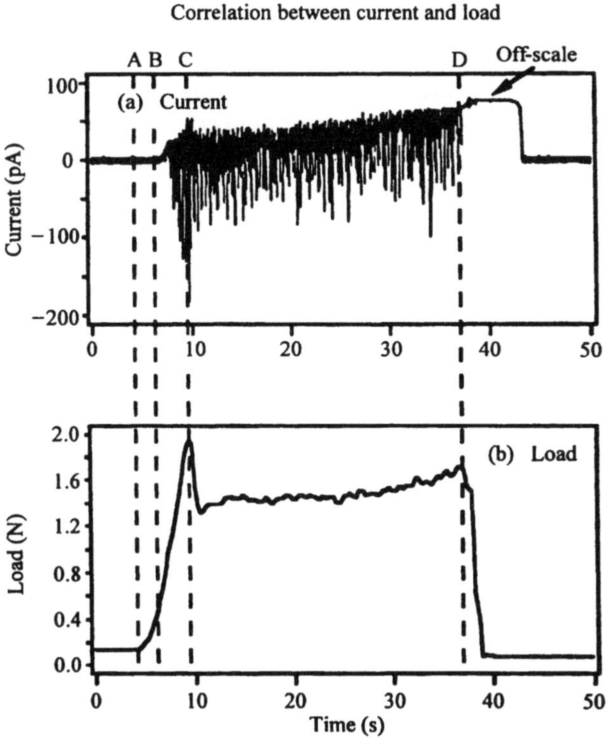

Figure 8. (a) Current signal and (b) applied load accompanying peel at a speed of 0.5 mm/s.

However, the presence of strong negative current spikes suggests that this average charge density is a small fraction of the total charge produced by interfacial failure.

Insight into the negative current spikes can be obtained from simultaneous photon emission (phE) measurements and long wavelength electromagnetic signals (RE: radio emission). Simultaneous phE bursts and RE signals are strong indicators of gaseous discharge events, which have been shown to be responsible for the phE accompanying peeling in air [17–19]. Similar measurements confirm that the negative current spikes in Fig. 8 also accompany gaseous discharge events, which transfer a net negative charge from the peeled adhesive to the copper substrate. At low peel speeds ($\leqslant$ 0.02 mm/s), the negative current spikes and phE bursts virtually always occur in coincidence. However, at higher peel speeds (>1 mm/s), the number of current spikes observed at this level of sensitivity is much less than the number of photon bursts. This suggests that undetected mini-breakdown events are frequent even at high peel speeds. More sensitive current measurements in the 180° peel geometry do show small negative current spikes which correlate quite well with correspondingly small RE signals, confirming this interpretation.

Some very interesting effects are observed when the peeling is performed in different atmospheres. These experiments were performed in the T-peel geometry to simplify light collection for spectra measurements. As shown in Fig. 9(a), peeling in the T-peel geometry in air yields a current signal similar to that in Fig. 8, except for some slow variations in current accompanying variations in the speed of the peeling apparatus used in vacuum. In contrast, peeling in high vacuum (< 10^{-4} Pa) produces extremely strong *positive* current spikes, shown in Fig. 9(b). These spikes are at least three orders of magnitude more intense than the *negative* spikes observed in air at comparable peel speeds. However, the continuum component of current is comparable to that observed in air at the same peel speed. Figure 9(c) shows the current during peeling in SF_6, a highly electronegative gas with a high ionization potential; the number and magnitude of the current spikes are about half those observed in air, although their sign is negative, as in air.

Additional insight into the electrical events during peeling in these three atmospheres can be obtained from spectral observations of the accompanying phE, shown in Fig. 10. In air, the spectrum is dominated by the N_2 lines, confirming that gaseous discharges are the dominant source of phE. In vacuum, weak N_2 lines are observed, along with features due to CH, C_2, H, and H_2 species, which are plausible adhesive decomposition products. These lines indicate the presence of excited adhesive material, produced either by discharge events *along* the adhesive surface, i.e. surface flashover [20], or by intense electron bombardment of trace residual adhesive on the copper surface. The N_2 lines were always present in the spectra, unless great care was taken in excluding atmospheric nitrogen from the interface, e.g. by degassing the tape in vacuum before applying it to the copper substrate in vacuum. Degassing the tape in vacuum and applying the tape to the substrate in a SF_6-filled glove box yielded the spectra of Fig. 10(c). The phE spectrum shows no clear line features, and perhaps a very weak, broad-band luminescence.

The large size and number of current spikes in vacuum reflect the higher charge densities which can be attained in vacuum. The presence of air facilitates breakdown

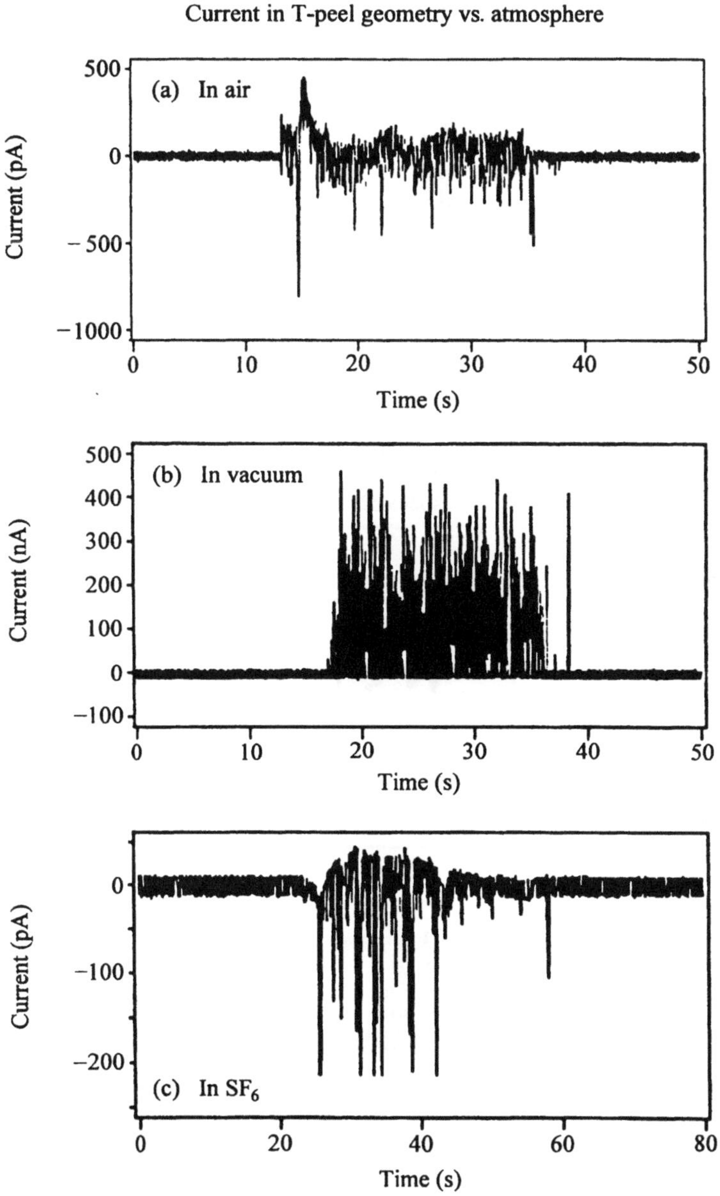

Figure 9. Current signals accompanying peel at a speed of 1 mm/s in the T-peel geometry: (a) in air, (b) in vacuum, and (c) in 1 atm of SF_6.

and increases 'leakage currents' along the adhesive surface. Both factors lead to more energetic and more frequent breakdown in vacuum. We attribute the intense positive current spikes to intense secondary electron emission during the discharge events. Energetic electrons (< 100 eV) striking the (oxidized) copper substrate can stimulate the emission of large numbers of electrons from the copper [21], resulting in a net flow

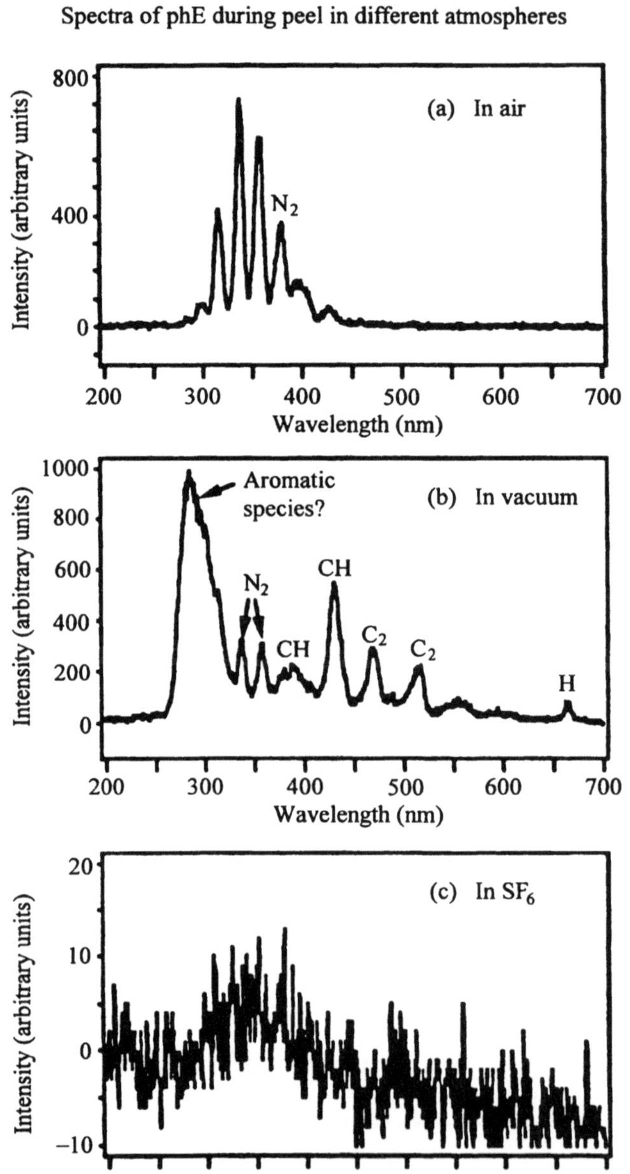

Figure 10. Time integrated spectrum of phE accompanying peel at a speed of 1 mm/s in the T-peel geometry: (a) in air, (b) in vacuum, and (c) in 1 atm of SF_6 (Integration time 40 s).

of electrons from the copper (i.e. a positive current spike). Atmospheric pressures of air or SF_6 hinder the escape of secondary electrons from the substrate, so that the net current is always a flow of electrons to the copper. Secondary electron emission in vacuum also tends to amplify the current signals accompanying breakdown, as several electrons can be emitted for every one that strikes the surface. Thus the intensity of the positive current spikes reflects both the high intensity of the breakdown event in vacuum, and the amplification of the secondary emission process.

The weakness of the negative current spikes in SF_6 is probably a consequence of electron scavenging by the highly electronegative and difficult to ionize SF_6 molecule. Electrons emitted in breakdown events tend to form rather stable SF_6^- ions before they are accelerated to energies capable of ionizing the SF_6 molecule. The absence of SF_6 line spectra (at this level of sensitivity) suggests that the negative current spikes are due to surface discharge events (i.e. charge flow via surface states created in a breakdown event), and that the absence of spectral features associated with hydrocarbons is due to collisional quenching by SF_6 molecules.

5.2. Current fluctuations vs. peel speed [22]

One way to improve the sensitivity of current measurements to small fluctuations along the peel front is to employ the 180° peel geometry, where the tape is drawn parallel to the substrate. Because the previously peeled adhesive is maintained at a fixed distance from the substrate, it does not induce a current in the copper. Thus the entire current is due to the removal of adhesive near the actual point of detachment.

Figure 11 shows typical current measurements during peeling in the 180° peel geometry in air at three peel speeds. At these peel speeds (2–6 mm/s), the number of large, negative current spikes (discussed above) is small, although a large number of smaller negative current spikes are observed in their place. Sensitive phE and RE measurements confirm that these current spikes accompany small gaseous discharge events. The behavior of the average current (between negative current spikes) increases with peel speed, as in the 90° peel geometry. Under these conditions a strong relation between load and low-frequency current fluctuations is observed, where the highest currents are observed during periods of falling load. This is consistent with the expectation that the current is a function of the rate of detachment. When the tape is drawn at a constant speed, transient increases in rate detachment decrease the applied load and increase the current.

The current fluctuations are readily analyzed by a variety of numerical techniques. For instance, the power spectra of the current signals at three different peel speeds are shown in Fig. 12. These spectra were computed by two distinct methods, once from the Fourier transform of the current signals and once by a maximum entropy technique [23]. Maximum entropy methods typically render broad band features in a smoother fashion, but are also capable of representing isolated, sharp features.

At the two lower peel speeds, the power spectra show a relatively constant level of 'power' at the low frequencies, a broad peak in the 50–60 Hz region, and a rapid drop in power as a function of frequency at the higher frequencies. (The contribution of 60 Hz electrical noise to the observed signals can be assessed from the sharp 60 Hz peak in the Fourier transform power spectra of Fig. 12(a). This peak is well below the 'signal' in the other two power spectra.) The total power, as measured by the area under the power spectra, is about five times greater at 4.5 mm/s than at 2.2 mm/s. The greater total power at 4.5 mm/s is consistent with the greater amplitude of the transient, negative-going spikes at 4.5 mm/s [Fig. 11(b)], which is about twice

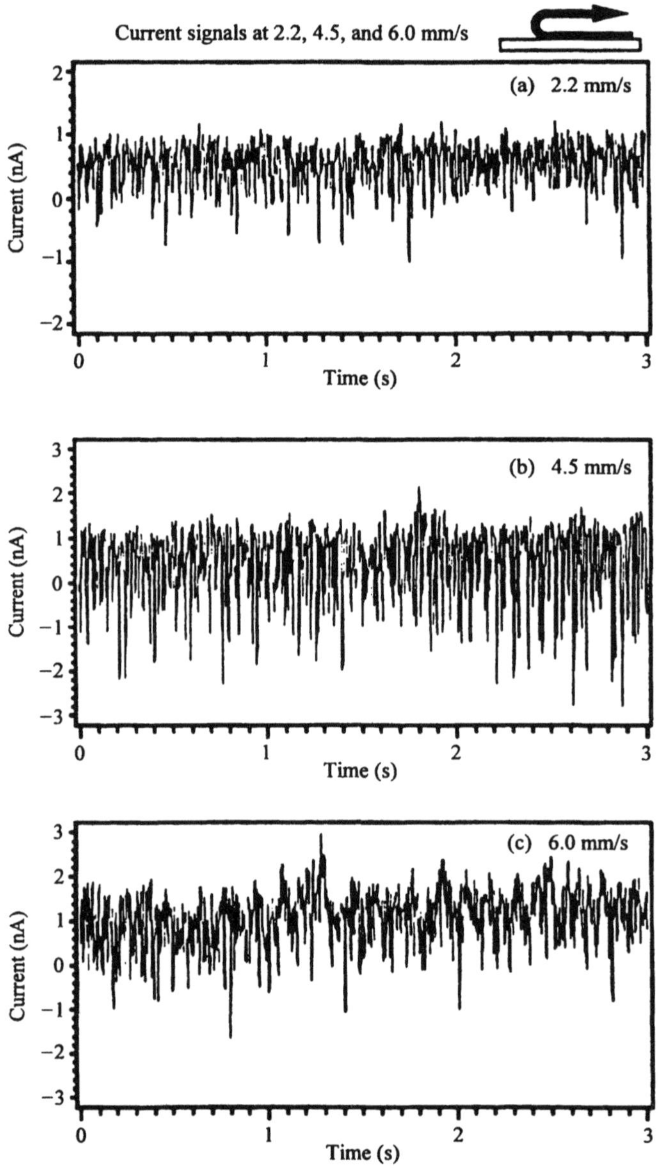

Figure 11. Current signals observed at peel speeds of (a) 2.2 mm/s, (b) 4.5 mm/s, and (c) 6.0 mm/s.

that at 2.2 mm/s [Fig. 11(a)]. Since the 'power' associated with a fluctuation varies as the square of its amplitude, a factor of two increase in amplitude would account for most of the increase in power.

The maximum entropy 'spectral estimates' follow the average behavior of the power spectra computed from the Fourier transforms quite well. At the lower two peel speeds, the general shapes of the spectral estimates are similar to the spectrum of an underdamped harmonic oscillator with a resonant frequency of 50–60 Hz. As noted

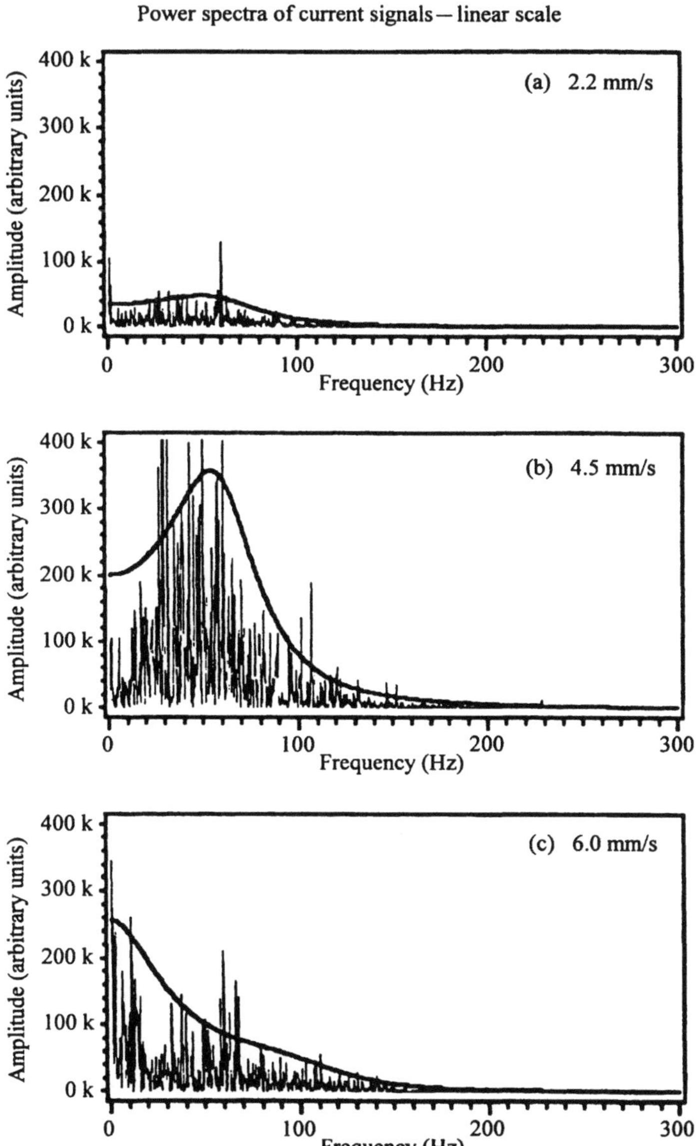

Figure 12. Power spectra of the current signals of Fig. 11 in the region of 0–300 Hz accompanying peel at (a) 2.2 mm/s, (b) 4.5 mm/s, and (c) 6.0 mm/s. The light lines show the power spectra as computed from the Fourier transforms of the current signals, while the smooth, dark lines show the power spectra estimates computed by the maximum entropy method.

below, this comparison can be made quantitative. It is important to note that features in this frequency range are too low to be attributed to individual breakdown events (which produce the negative current spikes) and too broad to be attributed to electrical noise (which produces a distinctive, but much smaller feature). Thus they appear to reflect a mechanical resonance along the peel front.

As the peel speed is increased to 6.0 mm/s, the power spectra of the current fluctuations change dramatically. A strong peak centered at zero frequency appears, suggesting a strong overdamped mechanical response, although a shoulder at about 75 Hz suggests that a weak underdamped response remains. The power spectrum at 6 mm/s is well described by two harmonic oscillators in parallel, one overdamped and one underdamped. This behavior may reflect an inhomogeneity in the mechanical response. For instance, interfacial failure *ahead of* the nominal peel front could yield current fluctuations with an entirely different (highly damped) character than interfacial failure along the peel front itself. Alternatively, the peel front as a whole may be oscillating between two different mechanical behaviors (intermittency).

The maximum entropy spectra (bold lines) of Fig. 12 can be quantitatively modeled in terms of underdamped (at the lower peel speeds) and underdamped (at the higher peel speeds) harmonic oscillators. The power spectrum, S, of a perturbed harmonic oscillator has the form [24, 25]

$$S(f) = \frac{N}{(f^2 + \delta)^2 + \varepsilon}, \tag{9a}$$

where N is the (constant) spectral amplitude of the random perturbations, f is the frequency,

$$\delta = \frac{\gamma^2}{2m^2} - \frac{k}{m}, \qquad \text{and} \qquad \varepsilon = \frac{\gamma^2 k}{m^3} - \frac{\gamma^4}{4m^4}. \tag{9b}$$

In a simple mass-spring-dashpot system, γ is the damping constant, k is the spring constant, and m is the effective mass. Underdamping prevails when $\delta < 0$ and overdamping when $\delta > 0$. The least squares fit of the maximum entropy spectral estimate to equation (9a) are displayed in Fig. 13. As suggested above, the power spectrum at 6.0 mm/s is modeled by two mass-spring-dashpot elements (arranged in parallel), one overdamped and one underdamped. Given the simplicity of these models, the fits to the experimental power spectra are excellent.

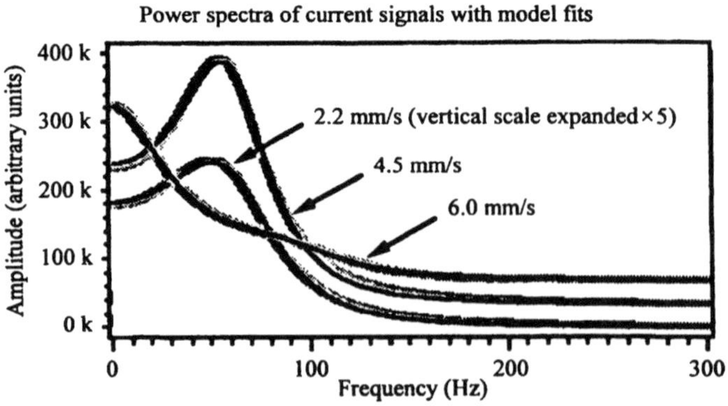

Figure 13. The maximum entropy power spectra of Fig. 12 (dark lines) and the least squares fit to the power spectra from equation (9) (light lines).

Peeling at lower relative humidities increases the amplitude of the fluctuations at all peel speeds, and raises the peel speed for the transition from underdamped to overdamped behavior. We attribute the greater fluctuations, in large part, to the greater adhesive-substrate adhesion at lower humidities. This produces greater strains in the adhesive along the peel front, and thus more rapid and extensive fibril movements. Assuming a rubber-like response in the adhesive, these greater strains are also associated with higher stiffness (higher k in equation (9b)). Overdamping is more difficult with 'stiff' springs (smaller δ in equations (9a) and (9b)), which would account for the higher velocities required for the transition.

The origin of the dramatic transition to largely overdamped behavior between 4 and 6 mm/s is not entirely clear. It is especially intriguing in that overdamped behavior is typical of highly dissipative systems — and is thus desired in adhesive systems. Since the performance of a pressure sensitive adhesive depends on dissipation in a large area adjacent to the region of actual detachment, this transition may not be evident in gross measures of peel behavior, such as the peel force. (The power spectra of fluctuations in the peel force become increasing broad in this region of peel speeds, but is otherwise not remarkable.) Nevertheless, the highest strains in the adhesive system occur along the peel front, which is thus the 'weak link' in the peel system. Thus any change in the behavior of the peel front is of considerable interest.

The broad, low frequency peaks in the power spectra of the current fluctuations can be described in terms of a simple mechanical response to a perturbation. However, the perturbation appears to be quite complex. For instance, chaotic systems typically display broad band noise similar to that observed here [26]. The definitive test for chaos is an extreme sensitivity to initial conditions, which can be quantified in terms of the Lyapunov exponent. Chaotic systems have at least one positive Lyapunov exponent. Unfortunately, the available algorithms do not yield reliable Lyapunov exponent estimates in many experimental systems [27], including this one. An important alternative analysis involves testing for fractal behavior (the search for a 'strange attractor'). Although fractal character does not mathematically require chaos, the number of known fractal, nonchaotic systems is small [28]. [The only exceptions (non-chaotic systems with non-integral dimensions) of which we are aware are dissipative systems driven simultaneously at two incommensurate frequencies. It does not seem likely that the peeling process (which can be viewed as a dissipative process driven at zero frequency) would constitute another exception.] This analysis confirms that the current signals are indeed fractal ($5.5 < D < 5.6$) [22], consistent with a chaotic origin, where D represents dimensionality. These dimension measurements exclude the possibility that the corresponding current fluctuations (reflecting mechanical and/or electrical perturbations along the peel front) are random. Random signals are inherently infinite dimensional [29], so that dimension estimates on random signals fail to converge altogether.

6. CONCLUSION

We have shown that for a variety of geometries, photon and electron emission are sensitive, time-resolved probes of interfacial failure. Photon emission and current signals can be employed to study 'embedded' interfaces, surrounded by the matrix material. These signals, along with other time resolved measurements (e.g. strain, radiofrequency emission, and electron emission), can provide information on the sequence of events with submicrosecond resolution, indicating the direction and speed of debond crack propagation, the onset of pullout, and the arrival of cracks at the matrix surface. Intense fracto-emission signals also accompany the subsequent pullout process. These probes require minimal contact with the sample, facilitating routine testing of fragile specimens (e.g. single fibers) and testing at high temperatures. Thus measurements of this kind have considerable potential to improve our understanding of debonding and pullout.

Fracto-emission and current measurements are also useful in adhesive systems where viscous flow is an important dissipation mechanism, such as the peeling of pressure sensitive adhesives. Visual observations of adhesive detachment reveal a highly branched structure of adhesive 'fingers' in a constant state of flux — branching, merging, extending, and retracting. However, these visual observations are very difficult to quantify in a time resolved fashion. Current measurements provide a time-resolved picture of the collective behavior of this very complex dynamical system. Numerical analysis of the current signal yields direct information on the mechanical response of the peel front, and indicates that the peel front is perturbed in a nonrandom, probably chaotic manner. Our present efforts are focused on improving the sensitivity and frequency range of the electrical measurements, as well as improving observational techniques for characterizing the events occuring in the peel zone.

Current and particle emission processes ultimately depend on atomic-level events along the interface, which is the source of the potentially very high spatial and temporal resolution of these measurements. Departures from *adiabatic* separation of atomic planes or molecular bonds are reflected in the topographical, electrical, and chemical states of the resulting fracture surface. By correlating and analyzing these properties, we are striving to probe the dynamics of the most dangerous type of fracture, i.e. catastrophic crack growth. The details provided by these and similar studies go hand-in-hand with developments in the fracture mechanics field. As computer performances improve, the mechanical description will naturally extend to nanometer scales, and perhaps down to atomic levels. Microscopic, dynamic descriptions of crack growth and related processes may require new syntheses of mathematical formalisms, where fractal geometry and chaos may play key roles. We expect the interdisciplinary nature of the fracture problem to spawn a host of fruitful collaborations among a wide range of engineering and scientific fields.

Acknowledgements

This work was supported by the Ceramics and Electronics Materials Division of the National Science Foundation Grant No. DMR 8912179, The Air Force Office of Scientific Research Contract No. AFOSR-F49620-91-C-0093, the 3M Company, and the Washington Technology Center.

REFERENCES

1. J. Lowell and A. C. Rose-Innes, *Adv. Phys.* **29**, 947 (1980).
2. J. T. Dickinson and L. C. Jensen, *J. Polym. Sci. Polym. Phys. Ed.* **23**, 873 (1985).
3. G. H. Campbell, M. Rühle, B. J. Dalgleish and A. G. Evans, *J. Am. Ceram. Soc.* **73**, 521 (1990).
4. Ma Zhenyi and J. T. Dickinson, *J. Appl. Phys.* **70**, 4797 (1991).
5. A. S. Crasto, R. Corey, J. T. Dickinson, R. V. Subramanian and Y. Eckstein, *Composites Sci. & Technol.* **30**, 35 (1987).
6. C. Atkinson, J. Avila, F. Betz and R. E. Smelser, *Mech. Phys. Solids* **30**, 97 (1982).
7. V. M. Karbhari and D. J. Wilkins, *Scripta Metall. Mater.* **24**, 1197 (1990).
8. K. A. Zimmermann, S. C. Langford and J. T. Dickinson, *J. Appl. Phys.* **70**, 4806 (1991).
9. K. Ohara, *Wear* **50**, 333 (1978).
10. J. T. Dickinson, L. C. Jensen and S. K. Bhattacharya, *Makromol. Chem. Macromol. Symp.* **7**, 129–152 (1987).
11. J. T. Dickinson and L. C. Jensen, *J. Polym. Sci. Polym. Phys.* **20**, 1925 (1982).
12. J. T. Dickinson, L. C. Jensen and A. Jahan-Latibari, *Rubber Chem. Technol.* **56**, 927 (1984).
13. J. T. Dickinson, L. C. Jensen and S. K. Bhattacharya, *J. Vac. Sci. Technol. A* **3**, 1398 (1985).
14. K. A. Zimmerman, S. C. Langford, J. T. Dickinson and R. P. Dion, *J. Polym. Sci. Polym. Phys.* **31**, 1229 (1993).
15. R. J. Good and R. K. Gupta, *J. Adhesion* **26** (1), 13 (1988).
16. Sunkyo Lee, L. C. Jensen, S. C. Langford and J. T. Dickinson, *J. Adhesion* (submitted).
17. E. E. Donaldson, J. T. Dickinson and X. A. Shen, *J. Adhesion* **19**, 267 (1986).
18. Ma Zhenyi, Fan Jiawen and J. T. Dickinson, *J. Adhesion* **25**, 63–77 (1988).
19. J. T. Dickinson, in: *Adhesive Bonding*, L. H. Lee (Ed.), pp. 395–423. Plenum Press, New York (1991).
20. H. C. Miller, *IEEE Trans. Electrical Insulation* **24**, 765 (1989).
21. W. H. Kohl, *Handbook of Materials and Techniques for Vacuum Devices*, pp. 564–568. Reinhold Publishing, New York (1967).
22. L. Scudiero, J. T. Dickinson, L. C. Jensen and S. C. Langford, *J. Adhesion* (submitted).
23. W. H. Press, B. P. Flannery, S. A. Teukolsky and W. T. Vetterling, *Numerical Recipes in Pascal: The Art of Scientific Computing*, pp. 473–479. Cambridge University Press, Cambridge (1989).
24. K. R. Symon, *Mechanics*, 3rd edn., p. 50ff. Addison Wesley, Reading, Mass (1971).
25. A. van der Ziel, *Noise: Sources, Characterization, Measurement*, pp. 16–18. Prientice-Hall, Englewood Cliffs, NJ (1970).
26. H. G. Schuster, *Deterministic Chaos: An Introduction*, 2nd edn, pp. 1–16. VCH, Weinheim, Germany (1988).
27. F. C. Moon, *Chaotic and Fractal Dynamics*, pp. 78–82, 307–310. John Wiley, New York (1992).
28. W. L. Ditto, M. L. Spano, H. T. Savage, S. N. Rauseo, J. Heagy and E. Ott, *Phys. Rev. Lett.* **65**, 533–536 (1990).
29. A. Ben-Mizrachi and I. Procaccia, *Phys. Rev. A* **29** (2), 975–977 (1984).

Fundamentals of Adhesion and Interfaces, pp. 205–224
D. S. Rimai, L. P. DeMejo and K. L. Mittal (Eds)

Direct measurement of molecular level adhesion between poly(ethylene terephthalate) and polyethylene films: determination of surface and interfacial energies

V. MANGIPUDI,[1] M. TIRRELL[1,*] and A. V. POCIUS[2]

[1] *Department of Chemical Engineering and Materials Science, University of Minnesota, Minneapolis, MN 55455, USA*

[2] *Adhesive Technologies Center, 3M Center, St. Paul, MN 55144, USA*

Revised version received 26 May 1994

Abstract—The strength of an adhesive bond depends on the thermodynamic work of adhesion, among other properties. In this paper, we report the direct measurement of the thermodynamic work of cohesion and adhesion between poly(ethylene terephthalate) (PET) and polyethylene (PE) films. The pull-off force between polymer surfaces was measured using the surface forces apparatus (SFA). Thermodynamic work of adhesion was determined from pull-off force measurements using the theory of contact mechanics developed by Johnson, Kendall, and Roberts (JKR theory). The values of the surface energies of PET and PE, and the interfacial energy between PET and PE were obtained from these measurements. The dependence of the measured values of the work of adhesion on the rate of separation, time in contact, and other variables that could reflect an irreversible contribution to the measured adhesion was found to be negligible. The critical surface tensions of PET and PE were determined from contact angle measurements. The critical surface tension of wetting depends on the characteristics of the probe liquids. The surface energy of PET determined by the direct force measurements is higher than the critical surface tension of wetting. These values are 61.2 mJ/m^2 and about 43 mJ/m^2, respectively. However, in the case of PE the surface energy determined using the SFA and the critical surface tension of wetting are about the same, 33 mJ/m^2. The interfacial energy between PET and PE, obtained from direct measurements, is about 17.1 mJ/m^2.

Keywords: Adhesion; direct measurement; surface energy; interfacial energy; surface forces apparatus.

1. INTRODUCTION

The determination of the surface energy of a solid is of interest in the fields of adhesion, adsorption, and wetting, where interactions at and across the interface operate. The adhesion between polymeric surfaces can be caused at a molecular level by any one or a combination of three possible mechanisms: namely, (i) van der Waals or

*To whom correspondence should be addressed.

other interactions that form secondary bonds across the interface; (ii) interdiffusion of polymer chains across the interface and coupling by entanglement with the bulk polymer; and (iii) the formation of primary chemical bonds between chains that cross the interface [1–3]. In the absence of mechanisms (ii) and (iii), the work of adhesion between two surfaces depends on the surface energies and the interfacial energy between them. This work is the energy required to debond or separate unit area of the surfaces in a reversible manner, and is termed the 'thermodynamic work of adhesion'. For two surfaces of the same material in contact, the thermodynamic work of cohesion is given by

$$W_c = 2\gamma, \tag{1}$$

where γ is the surface energy of the material. In the case of dissimilar materials in contact, the thermodynamic work of adhesion is

$$W_a = \gamma_1 + \gamma_2 - \gamma_{12}, \tag{2}$$

where γ_1 and γ_2 are surface energies of materials 1 and 2, and γ_{12} is the interfacial energy between materials 1 and 2. These equations are called the Dupré equations. Hereafter, we use W to denote the thermodynamic work of cohesion or adhesion. The surface and interfacial energies of solid polymers can be determined directly if it is possible to measure the work of adhesion between them in a reversible manner. Since it is difficult to achieve this in solids, traditionally, the surface and interfacial energies of solid polymers have been determined using indirect methods such as contact angle measurements.

The strength of an adhesive bond measured by typical mechanical tests like peeling and shearing gives an apparent work of adhesion, often termed the 'practical work of adhesion', W_P. W_P is usually significantly higher than W since real materials, especially polymers, are not perfectly elastic and dissipation of energy due to inelastic deformations occurs during the debonding or fracture process. For this reason, W_P is orders of magnitude greater than the thermodynamic work, W. However, it is thought that W_P and W are related. The studies of Gent and Schultz [4] and Andrews and Kinloch [5] suggest the following multiplicative dependence:

$$W_P = W(1 + \Phi), \tag{3}$$

where Φ is a dissipation function which depends on several bulk properties of the adhering materials as well as certain test parameters. The bulk properties that govern the dissipative effects include thickness, viscoelastic characteristics, and interdiffusivity of the adhering materials [2, 6–11]. The rate of crack propagation, temperature, and mode of crack propagation are the important test parameters that influence the contributions of various modes of energy dissipation to the practical work of adhesion [12–16]. The multiplicative effect of the thermodynamic work on the practical work of adhesion, as in equation (3), suggests that W is an important parameter in the design of adhesive bonds. If it is possible to measure W in a reversible manner, by minimizing or eliminating the dissipative effects, then it is possible to gain an insight into the relationship between W_P and Φ. The surface forces apparatus (SFA) has the potential to provide a means of measuring W reversibly, and hence determine the surface and interfacial energies. Originally developed by Israelachvili and co-workers, the SFA combines a direct and precise measure of forces acting between the surfaces with an accurate measure of their separation [17, 18]. Traditionally, the SFA has been used to measure interactions between several kinds of molecules and polymers which were adsorbed or deposited on mica surfaces under various conditions [17–21]. There have also been some reported measurements of adhesion between polymers adsorbed on mica and then dried [22]. In all these studies the base substrate has been mica. The primary reason for the use of mica is that it can be cleaved into smooth, thin sheets, making it suitable for the optics in the SFA. However, in recent years some studies on glass, sapphire, and metal substrates have been reported [23–26]. Merrill *et al.* [27] were the first to carry out adhesion measurements on thin polymer films. They measured the surface energy of biaxially oriented poly(ethylene terephthalate) (PET).

The objective of this work was to employ the SFA to directly measure the thermodynamic work of adhesion between polymers, and to compare the results with indirect measurements. The SFA makes it possible to measure W because of the small area of contact between the surfaces and the slow rate of separation available in the experiment. In this paper, we report the direct measurement of the surface energy of polyethylene (PE) and the interfacial energy between PET and PE. We also present the results of indirect estimates of the surface and interfacial energies of PET and PE by measuring contact angles of liquids of varying chemical constitution. The earlier work of Merrill *et al.* [27] on PET has been extended to study the influence of various bulk properties and test parameters on the adhesion strength measured using the SFA. We used PET and PE films for these studies because of their numerous applications. The PET and PE films used in these studies are semicrystalline. However, under the experimental conditions the amorphous domains of PET are glassy, while PE is far above its glass transition temperature. PE chain mobility, therefore, play a role in the measured strength of adhesion.

2. EXPERIMENTAL

2.1. Samples for the surface forces apparatus

The SFA combines a mechanical means of measuring forces with an optical means of determining the distance between the surfaces. It has been primarily used to measure force versus distance profiles between several kinds of surfaces [17–22]. These profiles can be obtained with a force sensitivity of 10 nN and a distance resolution of better than 5 Å. The distance is determined from the interferometric fringes, called fringes of equal chromatic order (FECO). The substrates of interest are back-silvered for interferometry purposes and are adhered to cylindrical quartz lenses using an adhesive. One of these lenses is mounted on a weak double cantilever leaf spring and the other on a rigid support. The surfaces are brought closer, and the deflection of the leaf spring determines the force between the substrates. The local radius of curvature of the contact geometry can also be determined from the FECO. (For details, see [17] and [27].) To be used in the SFA, the films need to be transparent to white light and smooth. Smoothness is a requirement for the characterization of the local geometry and to obtain a well-defined contact between the surfaces. To obtain useful interference fringes that allow easy optical separation, the substrates need to be thin and self-supporting for ease of handling and mounting. The optimum thickness of polymer films meeting these criteria is about 3–6 μm. As mentioned in the previous section, freshly cleaved mica sheets satisfy these requirements easily, thus making them the most widely used substrates in various studies using the SFA. In our studies, the generation of suitable polymer films demanded considerable effort.

2.2. *Sample preparation*

The PET films used in these experiments were extruded and then biaxially stretched to generate thinner films. To be able to stretch successfully and get smooth films, the slip agents normally added in commercial films were avoided. The unfilled PET resin was extruded to a 50 μm thick film. This film was then cut into 10 cm $\times$ 10 cm squares and biaxially stretched to a draw ratio of 4:1$\times$4:1 in a T. M. Long stretcher. The resultant films were about 3–5 μm thick. The samples were free from dust and static charge. The stretched film was mounted on a circular ring. On one side of the film, a silver layer of 530 Å thickness was deposited, and then a suitable adhesive was applied by spin-coating. This composite was then adhered to a cylindrical quartz lens of 1 cm diameter and 2 cm radius of curvature, exposing the PET surface. When applying the film to the lens, it was made certain that no air was entrapped between the lens and the substrate. Care was taken all along to keep the samples clean and any contamination was avoided.

The PE films were co-extruded with PET to a thickness of 2–4 μm. The PET/PE bilayer was mounted on a ring, silvered, and then coated with an adhesive on the back side of PE. The bilayer composite was mounted on the quartz lens and then the PET layer was peeled off, very slowly and carefully, exposing the PE surface.

The adhesive that was used to glue the film to the quartz lens forms an important component of the composite and, therefore, of the experimental method. Initially, we used two different pressure-sensitive adhesives (PSAs): an acrylic-based adhesive supplied by the 3M Co. and a silicone-based PSA (No. 355, Dow Chemical Co.). The use of a PSA sometimes resulted in time-dependent deformations. It was necessary that the adhesive be elastic under the test conditions. Kraton G1652 (Shell Chemicals) was found to be suitable for this purpose. A solution of Kraton in toluene was spin-coated on the backside of the film.

2.3. *Surface analysis*

The cleanliness of PET and PE used in these studies was determined by the analysis of their surface chemistry using X-ray photoelectron spectroscopy and ($\pm$) static secondary ion mass spectroscopy (SSIMS). In the case of the PE/PET bilayer, clean separation of these films was demonstrated by the attainment of (+) and (−) SIMS spectra that were identical to the surface of pure PE.

2.4. Contact angle measurements

The contact angles of various liquids on the PET and PE film samples were measured using a Rame-Hart goniometer. The liquids used for this purpose and their respective surface tensions are listed in Table 1. The liquids were doubly distilled and their surface tensions were measured using a Fisher Surface Tensiomat (Model 21). In contact angle measurements, a small drop of each liquid was placed on the surface using a clean syringe and was allowed to equilibrate for about 60 s. The advancing and receding contact angles were measured. The receding angles were measured by withdrawing liquid from the drop. In the case of glycerol, owing to its high viscosity, the equilibration time was longer than 3 min. The critical surface tensions of PET and PE were estimated from contact angle measurements using several different schemes suggested in the literature. These schemes are described briefly in Section 3.2.

3. THEORY

3.1. Surface forces, surface energy, and deformation of solids

When two smooth solid surfaces are brought into contact in the absence of any applied load, they deform due to the action of surface forces between them. Further, owing to the action of these attractive forces, a finite tensile load is required to separate the surfaces from contact. This tensile load is called the pull-off force, P_s. This pull-off

Table 1.
Surface tensions of contact angle liquids (at 25 °C, in mJ/m^2)

Methylene iodide	50.8
Dimethyl sulfoxide	44.2
Tricresyl phosphate	40.9
1-Bromonaphthalene	44.6
Formamide	58.2
Water	72.2
Glycerol	63.0
Ethylene glycol	45.8
Diethylene glycol	34.6
1,3-Butane diol	37.4
1,3-Propane diol	46.1
2-Methyl pentane diol	28.5
1,2,6-Trihydroxyhexane	51.1

force is related to the thermodynamic work of adhesion, W, and the geometry of the surfaces according to the theoretical relations developed by Johnson, Kendall, and Roberts (JKR theory) [28] and Derjaguin, Muller, and Toporov (DMT theory) [29]. If the pull-off force could be measured reversibly, the thermodynamic work of adhesion could be determined directly. From the work of adhesion, the surface and interfacial energies can be determined. The predictions of the JKR and DMT theories of contact mechanics are significantly different [28, 29]. However, in the case of low-energy surfaces like those of polymers it is found that the JKR theory correctly predicts the contact mechanics [27, 30–36]. These predictions are summarized in the next section. A critical comparison of the JKR and DMT theories can be found in the literature [27, 32, 36, 37].

3.1.1. JKR theory. The theory of Johnson, Kendall, and Roberts [28] is an extension of the Hertzian theory that accounts for deformations of smooth solid bodies in contact that are due to surface forces. This theory predicts a higher contact area under a given load than the Hertzian theory. The JKR theory considers the adhesion between two spheres simply as a change in surface energy only where they are in contact, i.e. the attractive forces are of infinitely short range and act only in the region of contact. The basis for the JKR theory is that the energy stored due to adhesion results in the deformation of the contact zone and an equilibrium is achieved. According to the JKR theory, at equilibrium, the radius of the deformed contact zone, a, at a given applied load P is given by

$$a^3 = \frac{R}{K}\left(P + 3\pi WR + \sqrt{(6\pi WRP + (3\pi WR)^2)}\right), \tag{4}$$

where W is the work of adhesion, R is the radius of curvature of the geometry, and K is the effective bulk modulus of the two adhering materials. This notation is de-

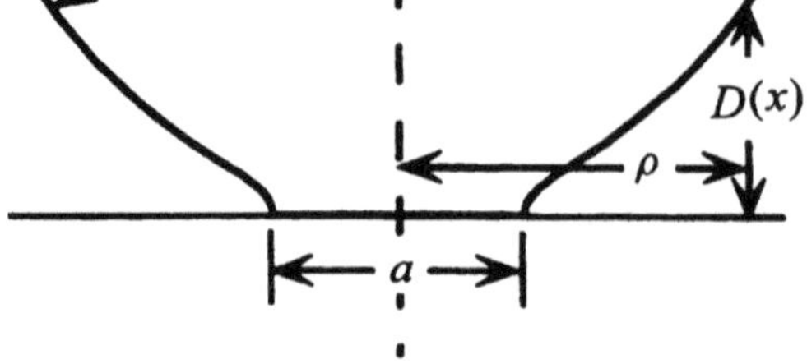

Figure 1. Notation used to describe the deformed shape of a sphere and a flat surface in contact. We can measure the radius of curvature, R; the contact radius, a; and the separation profile, $D(x)$ versus $x = (\rho/a)$, using the fringes in the SFA. ρ is the radial distance and P is applied load.

picted in Fig. 1. The theory also calculates the surface separation profile outside the contact zone which is given by the following equation:

$$D(x) = \frac{a^2}{\pi R}\left((x^2-1) + \left(x^2 - 2 + \frac{4}{3}\left(\frac{a_0}{a}\right)^{3/2}\right)\tan^{-1}(x^2-1)^{1/2}\right), \tag{5}$$

where $x = \rho/a$, ρ is the radial distance, a_0 is the contact radius at no applied load, and $D(x)$ is the distance between the two surfaces. Further, according to the JKR theory the tensile load required to separate the surfaces from contact is given by

$$P_s = \frac{3\pi WR}{2}. \tag{6}$$

This theory also predicts that at the instant of separation, the contact radius is about 63% of the initial no-load contact radius, a_0.

3.2. Contact angle analysis

Measurements of the contact angles of nonwetting liquids on a solid surface are very widely used to measure its critical surface tension of wetting, γ_c, which some consider to be a measure of the surface free energy, γ_s. This section briefly reviews the theory of contact angles. There are several models proposed for the determination of γ_c from contact angle data. In this paper, we denote the estimated critical tension by $\gamma_{c,x}$, where 'x' indicates the model used to analyze contact angle data. All these models involve the basic Young's equation

$$\gamma_{lv}\cos\theta = \gamma_{sv} - \gamma_{sl}, \tag{7}$$

where s, l, and v denote the solid, liquid, and vapor phases, and γ_{sv}, γ_{lv}, γ_{sl} are the interfacial energies. Solid surface energy γ_s and solid–vapor interfacial energy γ_{sv} are related through the spreading pressure, π_e. Since γ_{sl} is not known *a priori*, the above equation cannot be used for determining γ_{sv} and hence approximate models are used. The first approach was developed by Zisman and co-workers [38, 39]. According to Zisman, the cosines of the contact angles of several test liquids on a solid surface are plotted against their surface tensions and the data are extrapolated to $\cos\theta = 1$. The liquid surface tension corresponding to $\cos\theta = 1$ is termed the critical surface tension of wetting, γ_c, of the solid. In other words, γ_c equals the surface tension of the liquid which just exhibits a zero contact angle on the solid. Therefore, mathematically

$$\gamma_{c,\,Zisman} = \lim_{\theta\to 0}\gamma_{lv}. \tag{8}$$

From equations (7) and (8), one obtains

$$\gamma_{c,\,Zisman} = \gamma_{sv} - \gamma_{sl}. \tag{9}$$

$\gamma_{c,\,Zisman}$ may not be equal to γ_{sv}, because γ_{sl} need not be equal to zero, though in most cases it is neglected. γ_{sv} may not necessarily be equal to γ_s. Due to these limitations, Zisman's approach has been revised by several researchers. Girifalco and Good [40, 41] derived an expression for γ_{sl} in terms of γ_{sv} and γ_{lv} using the Berthelot relation for attractive constants. Using this in Young's equation results in the Good–Girifalco–Fowkes–Young [equation (10)]. A detailed derivation can be obtained from Good and Girifalco [40, 41], Fowkes [42], Wu [43], and Chaudhury and Whitesides [30].

$$\frac{(1+\cos\theta)}{2} = \left(\frac{\gamma_{sv}}{\gamma_{lv}}\right)^{1/2}. \tag{10}$$

Using this relation, the critical surface tension of a solid, $\gamma_{c,\,GGF}$, can be obtained from the plot of $(1+\cos\theta)/2$ versus $\gamma_{lv}^{-1/2}$.

Wu [43, 44] derived an equation of state for γ_c that accounts for interactions between the solid surface and the test liquid in a qualitative manner. γ_c is expressed in terms of ϕ to obtain a spectrum of critical surface tensions of wetting given by

$$\gamma_{c,\phi} = \phi^2\gamma_s - \pi_e = \frac{(1+\cos\theta)^2\gamma_{lv}}{4}, \tag{11}$$

where ϕ is the solid–liquid interaction parameter and π_e is the spreading pressure. $\gamma_{c,\,\phi}$ is plotted against γ_{lv} to obtain a curve whose maximum, $\gamma_{c,\,\phi,\,max}$, equals the surface energy of the solid. In the limit of $\phi_{max} = 1$, the above equation gives $\gamma_{c,\,\phi,\,max} = \gamma_{c,\,Wu} \approx \gamma_s$. For a detailed derivation one may refer to the original work.

In our studies, we evaluated the critical surface tension of wetting of PET and PE using all the above schemes, and compared the results with the values of surface energies obtained from direct force measurements.

4. RESULTS AND DISCUSSION

4.1. Adhesion measurements in the SFA

We carried out experiments to measure the pull-off force, P_s; the radius of curvature, R; the surface separation outside the contact zone, $D(x)$; and the contact radius at no applied load, a_0, using the surface forces apparatus (SFA). The radius of the contact zone, the surface separation profile outside the contact region, and the radius of curvature of the geometry were measured from the interferometric fringes. All these measurements were carried out in dry air at 25°C. As mentioned earlier, the dissipative effects in practical adhesion measurements are dependent on several parameters that include the rate of separation, thickness, and viscoelastic properties of the substrates forming the joint. We examined the influence of all these parameters on the pull-off force measurements. The velocity of crack propagation, v, was determined from the rate at which the contact radius changes. This velocity was varied by an order of magnitude, from 0.01 to 0.1 μm/s, by changing the rate at which the surfaces were pulled apart. The effect of viscoelastic and other time-dependent properties was examined by varying the time for which the surfaces were in contact under no applied load conditions. The films were adhered to the quartz lenses using Kraton G1652 in all cases, unless mentioned otherwise.

As shown by equation (6), the pull-off force is directly proportional to the radius of curvature. The normalized pull-off energy, P_n, is the ratio of the pull-off force to the radius of curvature:

$$P_n = \frac{P_s}{R} = \frac{3\pi W}{2}. \tag{12}$$

All the adhesion measurements are presented in terms of P_n, which should be a function of W alone if the measured pull-off force is free from dissipative effects.

4.1.1. PET–PET contact. Figure 2 is a plot of the normalized pull-off energy, P_n, for PET–PET cohesive contact as a function of the rate of crack propagation, v. As can be seen, the change in v does not have any significant effect on P_n. In a typical practical method, the measured adhesion strength increases as $(\text{rate})^{0.6-0.8}$ [5, 16]. Kendall [45] found that the practical work of adhesion of ethylene propylene rubber to glass, measured in a 90° peel test, varied as the 0.25 power of rate of crack propagation. Brown [2] reported a linear dependence of the fracture toughness on crack speed in the debonding of polyisoprene from a surface coated with polystyrene–polyisoprene diblock copolymer, in microscopic experiments. However, no rate dependence was observed in the SFA measurements on PET. Bulk dissipation effects, if any, also depend on the viscoelastic properties and thickness of the adhering substrates. P_n was measured as a function of the thickness of the PET films, changing the thickness from 4.5 to 7 μm, as shown in Fig. 3. P_n was independent of film thickness.

Figure 4 shows P_n measured after the surfaces remained in contact for different lengths of time. It clearly shows that there are no time-dependent effects in these measurements. Under the conditions of this measurement, PET is glassy and the molecular chain mobility is severely restricted. Little or no interdiffusion of polymer chains is expected while the surfaces are in contact. No time-dependent effect was observed. In all these measurements, the adhesive used to mount the film on the lens was Kraton G1652. At room temperature this copolymer is elastic. The deformation of the adhesive is reversible. The results clearly indicate that the measured pull-off force is free from any dissipative effects, and P_n can be used to determine the surface energy of PET. The average value of P_n obtained from all the above measurements is 576.8 ± 17.9 mJ/m^2. The error bars in Figs 2–4 represent the standard deviation of values obtained from multiple measurements.

The surface energy of PET can be determined from the value of P_n using the JKR theory [equation (12)]. To verify the validity of the JKR theory, the separation profile outside the contact region was measured under no load as well as under different compressive load conditions, and compared with the theoretical predictions. As shown in Fig. 5, the agreement between theory and experiments was good. It was also observed that the surfaces separate from contact at a finite contact radius, a_s. The ratio of a_s to a_0 is about 0.6, close to the JKR prediction of 0.63. These observations confirm that the JKR theory correctly predicts PET–PET contact mechanics. The surface energy of PET is determined from $P_n = 576.8 \pm 17.9$ mJ/m^2 using equation (12).

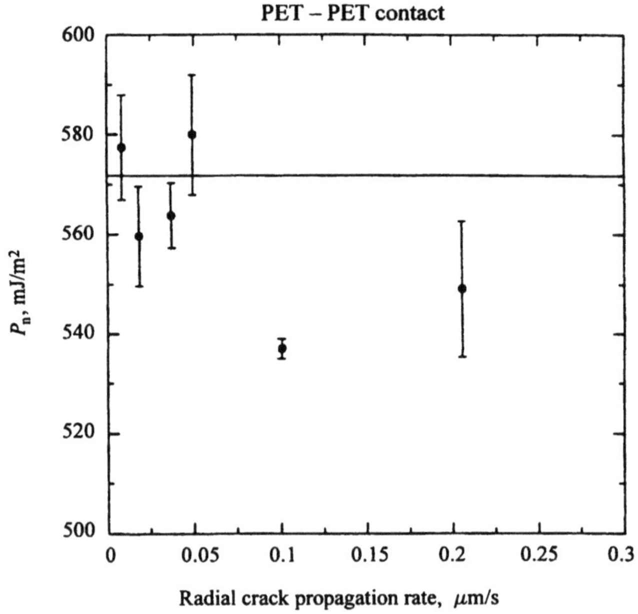

Figure 2. Normalized pull-off energy, P_n, versus the rate of crack propagation for PET–PET contact. The rate independence indicates the absence of dissipative contributions to the measured adhesion strength.

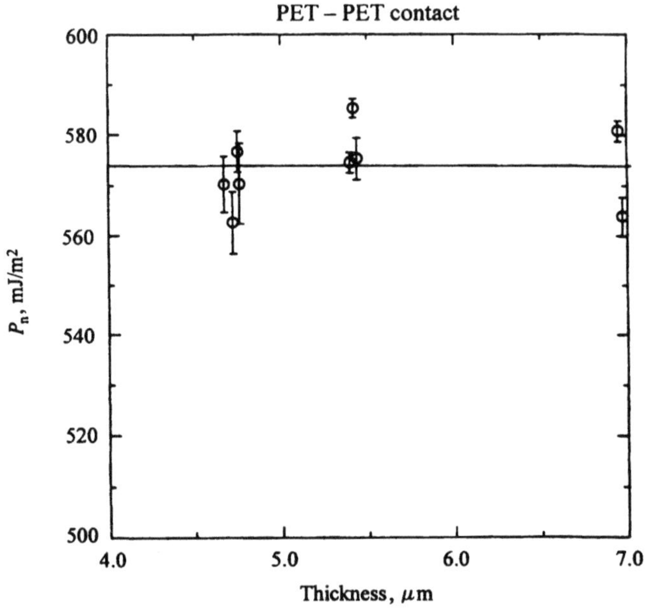

Figure 3. Normalized pull-off energy as a function of the thickness of films for PET–PET contact. Unlike in practical adhesion test methods, in SFA experiments on PET, the measured pull-off energy does not depend on the film thickness.

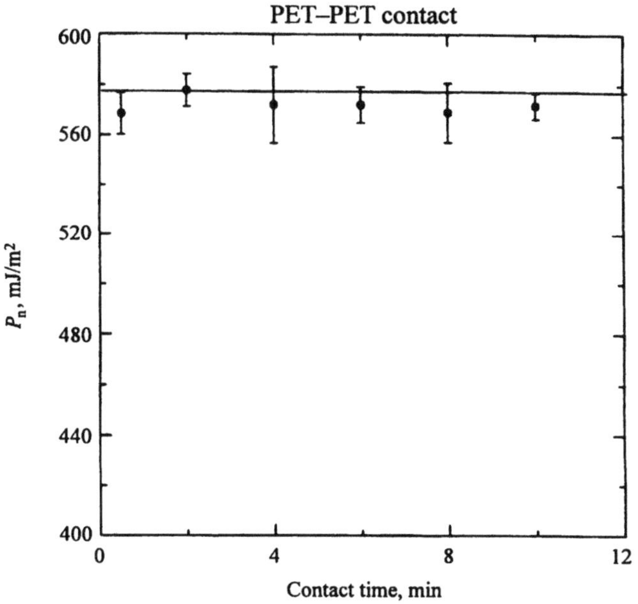

Figure 4. Normalized pull-off energy as a function of the contact time for PET–PET contact. The films are adhered to quartz lenses using Kraton G1652 as the backing adhesive, a thermoplastic elastomer which is elastic under experimental conditions. No significant time dependence is observed.

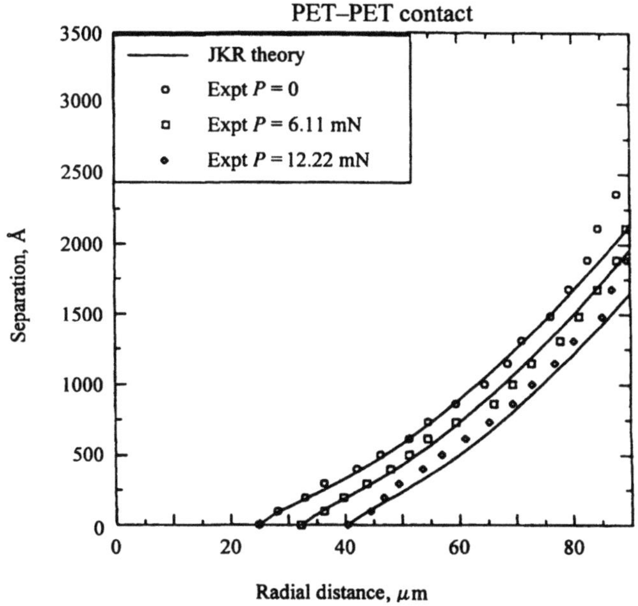

Figure 5. Surface separation profile outside the contact region for PET–PET contact under different loads. The data points represent the measured profiles and the solid lines are the predictions of the JKR theory. (○) No applied load; (□) 6.11 mN compressive load; (◇) 12.22 mN compressive load. The agreement between the experiments and theory is good. The DMT theory (not shown in this figure) predicts a smaller separation than the JKR theory at any given radial distance.

The surface energy of PET obtained from SFA measurements is 61.2 ± 1.9 mJ/m^2, in agreement with the measurements of Merrill *et al.* [27].

As mentioned earlier, Kraton G1652 adhesive responds elastically. This was established by the recovery to the undeformed shape by the surfaces immediately after they were separated from contact. The shape can be determined by the interferometric fringes. It should be noted that this was not the case when an acrylic or a silicone PSA was used. When a PSA was used, the deformations were found to be time-dependent even under no applied load conditions at longer times of contact, and the pull-off force increased with the contact time as shown in Fig. 6. The contact radius also increased with time, and the shape of the surfaces was recovered only after about 1–2 h had elapsed after the surfaces were separated from contact. Merrill *et al.* [27] used an acrylic PSA in their experiments and allowed the adhesive to relax for several hours between consecutive measurements. It is not necessary to do this when Kraton G1652 is used as it behaves elastically, especially under no applied load or when the applied load is low.

It should be pointed out here that the measurement of contact radius as a function of the load provides an independent means of determining W from equation (4) of the JKR theory. This route has been employed by several researchers including Johnson and co-workers [28, 30–32]. This method involves the application of significantly high compressive loads. Under such loads, we cannot ensure the elastic behavior of Kraton G1652. For this reason, we have not yet investigated this aspect of the JKR theory to determine the thermodynamic work of adhesion.

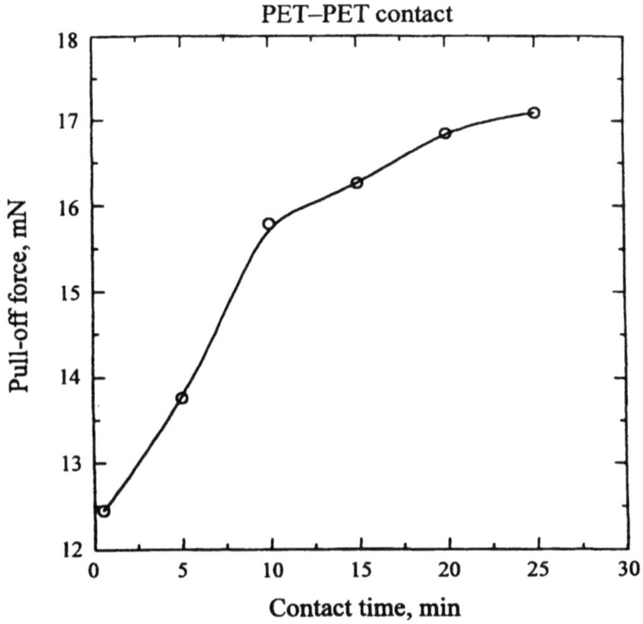

Figure 6. Pull-off force as a function of the contact time for PET–PET contact when the backing adhesive is a silicone-based pressure-sensitive adhesive. The adhesive deforms inelastically and the measured pull-off force increases with the time of contact.

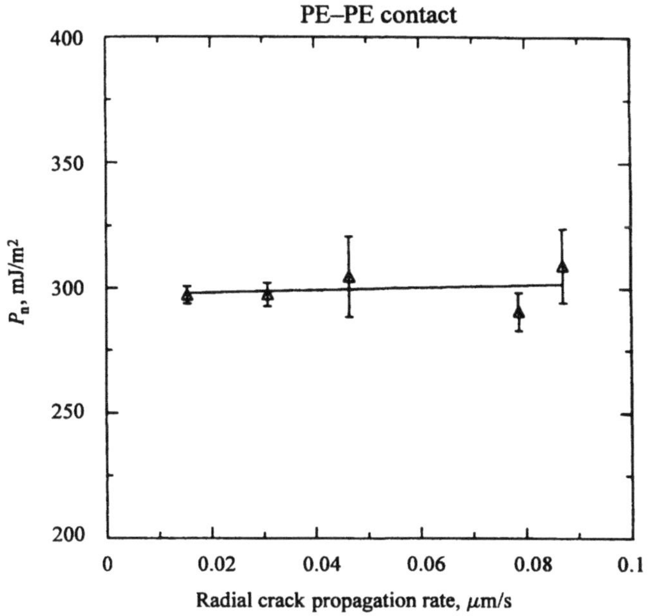

Figure 7. Normalized pull-off energy versus the rate of crack propagation for PE–PE contact. A change in rate does not affect the measured pull-off energy.

4.1.2. PE–PE contact. The normalized pull-off energy, P_n, was measured as a function of the rate of crack propagation and time of contact. Figure 7 is a plot of P_n versus the radial crack propagation rate, v. A change in v by a factor of 10, in the range of 0.01–0.1 μm/s does not affect the measured P_n. Figure 8 shows the variation of P_n with the time of contact. As the surfaces remain in contact for longer times, the pull-off force increases. It must be pointed out that no time-dependent deformation was observed over this time range. The contact radius and the shape of the contact zone remaind the same irrespective of the time of contact. Also, the deformation of the surfaces was observed to be reversible, i.e. the surfaces regained their undeformed shapes as soon as they were removed from contact, just as the PET surfaces did. However, unlike the case of PET–PET contact, for PE–PE contact the normalized pull-off energy increases with the contact time. As seen in Fig. 8, P_n does not change significantly initially but it increases after a certain time of contact. This behavior was observed in several different experimental runs on similar samples. These data represent the average values obtained from different measurements. As mentioned in the Introduction, the PE samples used in this study were semicrystalline and the amorphous domains are far above the glass transition temperature (T_g) at room temperature. Because of this, the polymer chains are mobile. This mobility could lead to some rearrangement or intermingling of chains across the interface, resulting in an increase in adhesion, due to the onset of mechanism (ii) (see Introduction). However, P_n does not vary significantly for shorter times in contact (for a contact time less than 4 min the variation is only about 5%) and is independent of the rate of crack propagation. This value of P_n is 313.8 ± 15.1 mJ/m^2, and is used to determine the

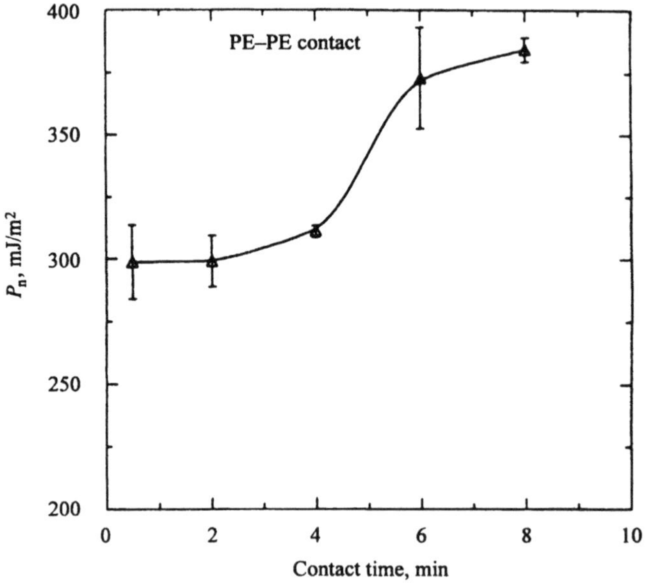

Figure 8. Normalized pull-off energy, P_n, as a function of the contact time for PE–PE contact. At shorter contact times, P_n does not significantly depend on the contact time. However, as the surfaces remain in contact for long times, the pull-off energy increases with time. This may be due to the interdiffusion of PE chains across the interface.

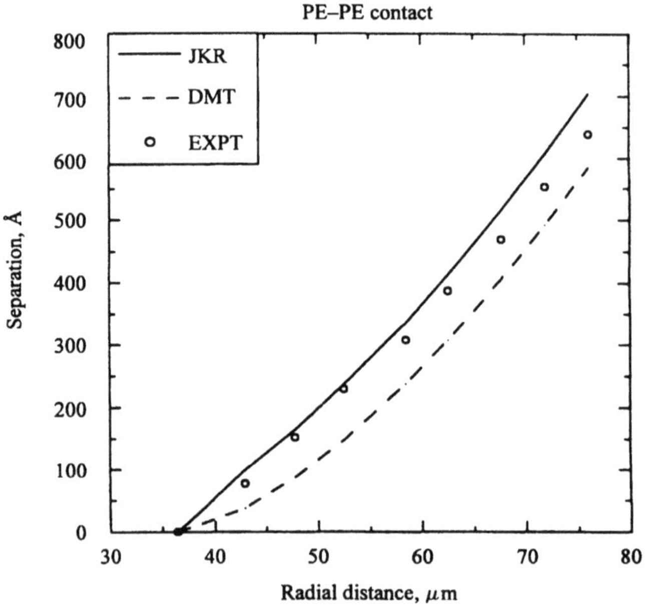

Figure 9. Surface separation profile outside the contact region for PE–PE contact at no applied load. The data points represent the measured profiles and the solid lines are the predictions of the JKR theory [28]. The agreement between the experiments and the theory is good. The dashed line is the DMT theory prediction, which assumes a Hertzian profile. The DMT theory predicts a smaller separation than the JKR theory at any given radial distance.

surface energy of PE. For PE–PE contact, the separation profile outside the contact zone (Fig. 9) and the ratio of a_s to a_0 were also measured, and these were found to be in agreement with the predictions of the JKR theory. Therefore, equation (12) is used to determine the surface energy of PE. This value is 33.3 ± 1.6 mJ/m^2.

4.1.3. PET–PE contact. The pull-off force between PET and PE surfaces was measured to determine the work of adhesion between them. Figure 10 shows a plot of P_n versus the rate of crack propagation. P_n increases slightly with an increase in the rate of crack propagation. However, at slower rates of separation P_n is independent of the rate. At higher rates (rate is increased by a factor of 8), there is an increase of about 6% in the measured value of P_n. The pull-off force was also measured as a function of the contact time. Figure 11 is a plot of P_n versus the contact time. There seems to be no significant variation in P_n with the contact time. This behavior is similar to that observed in the case of PET–PET contact, but unlike PE–PE experiments, in which case the pull-off force increases with the contact time. In PET–PET and PET–PE contacts, the PET surface is glassy, and the polymer chain mobility is restricted and there may not be any interpenetration of the polymer molecules across the interface. However, in the case of PE–PE contact, since both the materials are far above the T_g, there can be significant rearrangement on the surface and chain motion across the interface, leading to an increase in adhesion with time.

In PET–PE contact, for rates of crack propagation in the range of 0.008–0.06 μm/s, P_n is constant and its value is 364.3 ± 10.1 mJ/m^2. The work of adhesion, W, between

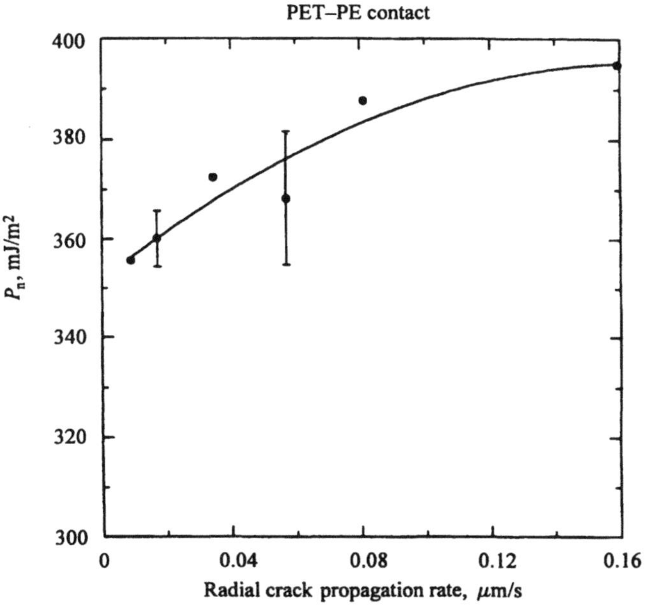

Figure 10. Normalized pull-off energy, P_n, versus the rate of crack propagation for PET–PE contact. At higher rates, P_n increases with the rate. However, at rates slower than 0.06 μm/s, P_n does not significantly depend on the rate; the variation in P_n is only ±2%. Hence, the average value of P_n at slower rates is taken to calculate the work of adhesion between PET and PE.

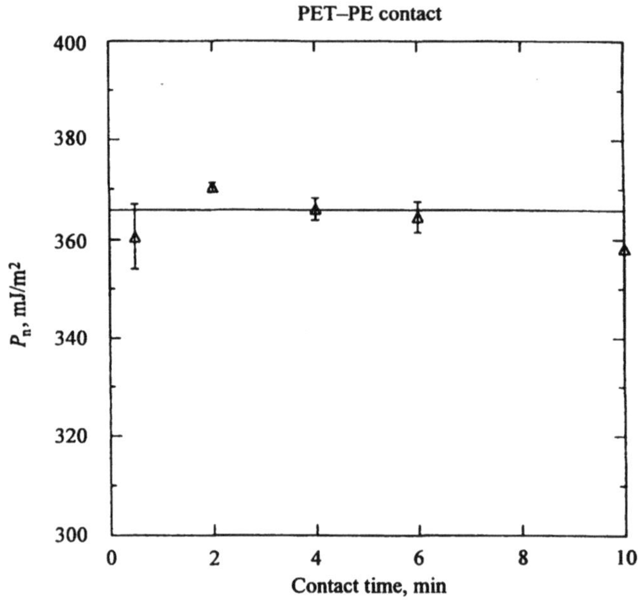

Figure 11. Normalized pull-off energy, P_n, as a function of the contact time for PET–PE contact. Unlike in the case of PE–PE contact (Fig. 8), P_n does not depend on the contact time. This behavior is similar to that observed for PET–PET contact (Fig. 4). Since PET is glassy, there is little or no chain mobility, and hence interdiffusion cannot occur.

PET and PE, obtained from these measurements, is $77.3 \pm 2.2\ \mathrm{mJ/m^2}$. From this value, the interfacial energy between PET and PE is determined using equation (2), and is found to be about $17.1\ \mathrm{mJ/m^2}$. It may be pointed out that this is considerably higher than the values of interfacial tension between polymers reported in the literature (see [43]).

4.2. Contact angle measurements

Advancing and receding contact angles were measured on PET and PE surfaces using the liquids listed in Table 1. Methylene iodide, dimethyl sulfoxide, tricresyl phosphate, and α-bromonaphthalene are predominantly dispersive in nature. For water, formamide, diols, and triols, the polar component of the surface tension is dominant due to H-bonding interactions. For this reason, the data obtained using these liquids were analyzed separately. In the contact angle measurements, some hysteresis was observed. The critical surface tension, $\gamma_{c,x}$, was calculated for both the advancing and the receding angles using the schemes discussed in Section 3.2. The results are listed in Tables 2 and 3. The hysteresis is higher when polar liquids are used, as can be seen from Tables 2 and 3. The surface energy obtained from pull-off force measurements is compared with γ_c calculated from receding contact angles as both represent debonding. The $\gamma_{c,\,\mathrm{Zisman}}$ values determined using nonpolar liquids and Zisman's plot are in agreement with the values reported in Zisman's original work, indicating that the samples used in this work are not significantly different from those studied by Zisman and co-workers.

Table 2.
Surface energy of PET (in mJ/m^2) measured using the SFA and the critical surface tension of wetting determined from contact angle data

1. SFA measurements 61.2 ± 1.9

2. Critical surface tension

	Dispersive liquids		Polar (H-bonding) liquids	
	Advancing	Receding	Advancing	Receding
Zisman's	38.2	38.8	30.7	36.4
Good–Girifalco–Fowkes–Young equation	39.0	39.4	36.8	38.4
Wu's equation	44.3	46.3	38.3	40.8

3. Literature values

Melt data[a]	44.6
Zisman's work[b]	43.0

[a] Solid surface energy can be estimated by extrapolating the data of the variation of surface tension in the liquid state with temperature to the solid state (from Wu [43]).
[b] Zisman [39].

Table 3.
Surface energy of PE (in mJ/m^2) measured using the SFA and the critical surface tension of wetting determined from contact angle data

1. SFA measurements 33.3 ± 1.6

2. Critical surface tension

	Dispersive liquids		Polar (H-bonding) liquids	
	Advancing	Receding	Advancing	Receding
Zisman's	24.8	32.1	16.2	19.2
Good–Girifalco–Fowkes–Young equation	30.3	34.5	25.4	26.8
Wu's equation	33.4	34.7	27.2	30.4

3. Literature values

Melt data[a]	31
Zisman's work[b]	31

[a] Solid surface energy can be estimated by extrapolating the data of the variation of surface tension in the liquid state with temperature to the solid state (from Wu [43]).
[b] Zisman [39].

The values of the surface energy and critical surface tension of PET determined by different methods are listed in Table 2. The surface energy measured using the SFA (61.2 mJ/m^2) and the critical surface tension of wetting (38.8–46.3 mJ/m^2) are significantly different. Also, the value of $\gamma_{c,x}$ depends critically on the characteristics of the liquids used for contact angle measurements, as well as on the model used to interpret the data. The contact angle hysteresis is significant when polar liquids are used. This can be because of specific dipole–dipole and H-bonding interactions between polar liquids and the PET surface. Use of polar liquids like diols and triols usually leads to erroneous predictions, which are significantly lower. γ_c depends on the validity of the model used. As discussed in Section 3.2, Zisman's plot is purely empirical and is based on Zisman's observation that $\cos\theta$ varies linearly with γ_{lv}. A truly linear variation of $\cos\theta$ with γ_{lv} was observed only when contact angles on dispersive solids such as poly(tetrafluoroethylene) were measured using dispersive test liquids like *n*-alkanes [38]. Significant deviations from linearity were observed when either the solid surface or the test liquids or both were nondispersive [46, 47]. The value of $\gamma_{c,GGF}$ obtained from the plot of the Good–Girifalco–Fowkes–Young [equation (10)] is 39.4 mJ/m^2 when nonpolar liquids are used and 38.4 mJ/m^2 when H-bonding liquids are used. These values are in the same range as those obtained from Zisman's plot. The difference in γ_c of PET for nonpolar liquids may be because of specific interactions between the liquids and the PET surface. Such interactions have been shown to predict lower values of γ_c [47, 48]. The equation of state [equation (11)] developed by Wu [44] takes into account, at least qualitatively, the interactions between the solid and the test liquid. The value of $\gamma_{c,Wu}$ for PET determined using this equation (46.3 mJ/m^2 with dispersive liquids and 40.3 mJ/m^2 with H-bonding liquids) is significantly higher than the predictions of Zisman's plot and the Good–Girifalco–Fowkes–Young equation. These results clearly indicate that the predictions of $\gamma_{c,x}$ can vary with the characteristics of the test liquids, and the validity of the models used.

In the case of PE, as shown in Table 3, the values of the surface energy obtained from SFA experiments and $\gamma_{c,x}$ obtained from contact angles of dispersive liquids are in agreement. It may be noted that the values determined from the contact angles of polar liquids are significantly lower than those obtained using dispersive liquids. The agreement between SFA measurements and contact angle measurements in the case of PE is unlike what is seen in the case of PET. This agreement can be because of the absence of any specific interactions between PE and the liquids. The dispersive forces acting between PE and the test liquids are similar to PE–PE interactions, which are also dispersive. However, in the case of PET, there can be specific interactions between the surface and the test liquids. These interactions may result in a lowering of the critical surface tension of wetting. In addition, the forces acting between two PET surfaces can be very different from those that come into play between PET and nonpolar liquids. However, no such significant differences exist in the case of PE. This may explain the difference that exists in the values of direct measurements and $\gamma_{c,x}$ of PET and the agreement between the direct measurements and $\gamma_{c,x}$ of PE. This aspect needs to be investigated further using other model polymer films.

5. CONCLUSIONS

In this paper, we have reported the direct measurement of the surface energies of PET and PE, and the interfacial energy between them. To the best of our knowledge, this is the first reported direct measurement of the interfacial energy between two different polymers in the solid state.

The surface forces apparatus (SFA) was used to measure molecular level adhesion between the polymers. The small area of contact, the slow rate of separation, and the use of thin polymer films (thickness $\approx$ 3–5 μm) make it possible to carry out dissipation-free measurements. In the SFA, the contact area between the surfaces is as small as 10^{-9} m^2, about 10^{-6} times smaller than that in a practical test of adhesion. The rate of crack propagation, 0.01 μm/s, is about 2–3 orders of magnitude slower. The reduction in the rate of separation and contact area results in minimization of dissipative effects that are otherwise associated with the methods used for strength measurement of a bond. Thus, the SFA offers a means of measuring surface and interfacial energies.

The surface energies, γ_s, of PET and PE measured using the SFA were compared with the critical surface tension determined from contact angle measurements. In the case of PET, the critical surface tension is about 43 mJ/m^2 and γ_s is equal to 61.9 mJ/m^2. Thus, γ_s is significantly higher than $\gamma_{c,x}$. However, in the case of PE, $\gamma_{c,x}$ (33.7 mJ/m^2) and γ_s (33.3 mJ/m^2) are about the same.

Both for PET and for PE, the value of $\gamma_{c,x}$ depends on the characteristics of the test liquids used for contact angle measurements; the values are lower when polar liquids are used. The dependence is more significant in the case of PET than in the case of PE. This is due to the specific interactions between the solid and the liquids in the case of PET. Dispersive liquids are normally more suitable for contact angle measurements.

The value of the critical surface tension depends on the model used to interpret the contact angle data. Zisman's plot may not give accurate results in all cases, especially when the test liquid or solid is nondispersive. In such cases, more rigorous models like the Good–Girifalco–Fowkes–Young equation, or Wu's equation of state may give better estimates of the critical surface tension.

REFERENCES

1. K. W. Allen, *J. Adhesion* **21**, 262 (1987).
2. H. R. Brown, *Macromolecules* **26**, 1666 (1993).
3. A. J. Kinloch, *J. Mater. Sci.* **15**, 2141 (1980).
4. A. N. Gent and J. Schultz, *J. Adhesion* **3**, 281 (1972).
5. E. H. Andrews and A. J. Kinloch, *Proc. R. Soc. London, Ser. A* **332**, 385 (1973).
6. E. H. Andrews and A. J. Kinloch, *Proc. R. Soc. London, Ser. A* **332**, 401 (1973).
7. H. R. Brown, *Annu. Rev. Mater. Sci.* **21**, 463 (1991).
8. A. N. Gent and A. J. Kinloch, *J. Polym. Sci. A* **29**, 659 (1971).
9. D. H. Kaelble, *J. Colloid Sci.* **19**, 413 (1964).
10. D. H. Kaelble, *J. Adhesion* **37**, 205 (1992).
11. M. J. Loukis and N. Aravas, *J. Adhesion* **35**, 7 (1991).
12. D. H. Kaelble, *J. Adhesion* **1**, 102 (1969).
13. K. Kendall, *J. Phys. D* **6**, 1782 (1973).
14. K. Kendall, *J. Phys. D* **8**, 512 (1975).
15. K. Kendall, *J. Phys. D* **8**, 1722 (1975).
16. D. Maugis and M. Barquins, *J. Phys. D* **11**, 1989 (1978).
17. J. N. Israelachvili and G. E. Adams, *J. Chem. Soc. Faraday Trans. 1* **74**, 975 (1978).
18. J. N. Israelachvili and D. Tabor, *Proc. R. Soc. London, Ser. A* **331**, 19 (1972).

19. G. Hadziioannou, S. Patel, S. Granick and M. Tirrell, *J. Am. Chem. Soc.* **108**, 2869 (1986).
20. J. N. Israelachvili, M. Tirrell, J. Klein and Y. Almog, *Macromolecules* **17**, 204 (1984).
21. P. F. Luckham and J. Klein, *J. Chem. Soc. Faraday Trans. 1* **80**, 865 (1984).
22. H. Watanabe and M. Tirrell, *Macromolecules* **26**, 6455 (1993).
23. R. G. Horn and D. T. Smith, *J. Non-Cryst. Solids* **120**, 72 (1991).
24. R. G. Horn, D. T. Smith and W. Haller, *Chem. Phys. Lett.* **162**, 404 (1989).
25. C. P. Smith, M. Maeda, L. Atanasoska, H. S. White and D. J. McClure, *J. Phys. Chem.* **92**, 199 (1988).
26. B. V. Thakkar, PhD Thesis. University of Minnesota (1991).
27. W. W. Merrill, A. V. Pocius, B. V. Thakkar and M. Tirrell, *Langmuir* **7**, 1975 (1991).
28. K. L. Johnson, K. Kendall and A. D. Roberts, *Proc. R. Soc. London, Ser. A* **324**, 301 (1971).
29. B. V. Derjaguin, V. M. Muller and Y. P. Toporov, *J. Colloid Interface Sci.* **53**, 314 (1975).
30. M. K. Chaudhury and G. M. Whitesides, *Langmuir* **7**, 1013 (1991).
31. Y. L. Chen, C. A. Helm and J. N. Israelachvili, *J. Phys. Chem.* **95**, 1036 (1991).
32. R. G. Horn, J. N. Israelachvili and F. Pribac, *J. Colloid Interface Sci.* **115**, 480 (1987).
33. D. Maugis and M. Barquins, *J. Phys. D* **16**, 1843 (1983).
34. V. M. Muller, V. S. Yushchenko and B. V. Derjaguin, *J. Colloid Interface Sci.* **77**, 91 (1980).
35. V. M. Muller, V. S. Yushchenko and B. V. Derjaguin, *J. Colloid Interface Sci.* **92**, 92 (1983).
36. D. Tabor, *J. Colloid Interface Sci.* **58**, 2 (1977).
37. D. Maugis, *J. Colloid Interface Sci.* **150**, 243 (1992).
38. H. W. Fox and W. A. Zisman, *J. Colloid Sci.* **5**, 514 (1950).
39. W. A. Zisman, in: *Contact Angle, Wettability, and Adhesion*, ACS Symp. Ser. No. 43, pp. 1–17. American Chemical Society, Washington, DC (1964).
40. L. A. Girifalco and R. J. Good, *J. Phys. Chem.* **61**, 904 (1957).
41. R. J. Good and L. A. Girifalco, *J. Phys. Chem.* **64**, 561 (1960).
42. F. M. Fowkes, *Ind. Eng. Chem.* **56**, 40 (1964).
43. S. Wu, *Polymer Interface and Adhesion*. Marcel Dekker, New York (1982).
44. S. Wu, *J. Colloid Interface Sci.* **71**, 605 (1979).
45. K. Kendall, *J. Phys. D* **8**, 115 (1975).
46. A. H. Ellison and W. A. Zisman, *J. Phys. Chem.* **58**, 260 (1954).
47. J. R. Dann, *J. Colloid Interface Sci.* **32**, 302 (1970).
48. J. R. Dann, *J. Colloid Interface Sci.* **32**, 321 (1970).

Fundamentals of Adhesion and Interfaces, pp. 225–234
D. S. Rimai, L. P. DeMejo and K. L. Mittal (Eds)

Wetting hysteresis: effects due to shadowing

M. E. R. SHANAHAN[1,*] and J. M. DI MEGLIO[2]

[1]*Centre National de la Recherche Scientifique, Ecole Nationale Supérieure des Mines de Paris, Centre des Matériaux P.M. Fourt, B.P. 87, 91003 Evry Cédex, France*

[2]*Laboratoire de Physique de la Matière Condensée (C.N.R.S. UA 792), Collège de France, 11 Place Marcelin Berthelot, 75231 Paris Cédex 05, France*

Revised version received 10 June 1994

Abstract—Wetting hysteresis due to isolated surface heterogeneities is now fairly well understood but when the solid presents a population of defects, complex cooperative effects between neighbours may exist. One such effect is that of 'shadowing', in which a proportion of the flaws near the triple line, and which would otherwise contribute to hysteresis, are masked by already existing deformations to the wetting front caused by neighbouring heterogeneities. This renders them inactive and, as a result, the hysteretic wetting force is only expected to be a linear function of density for sparse populations. Theoretical predictions are compared with experimental results obtained with model heterogeneous surfaces consisting of overhead projector transparencies bestrewn with circular ink spots — the defects. Agreement is found to be satisfactory when intrinsic angles on both the homogeneous solid and the flaws are finite, whereas the concordance is less satisfactory when the contact angle of the liquid on the homogeneous solid is zero.

Keywords: Defect (population); heterogeneity; hysteresis; model surfaces; shadowing; wettability; wetting.

1. INTRODUCTION

Although Young's equation predicts a unique value for the equilibrium contact angle at the triple line of a liquid in contact with an ideally homogeneous and rigid solid in the presence of a fluid (taken to be the vapour of the liquid or a second liquid immiscible with the first), it is well known that in virtually all practical conditions of wetting, the contact angle presents hysteretic properties. In particular, the advancing contact angle exceeds the receding contact angle. Several reasons for this hysteresis have been evoked, such as diffusion of the liquid into the solid leading to swelling, local adsorption, and the orientation of macromolecular chains in the case of polymeric substrates, but, for the most part, changes in the contact angle are attributed to the presence of heterogeneities on the solid surface, either chemical or physical, leading to 'anchoring' effects on the triple line [1–7]. Treatment of these anchoring effects due

*To whom correspondence should be addressed.

to isolated defects is now fairly well established [8–11] but when the population of heterogeneities is sufficiently dense, interaction between neighbouring defects renders the problem more complex. The modelling of such phenomena is delicate but anchoring occurs in analogous problems in other branches of condensed matter physics, such as the motion of vortex lines in supraconductors. Up to now, no general theory has been presented which is capable of accounting for the phenomenon of wetting hysteresis when defects interact. However, in a recent study, it has been shown that if the population of defects is sufficiently dense, neighbouring heterogeneities may be found within their correlation length and interaction may occur due to a phenomenon of 'shadowing'. Due to the proximity of certain 'active' defects giving rise to triple line deformation, some heterogeneities become 'inactive', lying within the 'shadow' or 'wake' of others [12]. In the present study, we show that the effects of shadowing can account for certain results obtained with experiments performed using model heterogeneous surfaces consisting of essentially homogeneous material bestrewn with less wettable defects of macroscopic dimensions [13].

2. THE PHENOMENON OF 'SHADOWING'

Consider an isolated, small, circular patch of surface free energy $\gamma_s - \varepsilon$ and radius r_d on an otherwise homogeneous flat, smooth solid substrate of surface free energy γ_s. The term ε is small compared with the liquid surface free energy, or tension, γ, but may be either positive or negative. However, for definiteness and consistency with the experimental details presented below, we will take ε as positive. This means that deformation to the wetting triple line will occur when the liquid front advances to contact the defect but that no significant disturbance will arise for the receding mode. It has been established elsewhere that the aforementioned deformation, $\delta(x)$, to the advancing triple line from its undisturbed, straight form is essentially logarithmic and given by [8, 14]:

$$\delta(x) \sim \frac{f}{\pi \gamma \theta^2}\left[\ln\left(\frac{r_0}{|x|}\right) - \frac{1}{2}\right]; \qquad |x| \gtrsim r_d, \tag{1}$$

where x is the distance from the defect centre measured parallel to the unperturbed wetting front; f corresponds to the local force 'pushing' the triple line (equal to $2r_d\varepsilon$ when the wetting line is extended to its maximum extent, just before it 'unhooks' and the liquid subsequently engulfs the defect); θ is the (small) intrinsic contact angle on the homogeneous solid; and r_0 is a cut-off distance, for present purposes taken as being of the same order as the capillary length, $\kappa^{-1} = [\gamma/(\rho g)]^{1/2}$ (ρ and g are, respectively, the liquid density and gravitational acceleration).

When a population of defects is present, the overall shape of the advancing triple line (which would be straight in the absence of defects) will resemble that shown in Fig. 1. Assuming all heterogeneities to be identical and their average density to be d (number per unit area of surface), there exists a band of width h near the triple line containing potentially active flaws: those further removed will either be entirely

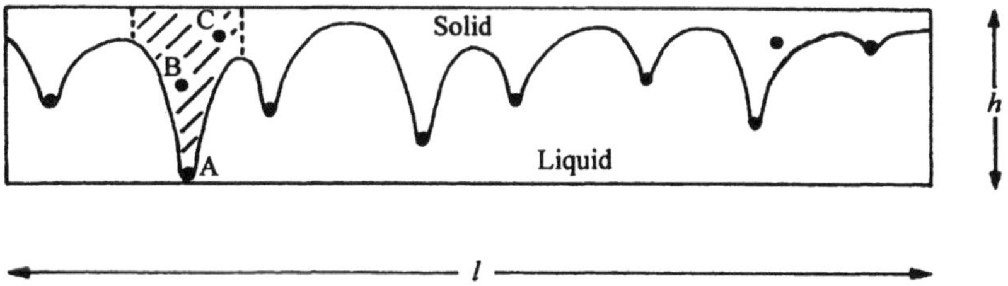

Figure 1. Sketch of a wetting front in advancing mode when the surface defects are less wettable than the surrounding solid. Heterogeneities B and C are inactive due to the shadowing phenomenon produced by A.

within the second fluid phase not yet having encountered the triple line (above the area shown schematically in Fig. 1) or entirely within the liquid after having been engulfed (below the area depicted in Fig. 1). If a single flaw were being considered, the value of h would be given by

$$h \sim fL/(\pi\gamma\theta^2), \tag{2}$$

where L is essentially the logarithm of the ratio of cut-off lengths; $L \sim \ln(r_0/r_d)+1/2$ [14]. In fact, there will be cooperative effects between heterogeneities increasing the value of h somewhat, but, provided that d is not too high, this effect may be ignored to a first approximation.

Considering Fig. 1, not all defects within the band of width, h, and length of triple line, l, will be active since some are in the 'shadow', or 'wake', of others. For example, in Fig. 1, the triple line having hooked onto flaw A, flaws B and C remain in the second fluid and do not therefore actively contribute to the deformation of the wetting front. This phenomenon effectively reduces the density of active defects from d to $\bar{d} = d - \Delta d$. A full appraisal of the situation will be complex and further work is currently in progress, but the following simplified argument may be used to assess the reduction in effective density, Δd, to a first approximation.

Average defect separation is of the order of $d^{-1/2}$ and a typical local deformation to the wetting front caused by a single defect will be given by equation (1) but with f given by $r_d\varepsilon$. The average fraction, ϕ, of the band of area hl *not* involved actively in triple line deformation will then be given simply by

$$\phi \sim \frac{2r_d d^{1/2}\varepsilon}{\pi\gamma\theta^2 h}\int_{r_d}^{(d^{-1/2}/2)}\left[\ln\left(\frac{r_0}{x}\right)-\frac{1}{2}\right]\mathrm{d}x. \tag{3}$$

Evaluation of the above integral and use of equation (2) leads to the following expression for the corrected value of the average density of active defects:

$$\bar{d} = d - \Delta d \sim d\left\{1 - \frac{[1+2\ln(2r_0 d^{1/2})]}{4L}\right\}. \tag{4}$$

3. HYSTERETIC WETTING FORCE

We are now in a position to be able to estimate the hysteretic wetting force, namely the difference in wetting force acting parallel to the solid surface and perpendicularly to the undisturbed triple line. This has been discussed previously and therefore will only be covered briefly here, but taking into account to above development [15].

When the triple line recedes, it experiences no retention from the population of defects since these, by hypothesis, are of slightly lower surface free energy than the surrounding solid. Dewetting will occur spontaneously upon arrival and passage of the triple line. However, an advancing triple line will meet resistance, as shown schematically in Fig. 1. On average, a defect will be at a distance $h/2$ from the undisturbed triple line and will therefore furnish a force of repulsion, $\bar{f}$, of $r_d\varepsilon$. For length, l, of triple line, a band of area hl contains, on average, $\overline{N}$ active, repulsive flaws with an average population density of $\bar{d}$. We may thus estimate the total average force of repulsion, F, from

$$F = \overline{N}\bar{f} \sim lh\bar{d}\bar{f} \sim \frac{l\bar{d}L(2r_d\varepsilon)^2}{2\pi\gamma\theta^2}. \tag{5}$$

The force F may be considered as the hysteretic wetting force for length l of (undisturbed) triple line and corresponds to the force difference between advancing and receding quasi-static motion of the wetting front. Clearly, in order to exploit equation (5), $\bar{d}$, as defined by equation (4), must be substituted. This demonstrates that the hysteretic wetting force would be expected to be a linearly increasing function of the actual density of surface heterogeneities for relatively sparse populations. Indeed, as d gets very large, F will pass through a maximum and then decrease. This is to be expected, since the solid surface then tends towards a homogeneously covered surface of 'defects'. The prior 'homogeneous solid' and the 'flaws' change roles.

4. EXPERIMENTAL

Experimental details are given in full in [13]. Model heterogeneous surfaces were made consisting of overhead projector transparencies onto which circular spots (corresponding to the flaws) were drawn using a plotter. Various densities, d, of flaws were produced. Two types of surfaces were used; those referred to as type A were bestrewn with randomly distributed spots of radius, r_d, equal to 0.75 mm, and type B, with a biased random distribution [13] of spots of radius 0.25 mm. For type A, the liquid used was n-hexadecane, for which γ is equal to 27.3 mN m^{-1} at ambient temperature and the capillary length, κ^{-1}, in air as the surrounding fluid, is 1.9 mm. The intrinsic contact angle (advancing) of n-hexadecane on the transparency surface is 13°, whilst on the ink constituting the flaws the equivalent value is 70°. In the case of the type B surface, the liquid employed was n-heptane with a surface tension of 20.2 mN m^{-1} and a capillary length of 1.7 mm. The intrinsic contact angle on the transparency surface is zero and that on the defect material is 13°. It should perhaps be noted that the flaw diameter for the type A surface is of the order of the capillary length for the system used. In addition, for both type A and type B surfaces, contact angle hysteresis is quite significant on the constitutive materials themselves, i.e. both on the transparency itself and the flaw material (except for *n*-heptane on the transparency, in which case both advancing and receding contact angles are zero) [13]. This is unfortunate and may perturb the results to some extent, but real systems without any hysteresis are extremely rare.

The transparencies were suspended above their respective liquids from a force sensor and dipped in and pulled out in order to measure wetting forces in the advancing and in the receding modes, respectively. Under quasi-static conditions, the difference between these forces (the receding value being greater than the advancing value) corresponds to the force F of equation (5).

5. RESULTS AND DISCUSSION

5.1. System using type A surface

We recall that for experiments with the type A surface, *n*-hexadecane wets but does not spread on either the homogeneous solid or the flaw material. Figure 2 represents the hysteretic wetting force difference, F, as a function of the the actual defect density, d, both being linear scales. The same results, but using logarithmic scales, are shown in Fig. 3. It may be seen that for low values of d, the hysteretic force difference is found to increase linearly with the defect density, as may be expected from a simple superposition of hysteretic effects from individual heterogeneities. However, for higher values of d, the gradient of F vs. d decreases and this has been attributed

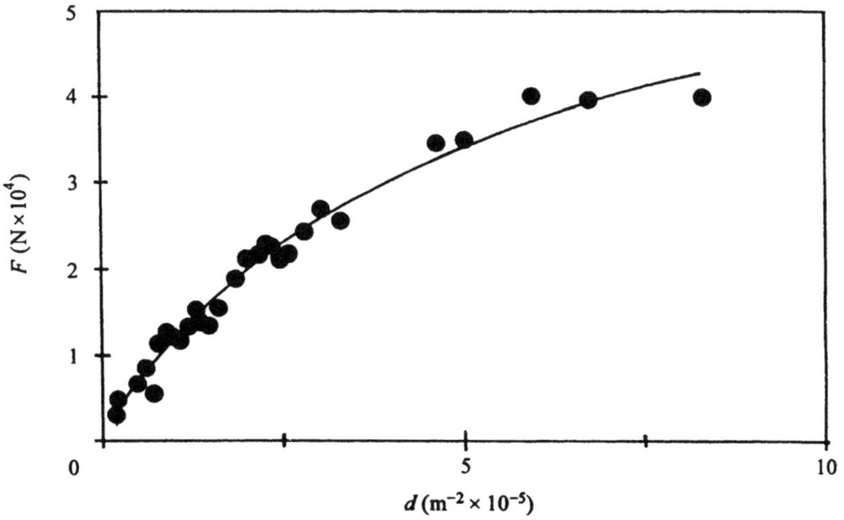

Figure 2. Hysteretic wetting force, F, vs. defect population density, d, for *n*-hexadecane on an overhead projector transparency bestrewn with circular ink spots (defects) of 0.75 mm radius.

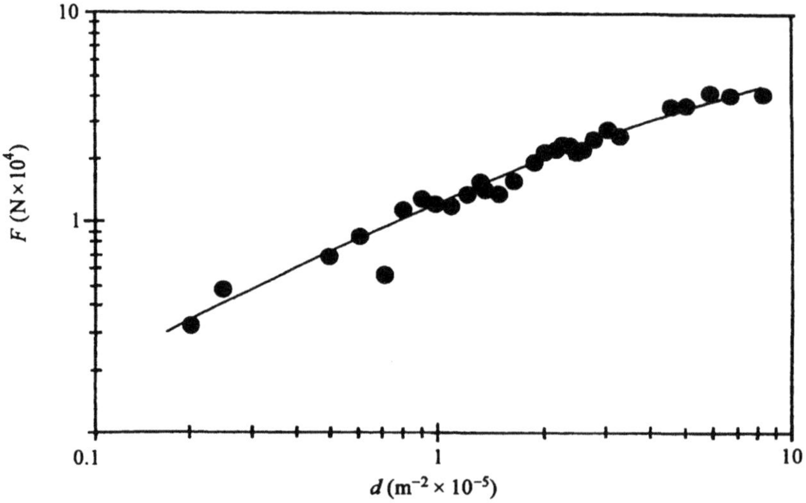

Figure 3. As for Fig. 2 but with logarithmic scales.

to the phenomenon of shadowing described above in Sections 2 and 3. The curves in Figs 2 and 3 correspond to a statistical adjustment to the experimental data in the form

$$F = d\,(A - B \ln d), \tag{6}$$

which corresponds to the mathematical form of equation (5) when equation (4) has been used to substitute for $\bar{d}$ (A and B are constants). It can be seen that the fit is most satisfactory, perhaps surprisingly so considering the simplifications used in the theoretical development and the already mentioned fact that the flaw diameter is comparable to the capillary length of the system, κ^{-1}. From the statistical fit, it was found that

$$B = \frac{l(2r_{\rm d}\varepsilon)^2}{8\pi\gamma\theta^2} = 3.2 \times 10^{-10}\ \mathrm{N\,m^2}. \tag{7}$$

Now the value of ε may be estimated from

$$\varepsilon = \gamma\left(\cos\theta - \cos\theta_{\rm d}\right), \tag{8}$$

where $\theta_{\rm d}$ is the intrinsic contact angle of the liquid on the flaw material. Using the above values of γ, θ, and $\theta_{\rm d}$, it is found that $\varepsilon \sim 17\ \mathrm{mN\,m^{-1}}$, corresponding to a maximum individual flaw force, $(2r_{\rm d}\varepsilon)$, of $\sim 2.6 \times 10^{-5}$ N. Taking the experimental value of l as 30 mm and the other values as given above, it is found that the calculated value of B as given in equation (7) is *ca.* $5.7 \times 10^{-10}\ \mathrm{N\,m^2}$, which is in acceptable agreement with that predicted from statistical treatment of the results.

It is deduced that the logarithmic coefficient, L,

$$L = \frac{A}{2B} + \ln\left(2r_{\rm d}\right) \qquad \text{with} \quad A = 1.93 \times 10^{-9}\ \mathrm{N\,m^2},$$

is of the order of 1.1, leading to an estimate of the cut-off distance, r_0, introduced in Section 2, of *ca.* 1.4 mm. Given that the capillary length, κ^{-1}, of this system is 1.9 mm, we obtain good agreement with theoretical predictions. (The cut-off distance will not necessarily be *exactly* equal to the capillary length.) Overall we can see that the shadowing phenomenon is capable of explaining experimental results obtained from the type A system quite adequately, although it would perhaps be more convincing to use defects smaller than the present radius, r_d, of 0.75 mm.

5.2. System using type B surface

In the case of the type B surface in which the liquid used was *n*-heptane, we note that the liquid presents a zero contact angle on the homogeneous solid surface and a small, but finite, value on the flaw material. We wished to see if the proposed shadowing model would apply when $\theta = 0$, as this case was also studied in [13].

Figures 4 and 5 present the hysteretic wetting force difference, F, as a function of the defect density, d, on linear and logarithmic scales, respectively. Again it appears that for low-density defect populations, F increases approximately linearly with d but that the gradient decreases for higher values of d. Qualitative agreement is present but a statistical fit of the experimental data using equation (6) leads to less satisfactory results than in the case of the type A surface. The expression analogous to equation (7) is

$$B = \frac{l(2r_d\varepsilon)^2}{8\pi\gamma\theta^2} = 1.3 \times 10^{-11}\ \mathrm{N\,m^2}. \tag{9}$$

Using the experimental value of l (20 mm) and ε from equation (8) (ε is equal to *ca.* 0.5 $\mathrm{mN\,m^{-1}}$), together with the values of r_d and γ given above, it is found that

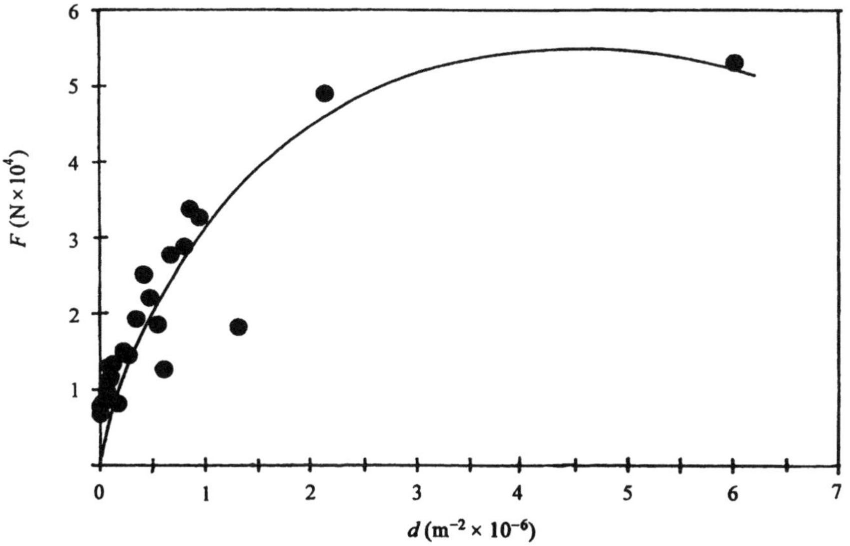

Figure 4. As for Fig. 2 but for *n*-heptane and with defects of 0.25 mm radius.

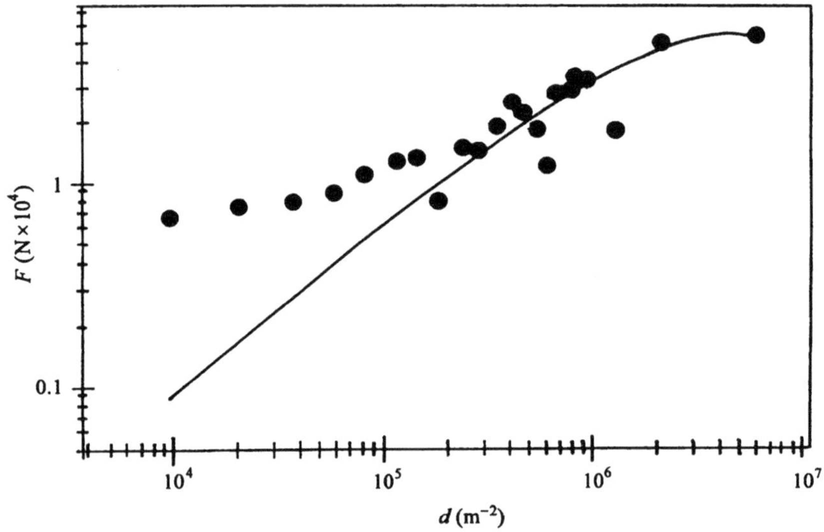

Figure 5. As for Fig. 4 but with logarithmic scales.

the intrinsic value of the contact angle on the homogeneous solid should be 0.8°, a value compatible with the directly observed zero angle. However, the value of $L(0.8)$ obtained (with $A = 9.6 \times 10^{-11}$ N m^2) leads to a cut-off, r_0, of 0.34 mm, which is a factor of 5 smaller than the capillary length, κ^{-1}, of the system (1.7 mm). The model would therefore seem to be inadequate to treat this case where the intrinsic contact angle of the liquid on the unperturbed solid surface is zero. Indeed, it can be seen in Fig. 5 that the best fit is relatively poor on a logarithmic scale, especially for low values of d. This could be at least partly due to some imprecision in the lower values of F corresponding to sparse heterogeneity populations. Using the value of ε of *ca.* 0.5 mN m^{-1} and the spot radius of 0.25 mm, we calculate the maximum force of repulsion of a single flaw to be *ca.* 2.6×10^{-7} N. Compared with the value of the equivalent parameter for the type A system, this is 100 times smaller. Nevertheless, we should not lose sight of the fact that the original work leading to equation (1) assumed a small, but non-zero intrinsic contact angle on the homogeneous solid, whereas we realize that the present system is unlikely to fit these requirements exactly.

Recently, a model has been proposed to describe the elasticity of a wetting triple line interacting with a defect when the intrinsic contact angle on the homogeneous solid is zero [16]. Instead of equation (1), the following expression is obtained:

$$\delta(x) \sim \left(\frac{f}{\pi\gamma\kappa^2}\right)^{1/3} \frac{\ln^{1/2}|\coth(\pi x/4H)|}{\ln^{1/6}|\coth(\pi r_d/2H)|}, \tag{10}$$

where $H = 2^{1/2}\kappa^{-1}$ and other symbols have been changed from those of the original article [16] in order to conform to the present nomenclature. Note the dependence of triple line deformation on $f^{1/3}$ instead of the previous linear behaviour. It has been shown that this equation satisfactorily describes deformation to the triple line in the case of an isolated defect [17]. We have therefore used a development along lines similar to those given in Section 2 to obtain the reduction in effective, or active, flaw density, Δd, due to the shadowing phenomenon. The result of the calculation only will be given for brevity:

$$\Delta d \sim \frac{8d^{3/2}}{2^{1/3}\pi\kappa\, \ln^{1/2}\left[\coth\left(\frac{\pi r_{\mathrm{d}}}{4H}\right)\right]}\left[\exp\left(\frac{-\pi r_{\mathrm{d}}}{4H}\right) - \exp\left(\frac{-\pi}{8Hd^{1/2}}\right)\right]. \tag{11}$$

Following Section 3, we may estimate the hysteretic force difference, F, by averaging force contributions from individual defects over the range of activity in the band of width h; we thus obtain

$$F \sim \frac{l\bar{d}\, \ln^{1/2}\left[\coth\left(\frac{\pi r_{\mathrm{d}}}{4H}\right)\right](2r_{\mathrm{d}}\varepsilon)^{4/3}}{4(\pi\gamma\kappa^2)^{1/3}\ln^{1/6}\left[\coth\left(\frac{\pi r_{\mathrm{d}}}{2H}\right)\right]}. \tag{12}$$

In this case, the overall hysteretic force difference depends on the force of an individual flaw to the power 4/3, compared with the earlier equation (5) for a small, finite contact angle on the homogeneous solid where the power is 2.

Equation (12) has been applied to the results obtained with the type B system but to little avail. In effect, the shadowing phenomenon is less significant using equation (12) [with $\bar{d} = d - \Delta d$ obtained from equation (11)] than with equation (5).

A virtually linear relationship between F and d is to be expected. We are thus led to conclude that our interpretation of the type B system experiments is inadequate and further work is required to understand the behaviour when the intrinsic contact angle of the liquid on the homogeneous solid surface is zero. A possibility presently being studied experimentally is that of coupling effects between flaws which could be a consequence of the propagation of 'unhooking waves' along the triple line.

6. CONCLUSIONS

The effects of populations of wetting flaws of different densities on wetting hysteresis have been studied both experimentally and theoretically. For sparse populations, a linear relationship between the hysteretic wetting force, i.e. the difference in force occurring as a result of advancing and of receding motion of the wetting front, and the flaw density should be obtained. However, for denser populations, a shadowing effect is expected in which a fraction of the defects near the triple line become inactive, being within the shadow, or wake, of other active heterogeneities. Using model surfaces consisting of overhead projector transparencies bestrewn with circular ink spots, it has been shown that the theory of shadowing satisfactorily explains changes in wetting hysteresis with population density, when intrinsic contact angles on both the homogeneous solid and the flaws are finite, even if the spot dimensions are close to the expected limits of applicability of the calculations. However, in the case when the intrinsic contact angle of the liquid on the homogeneous solid is zero, agreement between theory and experiment is less satisfactory. Even allowing for non-linear contact line elasticity, there is discrepancy. Further work is required in order to understand this case, and present experimental work aims at elucidating any possible coupling effects between defects due to the propagation of unhooking waves along the triple line.

Acknowledgement

We thank E. Raphaël for useful discussions.

REFERENCES

1. R. S. Hansen and M. Miotto, *J. Am. Chem. Soc.* **79**, 1765 (1957).
2. R. E. Johnson and R. H. Dettre, *J. Phys. Chem.* **68**, 1744 (1964).
3. J. D. Eick, R. J. Good and A. W. Neumann, *J. Colloid Interface Sci.* **53**, 235 (1975).
4. R. J. Good, *J. Colloid Interface Sci.* **59**, 398 (1977).
5. H. Yasuda, A. K. Sharma and T. Yasuda, *J. Polym. Sci., Polym. Phys.* **19**, 1285 (1981).
6. R. G. Cox, *J. Fluid Mech.* **131**, 1 (1983).
7. A. Carré, S. Moll, J. Schultz and M. E. R. Shanahan, in: *Adhesion 11*, K. W. Allen (Ed.), p. 82. Elsevier, London (1987).
8. J. F. Joanny and P. G. de Gennes, *J. Phys. Chem.* **81**, 522 (1984).
9. Y. Pomeau and J. Vannimenus, *J. Colloid Interface Sci.* **104**, 477 (1985).
10. L. W. Schwartz and S. Garoff, *J. Colloid Interface Sci.* **106**, 422 (1985).
11. M. E. R. Shanahan, *J. Phys. D, Appl. Phys.* **22**, 1128 (1989).
12. M. E. R. Shanahan, *C.R. Acad. Sci.* **313(II)**, 613 (1991).
13. J. M. di Meglio, *Europhys. Lett.* **17**, 607 (1992).
14. M. E. R. Shanahan, *J. Phys. D, Appl. Phys.* **23**, 703 (1990).
15. M. E. R. Shanahan, *J. Adhesion Sci. Technol.* **6**, 489 (1992).
16. E. Raphaël and J. F. Joanny, *Europhys. Lett.* **21**, 483 (1993).
17. J. Marsh and A. M. Cazabat, *Europhys. Lett.* **23**, 45 (1993).

Fundamentals of Adhesion and Interfaces, pp. 235–245
D. S. Rimai, L. P. DeMejo and K. L. Mittal (Eds)

Interaction forces between hydrophobic mica surfaces

JONATHAN WOOD* and RAVI SHARMA†

Materials Science and Engineering Division, Eastman Kodak Company, Rochester, NY 14650-2158, USA

Revised version received 5 December 1994

Abstract—The interactions between pairs of hydrophobic monolayers of dimethyldioctadecylammonium bromide (DMDOA) surfaces and pairs of octadecyltriethoxysilane (OTE) surfaces were measured directly using the surface force apparatus. In the case of DMDOA surfaces, a long-range attraction detected at about 100 nm was observed, in good agreement with published data for DMDOA. In contrast, for the OTE surfaces, no interaction was discernible until there was a jump-in at an estimated separation of 13 nm. Furthermore, for the DMDOA surfaces, cavitation was observed on separation, as seen previously, while for the OTE surfaces, cavitation occurred on contact (spontaneous cavitation). The nature or extent of interaction between the OTE surfaces did not change in the presence of electrolyte (up to 0.1 M KNO_3). Adhesion force measurements for these surfaces in air and when immersed in water are presented.

Keywords: Hydrophobic force; hydrophobic mica surface; adhesion; cavitation; surface energy; Johnson, Kendall, Roberts theory; surface reconstruction; dimethyldioctadecylammonium bromide; octadecyltriethoxysilane; prepolymerized Langmuir–Blodgett monolayer.

1. INTRODUCTION

Direct force measurements have revealed a long-range attraction (hydrophobic attraction) exists between two hydrophobic surfaces in water and in aqueous solutions. For example, attraction beginning from greater than 50 nm is commonly measured [1]. This attraction cannot be accounted for by continuum theories, such as the Lifshitz theory or the classical Derjaguin–Landau–Verwey–Overbeek (DLVO) theory [2, 3]. The experiments have utilized hydrophobic surfaces prepared by a variety of methods: (1) Langmuir–Blodgett (L–B) deposition of cationic surfactants [2, 4]; (2) chemical attachment of alkyl silanes on silica [5] and on mica which has been activated using a water plasma [6]; (3) adsorption of surfactants from solution [4, 7–12]; and (4) polymerized ammonium amphiphiles transferred by L–B

*Present address: Max-Planck-Institut für Polymerforschung, Ackermannweg 10, 55128 Mainz, Germany.

†To whom correspondence should be addressed.

deposition [13]. These surfaces have provided a puzzling and sometimes inconsistent array of data [1, 14, 15].

Previous work has been hampered by electrical neutrality and stability-related issues, particularly as surfaces which were stable were not necessarily electrically neutral [6, 7]. Recently, a method of depositing a robust hydrophobic monolayer on mica by L–B deposition of a prepolymerized monolayer of octadecyltriethoxysilane (OTE) has been described [16]. We present a comparison of direct force measurements, contact angles, and adhesion measurements obtained for OTE surfaces and dimethyldioctadecylammonium bromide (DMDOA) surfaces. In addition, we report on what we believe is the first *direct* evidence of spontaneous cavitation when two hydrocarbon surfaces are brought together in pure water.*

2. EXPERIMENTAL

2.1. Materials

Dimethyldioctadecylammonium bromide (DMDOA) was obtained from Kodak and recrystallized from ethanol before use. Octadecyltriethoxysilane (OTE) was purchased from Petrach and used without further purification. Hexadecane (Aldrich), chloroform (Baker), potassium nitrate (Baker), 0.1 N nitric acid (Baker), methylene iodide (Kodak), methanol (Baker), and ethanol (Kodak) were all used as received. Water was doubly distilled before being further purified by a Milli-Q water polishing unit to give a final resistivity of 18 MΩ cm. In addition, the water used in the surface force experiments was deaerated by heating under vacuum for about 30 min.

2.2. Surface pressure measurements

Surface pressure ($\pi - A$) isotherms were obtained for DMDOA and OTE monolayers, over pure water and pH 2 subphases, respectively, using a KSV 5000 Langmuir trough (maximum area 675 cm^2). Solutions of DMDOA (1 mg/ml) in chloroform–ethanol (95 : 5 by vol.) and 2 mg/ml solutions of OTE in chloroform–methanol (95 : 5 by vol.) were spread on the freshly aspirated subphase surface and allowed to stand for either 5 (DMDOA) or 30 (OTE) min before compressing the film at a rate of 15 mm/s. All isotherms were run at least twice in the direction of increasing pressure using a freshly spread film. Under optimum conditions, two isotherms run successively were completely superimposable; however, more typical error values were ± 0.2 mN/m in surface pressure and ± 0.005 nm^2/molecule in area. All isotherms were recorded at 24 ± 0.5°C.

*Spontaneous cavitation or cavitation on contact has been directly observed with fluorinated surfaces [2]. For hydrocarbon surfaces, however, cavitation has been directly observed only upon separation [2, 11], although cavitation on contact has been inferred in an experiment using silanated quartz filaments [5]. There has also been one unpublished report [1] of cavitation on contact between hydrocarbon surfaces.

2.3. Mica surface modification

Freshly cleaved mica surfaces — either large sheets for the evaluation of Langmuir–Blodgett (L–B) deposition transfer or thin (1–3 μm), partially silvered sheets glued with silver side down on quartz discs for use in the surface force apparatus (SFA) — were employed as substrates for modification.

Monolayers of DMDOA were transferred to these substrates by a standard L–B deposition technique (at 25 mN/m) and were allowed to dry in the airstream of a laminar flow clean bench for 15 min prior to use. Prepolymerized monolayers [16–18] of OTE prepared on a nitric acid subphase [17] (pH 2) were transferred to mica substrates that had been treated with radio-frequency-generated Ar/H_2O plasma [2 min at ≈30 W (low-power setting on a Harrick Plasma Cleaner, PDC-236), 500 mTorr] before L–B deposition was performed at 20 mN/m. After drying in a clean airstream for 15 min, these samples were baked in a vacuum oven (100°C, 100 mTorr) for 2 h before use.

2.4. Contact angle measurements

Advancing and receding contact angles were measured with a Ramé-Hart goniometer using pure water, aqueous potassium nitrate solutions (10^{-1}, 10^{-2}, and 10^{-3} M), hexadecane, and methylene iodide as test liquids. Advancing contact angles were obtained by increasing the volume of a drop — the maximum angle attained before advancement of the contact line was recorded as the advancing angle. Receding contact angle measurements were made by decreasing the volume of a drop and recording the minimum angle obtained before the contact line receded. All contact angle measurements were accurate to ±1°.

2.5. Direct force measurements

Force measurements were performed using the surface force apparatus designed by Israelachvili and Adams [19]. This apparatus employs fringes of equal chromatic order (FECO) to measure the separation of mica surfaces which are arranged in a crossed cylindrical configuration, with a resolution of 0.2 nm, whilst the force is measured by the deflection of a spring on which one of the surfaces is mounted, to an accuracy of about 10^{-7} N.

Force results, F, obtained at various distances of separation (D) were normalized by dividing them by the surface radius of curvature, R, and can be easily converted to the corresponding interaction energies (E) between flat plates by application of the Derjaguin approximation [20] for crossed cylinders:

$$E(D) = \frac{F(D)}{2\pi R}. \tag{1}$$

Adhesion forces, F_a, were measured by the pull-off force method [21, 22], in which the distance to which the two surfaces jump out on separation is recorded and multiplied by the spring constant. Surface energies can then be calculated by application of either the Johnson, Kendall and Roberts (JKR) [23] or Derjaguin, Muller and Toporov (DMT) [24] theory of adhesion mechanics. These limiting theories have recently been discussed in the context of a JKR–DMT transition by Maugis [25].

3. DISCUSSION

3.1. Characterization of contact angles on DMDOA and OTE monolayers on mica

Advancing and receding water contact angles on DMDOA and OTE surfaces are presented in Table 1. The contact angles obtained for these surfaces also agree closely with those from previous studies [2, 26, 27]. The water contact angle values of DMDOA surfaces are lower than those for OTE surfaces, indicating that the OTE surfaces are more hydrophobic.

Furthermore, the large contact angle difference between the advancing and receding values obtained for DMDOA surfaces is typical for physisorbed monolayer films and has been ascribed to a certain percentage of the molecules in the film 'flipping over' so that they present a hydrophilic head-group to the surface [22].

Table 1.
Comparison of the advancing (θ_a) and receding (θ_r) water contact angles on DMDOA and OTE monolayers deposited on mica

Surface	Water contact angles
DMDOA	$\theta_a = 95°$, $\theta_r = 56°$
OTE	$\theta_a = 112°$, $\theta_r = 93°$

Table 2.
Summary of advancing (θ_a) and receding (θ_r) contact angle data for OTE-coated mica surfaces

Test liquid	Contact angles
Water	$\theta_a = 112°$ $\theta_r = 93°$
10^{-1} M KNO_3	$\theta_a = 110°$ $\theta_r = 94°$
10^{-2} M KNO_3	$\theta_a = 110°$ $\theta_r = 94°$
10^{-3} M KNO_3	$\theta_a = 110°$ $\theta_r = 94°$
Hexadecane	$\theta_a = 45°$ $\theta_r = 39°$
Methylene iodide	$\theta_a = 72°$ $\theta_r = 68°$

Advancing and receding contact angles for various test liquids on an OTE monolayer deposited on flat mica sheets are presented in Table 2. The advancing contact angle values of 112° for water, 45° for hexadecane, and 72° for methylene iodide are in good agreement with literature data for a closed-packed layer of hydrocarbon chains [26, 27]. Furthermore, the monolayers were robust, as the advancing and receding contact angles did not change over several hours when tested with KNO_3 solutions. Potassium ions apparently compete for sites on the mica surface by an ion exchange mechanism [7]. The fact that the contact angles were stable when tested with KNO_3 solutions implies that the OTE monolayer underneath the drop was not displaced by K^+ ions. The contact angle hysteresis of 19° is larger than the 10° observed for self-assembled monolayers on silicon wafers [27c] and also larger than the 3° for OTE on mica prepared by Kessel and Granick [26]. We believe that this difference is partly due to the method used to measure the advancing and receding contact angles. Wasserman *et al.* [27c] and Kessel and Granick [26] measured the angle obtained after the drop had advanced and after the drop had receded, neither of which gives the maximum (θ_a) and minimum (θ_r) contact angles. According to Wasserman *et al.* [27c], the advancing contact angle was 2–3° smaller than the maximum angle and if this is used as a guide, the minimum angle would also be 2–3° smaller; then the hysteresis increases to 14–16°. This is only 3–5° smaller than the value obtained for the OTE surfaces prepared in this study. Contact angle hysteresis on our OTE surfaces was also measured by advancing and receding a drop. As expected, the 'advancing' contact angle reduced to 108° and the hysteresis decreased to 8°.

3.2. Surface force characterization of DMDOA

Force–distance curves obtained for the interaction between DMDOA surfaces in water are presented in Fig. 1. The attractive interactions begin at about 100 nm, which is slightly larger than that previously observed [2, 4]. The contact thickness corresponding to two monolayers of DMDOA was 3.3 and 3.5 nm in air and in water, respectively (Table 3). This suggests that the DMDOA monolayer was not as closely packed as C18 alkyl thiol on silver [28] and as a result, the layer was more compressible on contact. Adhesion forces in air and water are also comparable to previous DMDOA data [2] and to hydrocarbon chains packed in a crystalline state [22]. Application of the JKR and DMT theories to the pull-off force yields surface energies (solid–vapor surface energies, Table 4) which are in good agreement with published values for hydrocarbon surfaces [22]. Solid–liquid surface energies of 41 and 56 mJ/m^2 are obtained from the pull-off force in water using the DMT and JKR models, respectively. These values are higher than those measured previously for dihexadecyldimethylammonium acetate monolayers (28–36 mJ/m^2) [11]. However, the pull-off force of 520 mN/m for DMDOA surfaces is in reasonable agreement with that obtained by Christenson and Claesson [2] and therefore the high value for the solid–liquid surface energy obtained in this study is not unprecedented. Also in agreement with previous observations, cavitation was observed on separating the surfaces — a dark band appeared in the middle of the FECO fringes, suggesting that a single gas bubble had

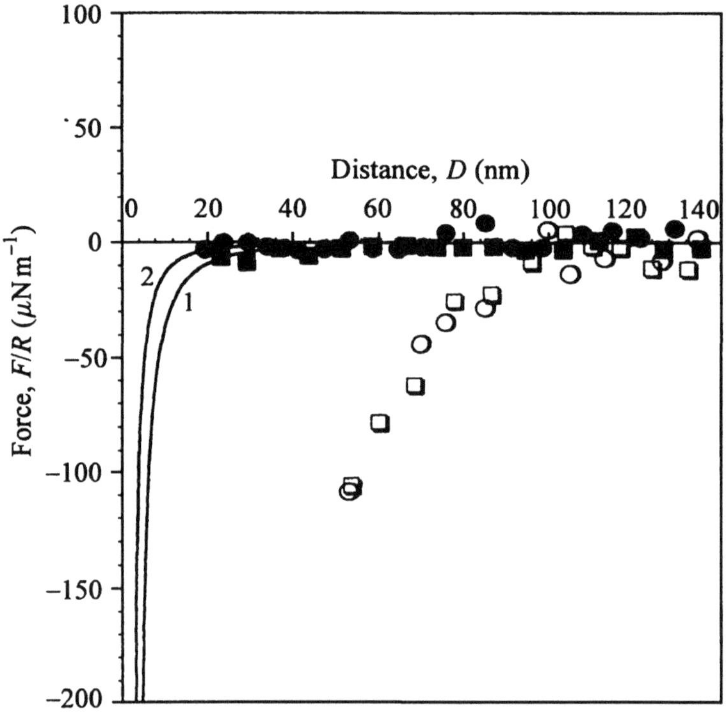

Figure 1. Normalized forces measured between mica surfaces coated with DMDOA (open symbols) and OTE (filled symbols) monolayers plotted as a function of the monolayer surface separation when immersed in deaerated water (pH 5.6). The solid lines represent theoretical van der Waals plots assuming Hamaker constants of 0.8×10^{-20} J (outer curve, 1) and 2.2×10^{-20} J (inner curve, 2). The square and circle symbols represent two separate experiments.

Table 3.

Contact thickness measurements and adhesion force (pull-off force) of L–B monolayers of DMDOA and OTE deposited on mica in air and when immersed in water using the SFA

Surface	Layer thickness (nm)		Adhesion force (mN/m)	
	In air	In water	In air	In water
DMDOA	3.3	3.5	290	520
OTE	4.2	4.2	270	480

Table 4.

Surface energies of L–B layers of DMDOA and prepolymerized OTE deposited on mica using the adhesion force data presented in Table 3 in the JKR and DMT equations

Surface	Surface energy (mJ/m^2)
DMDOA	31 (23 — DMT)
OTE	29 (22 — DMT)

formed [2]. Thus, we have measured a long-range force extending to about 100 nm which is in reasonable agreement with published data. Surface characteristics such as the water contact angles and the pull-off adhesion force are also in good agreement with previous data.

3.3. Surface force characterization of OTE

The OTE data obtained in this study show no discernible attraction ($F/R = 10^{-4}$ N/m or less) until about 17 ± 2.5 nm,* after which point there was a jump-in (Fig. 1). The exact position of the jump-in is closer than the last (stable) data point by a distance less than the step size used to obtain the next data point. Typically the surfaces were brought together at an average of 4 nm increments, so the jump-in position is estimated to be at 13 ± 2.5 nm monolayer surface-to-surface separation. The force profiles for OTE are distinctively different from those obtained for DMDOA in this study and by others [2]. The force profiles for OTE surfaces are also very different from those involving chemisorbed alkyl silanes on mica [6] or on quartz [5b]. Interestingly, for a polymerized ammonium amphiphile transferred in the up-stroke mode using an L–B trough, no force was detected until a jump-in was observed at 24 ± 6 nm [13]. This is qualitatively similar to the force profile for OTE obtained in this study, in that no force was also detected until a jump-in estimated to be at 13 ± 2.5 nm. However, there was no effect of electrolyte (up to 0.1 M KNO_3) on the force profiles for the OTE surfaces, while in the case of the polymerized ammonium amphiphile a weak repulsion was observed in the presence of 1 mM and 10 mM NaBr. The slight repulsion might indicate that the hydrophobic layer was unstable in the presence of NaBr. A trend on the effect of electrolyte on the force profiles for hydrophobic surfaces has been noted, in that it appears that electrolyte reduces the magnitude and not the decay length of the attractive force [15]. More recently, Tsao *et al.* [29] found that a 2 : 2 electrolyte ($MgCl_2$) was more effective than a 1 : 1 electrolyte (NaCl) in reducing the magnitude of the attractive force between hydrophobic mica surfaces. In this study, the hydrophobic mica might have been slightly negatively charged. Thus, it appears that there is an electrostatic contribution due to charging of the surface when being prepared [29] or by adsorption of ions [30], but apparently this was not the case with the OTE surfaces studied here as the force profiles did not change in the presence of electrolyte.

The contact distance, i.e. the thickness of the two hydrophobic OTE layers, was 4.2 nm (or 2.1 nm for each monolayer). This is slightly smaller than the transextended length of 2.6 nm for R–Si–O (R = C18) and that typically obtained by ellipsometry for R–Si–O on silica [31]. This difference is suggestive of compression of the layer on contact and is in excellent agreement with a similar trend observed by Levins and Vanderlick [28] for alkyl thiols on silver; they obtained a thickness of 2.15 nm for a C18 thiol from contact distance measurements using the surface force apparatus.

---

*The average number of 17 nm was obtained from over 15 measurements and the uncertainty of ± 2.5 nm is one standard deviation.

Another interesting distinction between the OTE surfaces studied here and previously studied hydrocarbon surfaces is that we have direct evidence of spontaneous cavitation on contact (see the discontinuous FECO in Fig. 2). By spontaneous cavitation or cavitation on contact we mean that once the surfaces had jumped into contact, discontinuities in the FECO [2] attributable to vapor bubbles were observed. In the OTE experiments the first discontinuities appeared at 0.5 s or less, but these continued to grow for the next 2–3 s, after which they stopped growing. Cavitation on contact has been directly observed for fluorocarbon-coated mica [2] and inferred to have occurred between hydrocarbon-coated silica [5a]. Cavitation has also been observed on separation using hydrocarbon-coated mica surfaces [11] including DMDOA-coated mica surfaces [2], but in this case the cavitation is not spontaneous. The cavitation on contact observed for the OTE surfaces is indicative of a strongly hydrophobic nature of the surfaces. The nonwetting condition ($\theta > 90°$) necessary for cavitation has been verified in experiments in which glass surfaces were immersed in mercury [32]. In addition, according to the theory of Yushchenko *et al.* [33], the receding contact angle of water on a hydrophobic surface would have to be greater than 90° in order for cavitation to occur [2], a condition which is easily satisfied for these OTE surfaces.

Adhesion forces for OTE surfaces in air and water are comparable to previous results (Table 3). Application of the JKR and DMT theories gives numbers which are in the appropriate range for hydrocarbon surfaces [22]. From the JKR theory we write

$$\frac{F}{R}(\text{air}) = 3\pi\gamma_{\text{SV}},$$

where $(F/R)(\text{air})$ is the maximum force per unit length required to separate two identical solid (S) surfaces in their vapor (V). Similarly,

$$\frac{F}{R}(\text{water}) = 3\pi\gamma_{\text{SL}}$$

is the maximum force per unit length required to separate two identical surfaces when immersed in a liquid (L). Using Young's equation, it follows that

$$\frac{F}{R}(\text{air}) - \frac{F}{R}(\text{water}) = 3\pi\gamma_{\text{LV}}\cos\theta.$$

Inserting values from Table 3 and using $\gamma_{\text{LV}} = 72.2$ mN/m for water, a contact angle of 108° is obtained. This is in reasonable agreement with the measured contact angle of 112°, as shown in Tables 1 and 2. This procedure, which verifies Young's equation, was first used by Johnson *et al.* [23] and has since been used successfully by others [34–37].

It is interesting to note that the pull-off force was not affected in a significant manner by the vapor cavities that had formed on contact as the pull-off force is comparable to that obtained for DMDOA. In contrast, a substantially higher pull-off force was measured in the case of hydrophobized quartz filaments and this was attributed to the formation of cavities that had bridged the two surfaces [5].

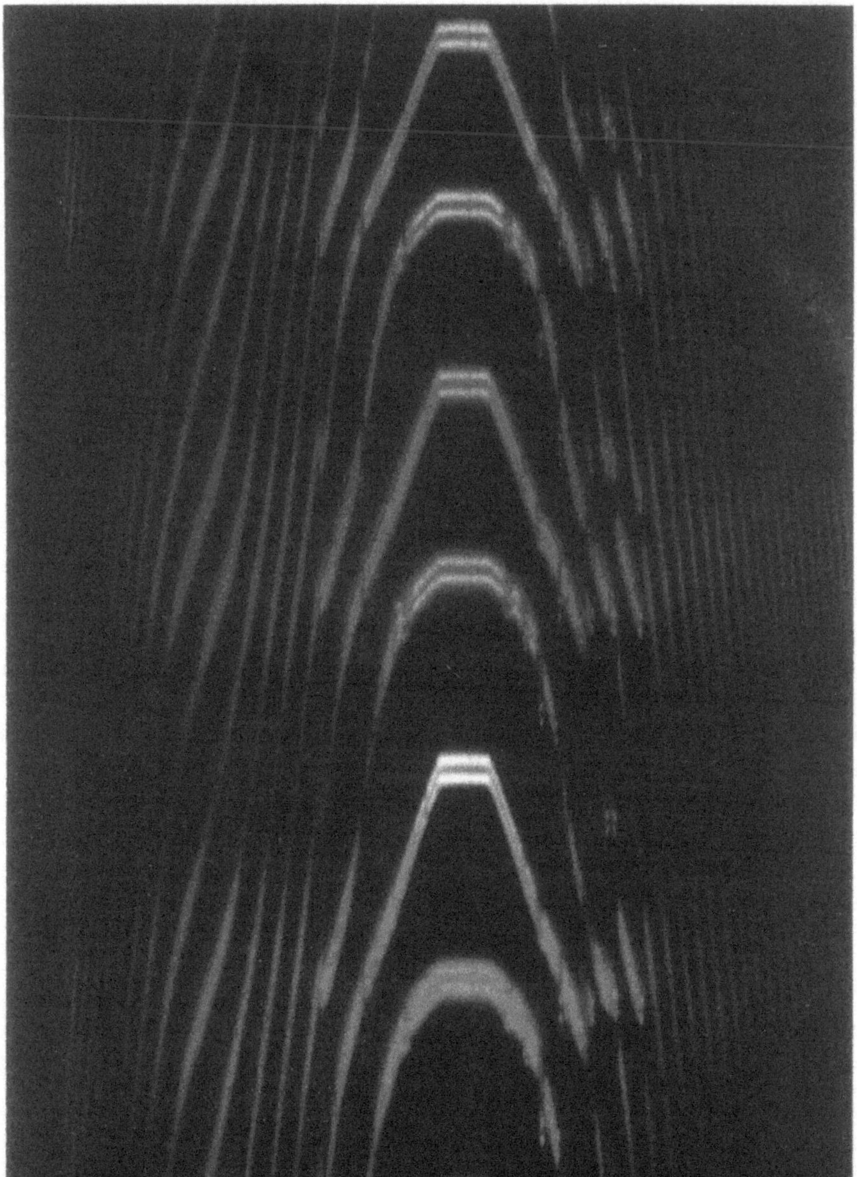

Figure 2. Discontinuities in the FECO (fringes of equal chromatic order) appear due to vapor cavities that form spontaneously when two OTE surfaces jump into contact in water.

4. CONCLUSION

Direct measurements of the interaction force between two DMDOA surfaces and two OTE surfaces in water have been made using the SFA. In the case of DMDOA surfaces, a very-long-range attraction first detected at about 100 nm is evident and is in reasonable agreement with previous work. Also in agreement with previous work is the cavitation that is observed on separating the two DMDOA surfaces. In contrast, the interaction profile between two OTE surfaces is distinctly different; no discernible interaction is evident until a jump-in is observed at an estimated surface separation of 13 nm. On contacting the surfaces, cavitation is observed, indicating that the surfaces are very hydrophobic (more hydrophobic than the DMDOA surfaces).

Adhesion forces between pairs of DMDOA and pairs of OTE surfaces were measured in air and in water. It is interesting to note that the adhesion force (pull-off force F_a) measurements in water for OTE and DMDOA surfaces indicate that the solid–liquid surface energies are approximately equal in both cases (the OTE surface energies are slightly lower), even though the OTE surfaces are more hydrophobic than the DMDOA surfaces. One possibility is that once the surfaces are in contact, the DMDOA monolayers restructure to form an interface equivalent to that formed between two OTE monolayers, and so the force required to separate them is also the same. OTE monolayers are not expected to undergo extensive restructuring as these layers are polymerized, but the physisorbed DMDOA molecules should be relatively mobile. Restructuring of monolayers has previously been discussed [22, 37].

The pull-off force values for two OTE surfaces and two DMDOA surfaces in air also indicate no significant difference in the surface energies between the DMDOA and OTE surfaces. However, these two surfaces have very different water contact angles. DMDOA has a higher contact angle hysteresis, which may indicate restructuring of the monolayer underneath the water drop. This could explain why adhesion force measurements, measurements which interrogate hydrocarbon–hydrocarbon contact strength, do not reflect the structural differences detected by water contact angles (water-induced restructuring of the DMDOA monolayer underneath the water drop).

REFERENCES

1. H. K. Christenson, in: *Modern Approaches to Wettability: Theory and Applications*, M. E. Schrader and G. Loeb (Eds). Plenum Press, New York (1992).
2. H. K. Christenson and P. M. Claesson, *Science* **239**, 390 (1988).
3. P. M. Claesson, P. C. Blom, P. C. Herder and B. W. Ninham, *J. Colloid Interface Sci.* **114**, 234 (1986).
4. P. M. Claesson and H. K. Christenson, *J. Phys. Chem.* **92**, 1650 (1988).
5. (a) Ya. I. Rabinovich, B. V. Derjaguin and N. V. Churaev, *Adv. Colloid Interface Sci.* **16**, 63 (1982). (b) Ya. I. Rabinovich and B. V. Derjaguin, *Colloids Surfaces* **30**, 243 (1988).
6. (a) J. L. Parker, D. L. Cho and P. M. Claesson, *J. Phys. Chem.* **93**, 6121 (1989). (b) J. L. Parker, P. M. Claesson, D. L. Cho, A. Ahlberg, J. Tidblad and E. Blomberg, *J. Colloid Interface Sci.* **134**, 449 (1990).
7. H. K. Christenson, P. M. Claesson, J. Berg and P. C. Herder, *J. Phys. Chem.* **93**, 1472 (1989).
8. Y.-H. Tsao, S. X. Yang, D. F. Evans and H. Wennerström, *Langmuir* **7**, 3154 (1991).
9. J. N. Israelachvili and R. M. Pashley, *Nature* **300**, 341 (1982).
10. J. N. Israelachvili and R. M. Pashley, *J. Colloid Interface Sci.* **98**, 500 (1984).

11. R. M. Pashley, P. M. McGuiggan, B. W. Ninham and D. F. Evans, *Science* **229**, 1088 (1985).
12. P. Kékicheff, H. K. Christenson and B. W. Ninham, *Colloids Surfaces* **40**, 31 (1989).
13. K. Kurihara and T. Kunitake, *J. Am. Chem. Soc.* **114**, 10927 (1992).
14. H. K. Christenson, J. L. Parker and V. V. Yaminsky, *Langmuir* **8**, 2080 (1992).
15. H. K. Christenson, P. M. Claesson and J. L. Parker, *J. Phys. Chem.* **96**, 6725 (1992).
16. J. Wood and R. Sharma, *Langmuir* **10**, 2307 (1994).
17. S. W. Barton, A. Goudot and F. Rondelez, *Langmuir* **7**, 1029 (1991).
18. Y. Okahata, M. Yokobori, Y. Ebara, H. Ebato and K. Ariga, *Langmuir* **6**, 1148 (1990).
19. J. N. Israelachvili and G. E. Adams, *J. Chem. Soc., Faraday Trans. 1* **74**, 975 (1978).
20. B. V. Derjaguin, *Kolloid-Z.* **69**, 155 (1934).
21. R. G. Horn, J. N. Israelachvili and F. Pribac, *J. Colloid Interface Sci.* **115**, 480 (1987).
22. Y. L. Chen, C. A. Helm and J. N. Israelachvili, *J. Phys. Chem.* **95**, 10736 (1991).
23. K. L. Johnson, K. Kendall and A. D. Roberts, *Proc. R. Soc. London, Ser. A* **324**, 301 (1971).
24. B. V. Derjaguin, V. M. Muller and Yu. Toporov, *J. Colloid Interface Sci.* **53**, 314 (1975).
25. D. Maugis, *J. Colloid Interface Sci.* **150**, 243 (1992).
26. C. R. Kessel and S. Granick, *Langmuir* **7**, 533 (1991).
27. (a) M. Pomerantz, A. Segmuller, L. Netzer and J. Sagiv, *Thin Solid Films* **132**, 153 (1985).
(b) N. Tillman, A. Ulman, J. S. Schildkraut and T. L. Penner, *J. Am. Chem. Soc.* **110**, 6136 (1988).
(c) S. Wasserman, Y.-T. Tao and G. M. Whitesides, *Langmuir* **5**, 1074 (1989).
28. J. M. Levins and T. K. Vanderlick, *J. Phys. Chem.* **96**, 10405 (1992).
29. Y.-H. Tsao, D. F. Evans and H. Wennerström, *Langmuir* **9**, 779 (1993).
30. H. K. Christenson, J. Fang, B. W. Ninham and J. L. Parker, *J. Phys. Chem.* **94**, 8004 (1990).
31. V. DePalma and N. Tillman, *Langmuir* **5**, 868 (1989).
32. V. V. Yaminsky, V. S. Yushchenko, E. A. Amelina and E. D. Shchukin, *J. Colloid Interface Sci.* **96**, 301 (1983).
33. V. S. Yushchenko, V. V. Yaminsky and E. D. Shchukin, *J. Colloid Interface Sci.* **96**, 307 (1983).
34. R. M. Pashely and J. N. Israelachvili, *Colloids Surfaces* **2**, 169 (1981).
35. R. M. Pashely, P. M. McGuiggan, R. G. Horn and B. W. Ninham, *J. Colloid Interface Sci.* **126**, 569 (1988).
36. J. N. Israelachvili, *Adv. Colloid Interface Sci.* **16**, 31 (1982).
37. (a) J. D. Andrade, S. M. Ma, R. N. King and D. E. Gregonis, *J. Colloid Interface Sci.* **72**, 488 (1979).
(b) H. Yasuda and A. K. Sharma, *J. Polym. Sci. Polym. Phys. Ed.* **19**, 1285 (1981).
(c) A. Carre, S. Moll, J. Schultz and M. E. R. Shanahan, in: *Adhesion*, K. W. Allen (Ed.), Vol. 11, p. 82. Elsevier Applied Science Publishers, London (1986).

Fundamentals of Adhesion and Interfaces, pp. 247–260
D. S. Rimai, L. P. DeMejo and K. L. Mittal (Eds)

Cracks at adhesive interfaces

KEVIN KENDALL
Chemistry Department, Keele University, Keele, Staffs ST5 5BG, UK

Revised version received 7 May 1994

Abstract—The application of Griffith's energy balance argument to cracks at adhesive interfaces is studied. Adhesive interfaces are generally brittle, representing the simplest form of fracture mechanics geometry because cracks are constrained to travel along the interface, giving a defined crack path which eases analysis. Experimentally, such cracks may be propagated along the interface between optically smooth rubber pieces, and measured through the transparent material. The development of adhesive fracture test-pieces since Griffith's time reveals difficulties in his reasoning, and allows improved understanding of the energy balance method. The most important conclusion is that stress does not normally enter the cracking criterion. It is demonstrated experimentally that stress may remain constant while the crack criterion changes. The strength of an adhesive interface is shown to be a meaningless parameter; instead, the work of adhesion, or adhesive energy, which is the work of adhesion together with energy losses, should be used to define the behaviour of cracks at interfaces.

Keywords: Cracks; work of adhesion; adhesion energy.

1. INTRODUCTION

This year marks the 100th anniversary of Griffith's birth [1, 2], so it is a timely occasion to discuss the influence of his energy balance theory on our understanding of cracks at adhesive interfaces. Here an interface is defined as a surface at which the molecules of two materials meet. It follows that an adhesive interface is one which requires mechanical force to separate the two materials. This mechanical force F is the adhesion force or, more simply, the adhesion.

Since adhesion is usually determined by cracking an interface, Griffith's theory provides the main link between adhesion measurements and our knowledge of interface structure and chemistry, which may be quantified by the thermodynamic work of adhesion W, describing the molecular bond energies across the interface. Our aim is therefore to find this link between adhesion force and work of adhesion.

The most obvious feature of an adhesive interface is brittle failure along the interface. It is a rare joint which exhibits good resistance to cracks. Consequently, once a flaw is initiated in a typical adhesive interface, a crack propagates quickly, often accelerating catastrophically under a small driving force. Such behaviour is similar to the fracture of glass or ceramic, displaying the characteristic sudden failure, the low strength, the unreliability, the invisible defects, the sensitivity to manufacturing conditions, and the ease of damage by impact or by environmental agents. Adhesive interfaces thus do not behave like the safe metals of common engineering construction. Instead, they must be designed like glass or ceramic devices, and are frequently shunned by mainstream engineers.

It is ironic that Griffith was a mainstream engineer, who made a major contribution to gas turbine technology, yet diverted his attention to glass as a model in explaining the influence of surface defects on strength. Appropriately, the theory which he used to describe the cracking of glass also describes cracks at adhesive interfaces with some accuracy. This paper first gives the background to the Griffith argument, and then points out the paradox in his paper, which seems to suggest that the stress criterion of fracture can be consistent with the energy balance. By studying several adhesion test geometries, such as peeling [3] or frictionless cleavage [4], it is demonstrated that stress does not normally figure in the fracture equation. It is then shown experimentally that stress can remain constant while the cracking criterion changes. The Griffith equation would be better recast in terms of F/b, the crack line force which appears in other adhesion equations, rather than stress. The crack resistance for the Griffith geometry is then $(WEd)^{1/2}$, as observed for other elastically linked fracture geometries like lap-shear or wedging, but with a scaling of $(d/c)^{1/2}$. In these equations, F is the force, W is the thermodynamic work of adhesion, b is the crack width, c is its length, d is the characteristic sample dimension, and E is Young's modulus. The nomenclature is given in Fig. 7.

Treating adhesive fracture in this way allows other secondary influences on cracks — such as stiffening, bridging, residual stress, delamination, friction, and healing — to be understood more rationally.

2. EXPERIMENTAL SYSTEM

An adhesive interface with molecular contact between two materials is readily formed from two pieces of rubber with optically smooth surfaces (Fig. 1a). Roberts [5]

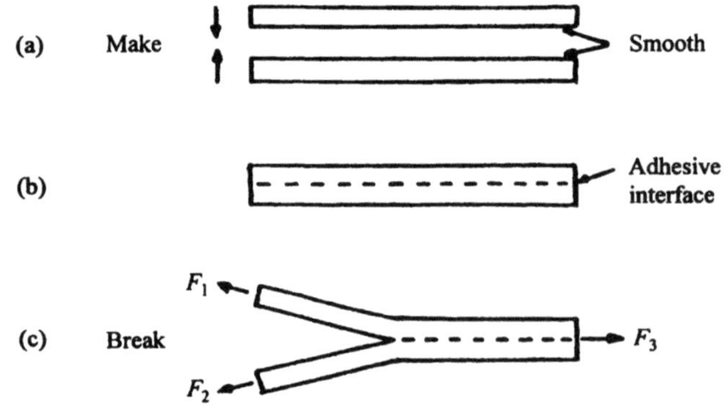

Figure 1. (a) Two optically smooth rubber samples; (b) adhesive interface formed between smooth surfaces; (c) forces applied to crack the adhesive interface.

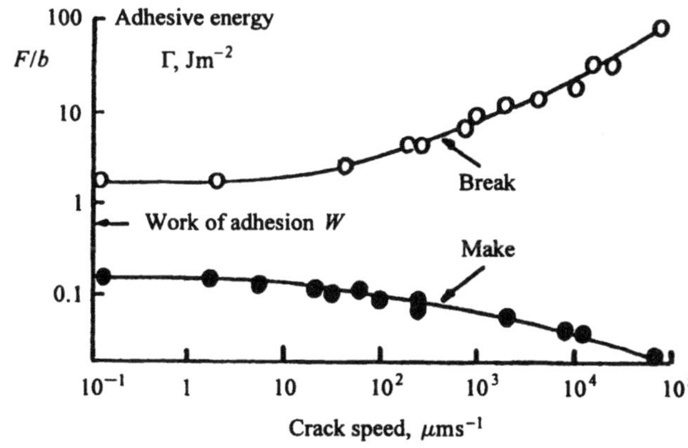

Figure 2. Dependence of the crack speed on the peel force/width (F/b) for smooth rubber in contact with glass. F/b is equivalent to the work of adhesion W at equilibrium or to the adhesive energy Γ when losses occur. The lower curve shows the crack speed during the formation of the adhesive interface. The upper curve shows the crack speed during breakage, after 300 s dwell time.

showed how to make such smooth rubber samples by moulding gum rubber against an optically smooth surface such as glass, crosslinking the polymer, and then peeling the rubber from the smooth mould. Alternatively, instead of gum rubber, low-temperature curing materials such as silastomer or polyurethane could be used to make the two smooth adherends, as demonstrated so elegantly by Barquins, Maugis, and colleagues in a number of papers [6]. The two smooth rubber surfaces could be brought into contact to form an adhesive interface (Fig. 2b), and then the adhesion could be measured by applying forces to crack the interface (Fig. 2c). The two materials forming the adhesive interface were varied. Rubber/rubber or rubber/glass or rubber/metal interfaces could be made in a similar way.

In the results reported earlier [7] and expanded here, ethylene propylene rubber (Enjay 404) was twin-roll milled with 0.32% sulphur and 2.7% dicumyl peroxide, and then pressed at 160°C against a glass plate for 1 h to crosslink the material before chilling to room temperature. The rubber sheets were then peeled from the glass to reveal optically smooth surfaces which could be brought together under modest pressure, as shown in Fig. 1, to give reproducible contact (Fig. 1b). This gave a well-defined interface which was tested by applying forces to propagate an interfacial crack (Fig. 1c), observed optically through the transparent materials.

Several experimental precautions had to be taken when measuring the fracture of these interfaces. The first was recognition of the dwell-time effect, in which the adhesion force increased with the time of contact of the rubber pieces. The adhesive interface was therefore made a fixed time before cracks were tested.

A second phenomenon was the connection of crack speed with applied force, as shown in Fig. 2, and with temperature. As the force or the temperature was raised, so the crack speed increased. Crack speed and temperature were therefore held constant when making adhesion force measurements. Adhesion forces were higher than expected from the equilibrium work of adhesion under these conditions, so 'adhesive energy' Γ was the term used to describe the work needed to fracture unit area of interface in these examples, where adhesive energy must include the reversible work of adhesion, but also must contain some irreversible work.

At slow crack speeds, the adhesion tended towards the equilibrium value, which could be determined by measuring crack healing as well as crack breaking. However, it was evident that equilibrium was not readily attainable even at the lowest speeds (Fig. 2), as a result of adhesive hysteresis.

To summarize these effects, interfacial cracking experiments on smooth rubber interfaces display kinetic and hysteresis phenomena much like those exhibited by liquid droplets wetting solid surfaces. Adhesion increases with the contact time and with the crack speed. Reproducible results are obtained by measuring at constant times and speeds, such that the equilibrium equations are applied some way from the equilibrium point, to measure the adhesive energy Γ. Equilibrium adhesion and therefore the true thermodynamic work of adhesion W may be difficult to attain because of hysteresis, but may be estimated by considering the energy gap between making and breaking the joint.

3. GRIFFITH THEORY

Griffith studied the cracking of glass bulbs under internal pressure and showed that his fracture results could be explained by an energy balance theory. This theory was based on the idea of reversible thermodynamic fracture, in which the work expended by applied loads and elastic strain was entirely converted to surface energy during cracking.

By balancing the three energy terms: potential energy in the loads, strain energy in the elastic material, and surface energy in the crack faces, Griffith derived the criterion for stability of a slit, length $2c$, in an infinite thin elastic sheet with tensile stress σ applied at the extremities (Fig. 3). The criterion was correctly written in his 1924 paper as

$$\sigma = (EW/\pi c)^{1/2}, \tag{1}$$

where E was the Young's modulus of the elastic sheet and W was the work required to break unit area of material. Griffith imagined W to be the thermodynamic value.

This was a radical theory of fracture at the time because it showed that the energy needed to separate molecules at the crack tip was the important factor preventing crack propagation, whereas stress, elastic modulus, and crack length merely provided a mechanism or linkage allowing the work to be expended by the applied force. Essentially, Griffith proved that fracture could occur over a wide range of stresses depending on this linkage. Therefore, the Galilean stress criterion of fracture, which states that materials break at a critical stress, is not valid for a brittle system. Thus, strength is not a useful parameter for describing a brittle material. Equally, interface strength is meaningless for a brittle interface.

Curiously, Griffith did not quite believe his own theory. He persisted in claiming that failure occurs 'at a certain maximum tension'. This is a restatement of the stress

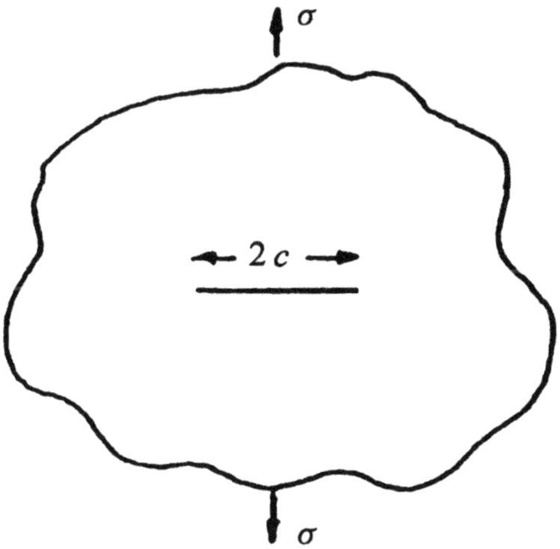

Figure 3. The Griffith crack geometry for an infinite elastic sheet.

criterion of fracture, whereas, in his mathematics, Griffith assumed an infinite stress at the crack tip. Later [8], Griffith reverted to an incorrect stress analysis of compression fracture.

A second issue was the deceptive agreement which Griffith found between his equilibrium work of fracture W, 1.1 J m^{-2}, obtained from surface tension measurements, and the value obtained for fracture tests, which turned out to be 3.5 J m^{-2}. Griffith, as a result of errors, thought agreement was within 10% but in fact there was a factor of 3 discrepancy. This difference must be explained by departure from equilibrium. The value of 3.5 J m^{-2} measured by cracking was an adhesive energy Γ rather than a thermodynamic work of adhesion W. Unidentified dissipative crack mechanisms were magnifying the thermodynamic energy.

Pragmatically, Irwin and his followers found much greater deviations for metals, where huge dissipations can occur in plastic flow. They avoided the equilibrium issue by balancing work rates [9]. Also they defined a parameter K_{IC} to replace $(EW)^{1/2}$ in equation (1). K_{IC} was called the critical stress intensity factor and was fitted empirically to fracture data, to include dissipation. This approach may have been useful for practical applications, but it has clouded the several mechanistic issues by lumping them in this single crude factor.

A final problem was the form of equation (1). Because it was framed in terms of stress, it has been viewed as a stress criterion of fracture. But, as shown above, fracture depends on energy and not on stress. Thus, it is not possible, in general, to predict cracking from stress measured at a point in a material. This became evident later, when studies of cracks at adhesive interfaces were reported.

4. FRACTURE AT ADHESIVE INTERFACES

A number of simpler crack geometries have been discovered since Griffith's time, illustrating the point that stress does not determine crack propagation. Many of these crack geometries involve adhesive interfaces, which simplify the analysis because cracks at adhesive interfaces are usually constrained to follow the interface, rather than wandering off unpredictably, as in the fracture of glass.

Two such geometries are shown in Fig. 4. The peel test was studied by Rivlin [3] using adhesive tapes. He showed the strong effects of crack speed and temperature, together with the remarkable stringing out of the adhesive, bridging the crack faces. He ignored elastic energy in the tape (Fig. 4a), because this should remain constant, and connected the applied peeling force F, divided by tape width b, to the work of adhesion W. In effect, he was applying the Griffith energy balance to the adhesive crack.

The same equation was found by Kendall and Fuller [4] to describe the cleavage force necessary to drive a frictionless wedge into an adhesive joint (Fig. 4b). Applying the Griffith energy theory to this geometry, there are three terms: the potential energy $-Fc$ in the applied force, the elastic strain energy U_{E} in the deformed elastic material, and the surface energy Wbc in the free crack faces. The total energy U_{T} is therefore

$$U_{\mathrm{T}} = U_{\mathrm{E}} - Fc + Wbc. \tag{2}$$

Applying the energy conservation criterion $\mathrm{d}U_{\mathrm{T}}/\mathrm{d}c = 0$, and noting that U_{E} is a constant during steady propagation, reveals the crack criterion

$$F/b = W. \tag{3}$$

This derivation is interesting because it shows that stress does not enter this Griffith-type analysis of the cracking condition. The crack is merely a mechanism for direct conversion of potential energy into work of adhesion and is not driven by stress. Nor is the cracking affected by elastic modulus or specimen thickness. The only important parameters are the crack line force F/b, which drives the crack, and the work of adhesion W, which resists fracture. This idea reinforces our physical intuition that stress *per se* is not needed to produce separation of interfaces. For example, surface separation can be brought about chemically, by wetting liquids or by dissolution. Also, we know that liquid interfaces can be separated despite their inability to support stress.

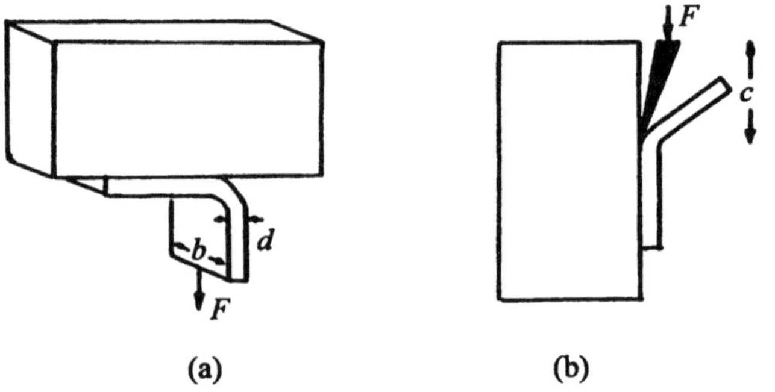

Figure 4. (a) Peeling of an elastic strip from a rigid substrate; (b) frictionless wedging of the elastic interface.

5. INFLUENCE OF ELASTIC ENERGY

In the above examples, the elastic energy term was significant, but remained constant and therefore could not contribute to the crack motion. If the elastic energy term is allowed to change with the crack length, as in Fig. 5b, which shows the Obreimoff cleavage test, then elastic energy can drive the crack and a different cracking criterion emerges.

Obreimoff [10] used the cleavage of mica to verify the Griffith energy balance concept. The crack was wedged open and held by friction, such that the bending displacement applied to the cleaved strip remained constant. The applied lateral force F causing the bending was therefore fixed in position and could not supply any work to the crack. Using the terminology of Fig. 5, the three energy terms are:

$$\text{potential energy} = 0 \qquad \text{because the displacement is fixed}$$

$$\text{elastic energy} = Ebd^3h^2/8c^3$$

$$\text{surface energy} = Wbc$$

and

$$\text{total energy } U_\mathrm{T} = Wbc + Ebd^3h^2/8c^3.$$

Taking $\mathrm{d}U_\mathrm{T}/\mathrm{d}c = 0$ and expressing $h = 4Fc^3/Ebd^3$ gives the crack criterion

$$F/b = \left(WEd \cdot d^2/6c^2\right)^{1/2}. \tag{4}$$

It is evident from this equation that the crack driving force is F/b and that the resistance to cracking is $(WEd)^{1/2}$, moderated by the scaling factor d/c.

Similarly, Fig. 5a shows how the peeling test can be modified so that elastic strain energy becomes important. The peel angle has been reduced from $\pi/2$ to θ, such that the effective line force is diminished, as Rivlin showed, to $(1 - \cos\theta)F/b$. To maintain peeling, the line force F/b must therefore be increased to $W/(1 - \cos\theta)$. Elastic stretching is then observed, and this makes peeling easier, as explained by Kendall [11] using the Griffith energy balance.

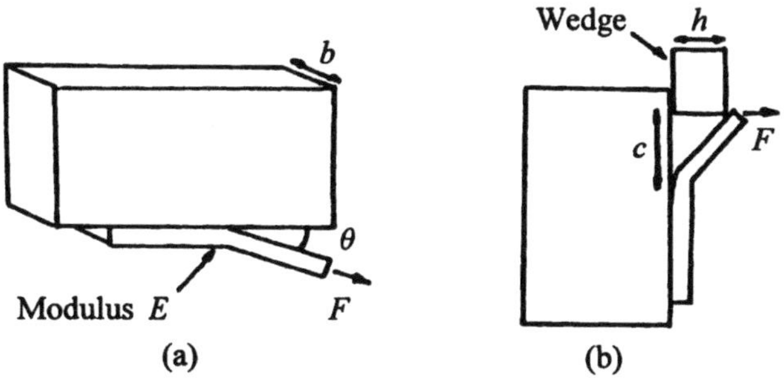

Figure 5. (a) Peeling at an angle θ; (b) Obreimoff crack geometry.

The three terms are:

$$\begin{aligned} \text{potential energy} &= -Fc\,(1-\cos\theta) - F^2c/bEd \\ \text{elastic energy} &= F^2c/bEd \\ \text{surface energy} &= Wbc. \end{aligned}$$

Summing these and differentiating with respect to crack length gives the fracture criterion

$$(F/b)^2/2Ed + (F/b)\,(1-\cos\theta) - W = 0. \tag{5}$$

In this example, the potential and elastic energy terms are distinct and separable, leading to a quadratic criterion of fracture. At large peel angles, the potential energy term is the main contribution and equation (5) simplifies to equation (3). At very small angles, the elastic term dominates and equation (5) simplifies to

$$F/b = (2WEd)^{1/2}, \tag{6}$$

which is the rigid/elastic lap-joint fracture criterion [12].

6. EXPERIMENTAL PROOF THAT STRESS DOES NOT DRIVE CRACKS

It is clear from equation (5) that stress in the elastic material can remain constant while the crack criterion may change. Thus, peel cracks can propagate while the stress is low and also while the stress is high. In other words, this equation supports the general conclusion from the Griffith theory that stress does not determine cracking.

A direct experimental verification of this idea can be obtained quite simply in the peel test geometry. Consider the peel test shown in Fig. 6a. Imagine applying a force F to this geometry and increasing the force until peeling almost occurs. This force creates a certain complex stress distribution in the rubber. The rubber is held onto a rigid substrate, but with a few molecules of rubber attached to its surface to provide a rubber/rubber adhesion condition.

The stress in the rubber of Fig. 6a is the same as that in Fig. 6b, in which the rigid substrate has been replaced by a mirror image of the rubber strip. If the stress

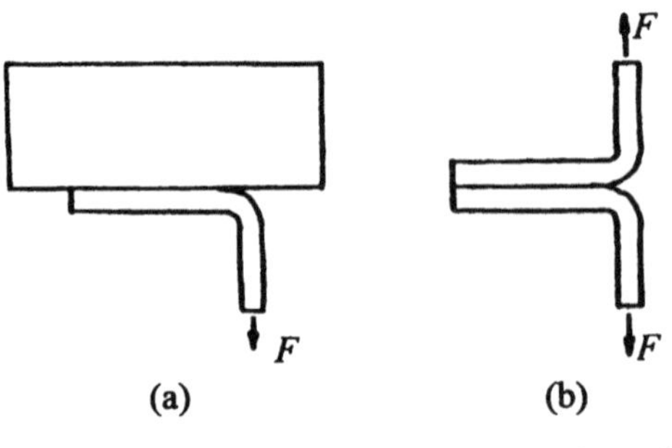

Figure 6. (a) Peel test with F just low enough to prevent cracking; (b) peel test at the same stress now fractures.

in the rubber dictated fracture, then peeling would also not occur in Fig. 6b because the crack did not move in Fig. 6a. However, peeling does occur in the geometry of Fig. 6b, and this is readily explained in terms of the energy balance theory, which shows that the criterion for cracking of Fig. 6b is

$$F/b = W/2, \tag{7}$$

a factor of 2 different from Fig. 6a, even though stress remains constant. This factor of 2 arises because the displacement is doubled in Fig. 6b, giving twice the energy input at constant stress.

This experiment was performed using ethylene propylene rubber crosslinked with dicumyl peroxide. The rigid substrate was polymethyl methacrylate (PMMA), with a polished smooth surface, covered by a thin film of the ethylene propylene rubber. A rubber strip was then brought into contact with the rubber-coated PMMA for 60 min before applying the peel force. The load F was set at 12 mN on a 10 mm wide strip of rubber and the crack was observed through the transparent rubber. No peeling was seen over a period of 30 min in the geometry of Fig. 6a.

Two rubber strips were held in contact for 60 min and then the same load of 12 mN was applied in the geometry of Fig. 6b. Peeling was then observed at a crack speed of 2 μm s^{-1}, as expected from the energy balance theory.

7. STRESS AND THE CRACKING CRITERION

The above examples indicate that stress does not generally enter the criterion for cracking. A list of some adhesive crack criteria is given in Table 1.

Linkage between driving force and crack resistance is direct in certain cases, for example, in peeling or in frictionless cleavage, the first two geometries of Table 1. In these cases, the crack driving force is of the form F/b and crack resistance is given by W, the work of adhesion. In other cases, like Obreimoff or lap-shear, the linkage is indirect, transmitted through elastic deformations, and then $(WEd)^{1/2}$ is the relevant crack resistance parameter. Both direct and elastic linkages may operate simultaneously, as in low angle peeling, where both peeling and stretching fracture

Table 1.
Fracture criteria obtained by the Griffith energy balance

$\pi/2$ peel	$F/b = W$	
Frictionless cleavage	$F/b = W$	
Lap-shear	$F/b = (2WEd)^{1/2}$	(rigid/elastic)
Griffith	$F/b = (WEd \cdot d/\pi c)^{1/2}$	
Obreimoff	$F/b = (2WEd \cdot d^2/12c^2)^{1/2}$	
Elastic peel	$0 = (F/b)^2/2Ed + (F/b)(1 - \cos\theta) - W$	

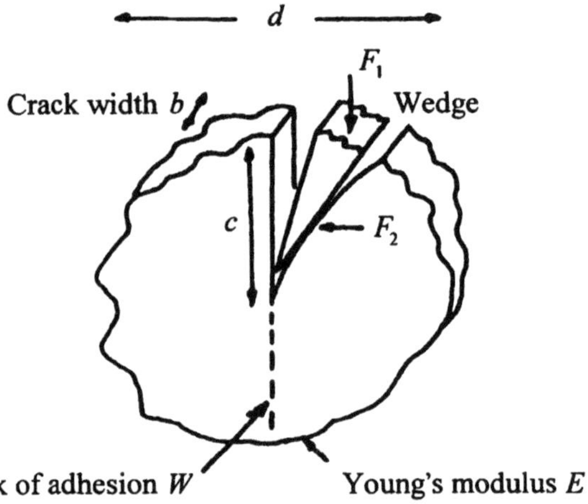

Figure 7. Crack in the vicinity of a wedge to illustrate the effects of friction. The wedge is driven by the force F_1 to open up the crack of length c in the interface of width b. However, the deformed material of elastic modulus E presses back on the wedge with a force F_2 which causes frictional resistance.

are observed depending on the film thickness, elastic modulus, and peel angle. Then the crack criterion is in the form of a quadratic equation.

It is clear from this discussion that Griffith cracking is little different from the lap-shear or the Obreimoff geometry and could better be written

$$F/b = (WEd \cdot d/\pi c)^{1/2}, \tag{8}$$

with the symbols shown in Fig. 7. Here the crack resistance scales with $(d/c)^{1/2}$, whereas the lap-shear crack resistance scales with $(d/c)^0$ and the Obreimoff crack scales with $(d/c)^1$.

8. SECONDARY INFLUENCES ON CRACKS

Having established these two simple linkages, direct and elastic, between the crack driving force and the work of adhesion, it is then straightforward to incorporate into the theory a range of secondary influences such as friction, bridging, biaxial loading, delamination, etc. The energy balance then reveals a rich variety of sub-linkages which influence adhesion enormously.

It has been common in practical applications to describe these secondary effects in terms of a change of interface toughness, lumping all the terms into an empirical K_{IC} parameter. A more sensible approach would identify the details of the specific mechanisms and make the necessary adjustments to the Griffith energy balance.

As an example, consider the effects of friction on the cracking of an adhesive interface. The overall effect of friction is to inhibit crack movement by making adhesion less reversible. Thus, the equilibrium equation of Griffith becomes less valid as heat is generated in the system, apparently increasing the toughness of the interface, as measured by a K_{IC} determination.

However, this apparent increase in toughness can arise at several locations in the mechanism, and these precise locations should be identified carefully. Figure 7 shows the possibilities around a wedge crack. A frictionless wedge with elastic material and reversible interface gives equation (3), $F/b = W$, by the Griffith argument. Friction can influence this equation in three ways: First, it can reduce the driving force to $F_1 - \mu F_2$, where μ is the friction coefficient between the wedge and the elastic material. Second, friction between molecules at the interface can increase the work W done at the interface to the non-equilibrium value Γ. Third, friction in the 'elastic' material can dissipate energy within the solid, giving a relaxation of the modulus E with time. This influences the elastic linkage.

The first two effects are easily incorporated into the energy balance by modifying the driving force and resistance terms in equation (3) to give

$$\left(F_1 - \mu F_2\right)/b = \Gamma, \tag{9}$$

and this itself is interesting because the friction force depends on the crack length and this provides a feed-back loop.

The third influence, the relaxation of elastic modulus, does not seem to enter equation (3), which has no modulus dependence. The effect of relaxation is therefore quite subtle and remained unexplained until recently [13, 14]. The relaxation was shown to give two effects: an energy loss term which raised the work of adhesion to WE_1/E_2, where E_1 was the initial modulus and E_2 was the relaxed modulus; and a shape change term depending on the crack length, crack speed, and $\mathrm{d}E/\mathrm{d}t$, the relaxation rate. This second effect, not recognized before, was shown to reduce the crack speed with time.

It is worth considering these three locations of crack inhibition in more detail.

8.1. Reduction of crack driving force

The crack driving force can be reduced in several ways besides friction. For example, bridging of the crack may occur, where ligaments of material hold the crack faces together and inhibit crack opening. Or additional crack closing forces may be applied to the specimen, or healing of the crack may reduce the driving force.

Bridging of the crack is particularly simple if a uniform pressure acts to hold the Griffith crack faces together [15]. It has been demonstrated that this pressure directly subtracts from the applied force in the Griffith criterion to give

$$\left(F - F_y\right)/b = (WEd \cdot d/\pi c)^{1/2}, \tag{10}$$

where F_y/bd is the bridging pressure.

Additional crack closing forces are more complex [16]. Consider the example of Fig. 8, in which a force F_1 stretches a lap-shear joint while an Obreimoff force F_2 is applied to the free arm of the crack. Adding the two energy terms leads to the crack criterion

$$\left(F_1^2/b^2 + F_2^2 \cdot 24c^2/d^2\right)^{1/2} = (2WEd)^{1/2}. \tag{11}$$

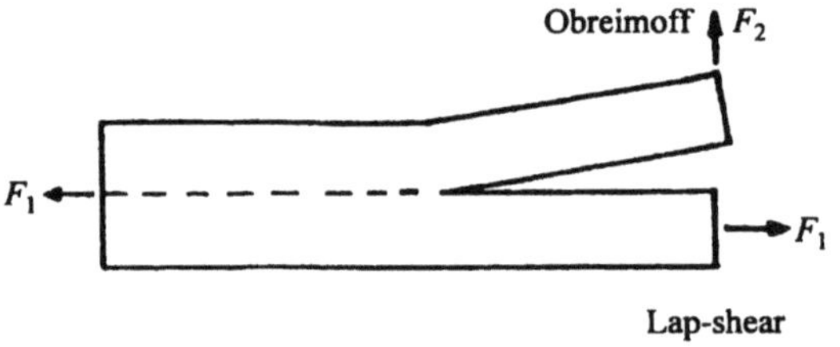

Figure 8. Lap-shear crack with an Obreimoff force applied to the free arm of the crack.

This equation is surprising to many engineers because it demonstrates once again that stress does not drive cracks. The crack speeds up whether the Obreimoff force is opening or closing the crack. Experimentally, this was demonstrated using adhering rubber strips 10 mm wide. F_1 was set at 2 N, driving the crack at 60 μm s^{-1}. When an Obreimoff force of 0.021 N was applied, the crack speed rose to 170 μm s^{-1}. If the Obreimoff force was then reversed to −0.021 N, the crack speed again rose to 170 μm s^{-1}. This experiment only works if the crack surfaces do not touch and heal. When healing occurs, then the crack slows down [16].

8.2. Increase in adhesive energy

Gent and his colleagues [17, 18] studied the problem of adhesive energy, showing the strong crack speed and temperature dependence. Results like those in Fig. 2 were explained in terms of viscoelastic losses in the rubber. However, it seems likely that similar effects could also stem from molecular dissipation at the interface [13]. The precise mix of interface and bulk losses is often difficult to estimate.

An interesting mechanism for increasing crack resistance is crack deflection by weak interfaces. A crack can be deflected along a different path if the adhesive energy is sufficiently low. To restart the original crack then requires further energy, thereby increasing resistance to fracture. This deflection mechanism was first suggested by Cook and Gordon [19], who proposed that interface strength was the important parameter. Kendall [20] showed that this was incorrect and demonstrated that an energy criterion was necessary. This was

$$W_1/W_2 > K, \tag{12}$$

where W_1 was the work of adhesion for the original interface, W_2 was the work of adhesion for the deflecting interface, and K was a constant, around 10, depending on the precise geometry.

8.3. Increase in elastic resistance

It is evident from the criteria of Table 1 that the elastic linkage is easier if the elastic modulus is low. Thus, making the Griffith or Obreimoff samples stiffer makes them more resistant to cracks. This is not true of the direct linkage, such as peeling, where stiffening has no effect.

By the same token, elastic energy originating from shrinkage or pre-stressing can also affect cracks. This is rather easy to demonstrate with the lap joint, which may be strengthened by up to a factor of 3 by properly pre-stretching the components before adhering them [21]. In contrast, peel joints are always weakened by residual strain energy.

Perhaps the most intriguing effect of elasticity on cracks is demonstrated by increasing the stiffness of a peeling film. Although the elastic modulus does not enter the cracking criterion $F/b = W$ for a uniform peeling strip, a change in elastic modulus does affect cracking. This was demonstrated by stiffening part of a rubber strip peeled from a glass plate [22]. The peeling speed was constant for the stiffened strip and was the same for the unstiffened film, but where the stiffness changed, the crack speed also changed. Using the energy balance theory it was shown that the change in elastic modulus from E_1 to E_2 at the crack tip produced an effective cracking criterion of

$$F/b = WE_2/E_1. \qquad (13)$$

9. CONCLUSIONS

The energy balance theory of fracture gives a good description of cracks at adhesive interfaces formed by bringing optically smooth rubber surfaces into contact. Cracking criteria give the adhesion force F in terms of W, the work of adhesion at the interface. In general, the criterion is a quadratic equation because the potential energy linkage depends on F, whereas the elastic linkage depends on F^2. Stress does not normally enter the cracking criterion. This theory allows many fascinating secondary mechanisms to be evaluated, including friction effects, viscoelastic relaxation, crack bridging, combined loading, crack deflection, residual stress, and change in stiffness.

REFERENCES

1. A. A. Griffith, *Philos. Trans. R. Soc. London, Ser. A* **221**, 163–198 (1920).
2. A. A. Atkins, in: *The Energetic of Fracture*, A. A. Atkins (Ed.), pp. 1–3. The Griffith Centenary Meeting, Inst. Materials. London (1993).
3. R. S. Rivlin, *Paint Technol.* **9**, 215–216 (1944).
4. K. Kendall and K. N. G. Fuller, *J. Phys. D* **20**, 1596–1600 (1987).
5. A. D. Roberts, *Eng. Mater. Des.* **11**, 579–582 (1968).
6. M. Barquins and D. Maugis, *J. Adhesion* **13**, 53–65 (1981).
7. K. Kendall, *Wear* **33**, 351–358 (1975).
8. A. A. Griffith, in: *Proc. 1st Int. Congr. Appl. Phys. Delft*, pp. 55–63. Delft Technische Boekhandel (1924).
9. G. R. Irwin, *J. Appl. Mech.* **24**, 361–364 (1957).
10. J. W. Obreimoff, *Proc. R. Soc. London, Ser. A* **127**, 290–297 (1930).
11. K. Kendall, *J. Phys. D* **8**, 1449–1452 (1975).
12. K. Kendall, *J. Phys. D* **8**, 512–522 (1975).
13. K. Kendall, *J. Adhesion* **5**, 179–202 (1973).
14. K. Kendall, *J. Adhesion* **7**, 55–72 (1974).
15. K. Kendall, *J. Mater. Sci. Lett.* **10**, 671–674 (1991).
16. K. Kendall, *Philos. Mag.* **A43**, 713–729 (1981).

17. A. N. Gent and R. P. Petrich, *Proc. R. Soc. London, Ser. A* **310**, 433–448 (1969).
18. A. N. Gent and A. J. Kinloch, *J. Polym. Sci.* **9**, 659–668 (1971).
19. J. Cook and J. E. Gordon, *Proc. R. Soc. London, Ser. A* **282**, 508–520 (1964).
20. K. Kendall, *Proc. R. Soc. London, Ser. A* **344**, 287–302 (1975).
21. K. Kendall, *J. Phys. D* **8**, 1722–1732 (1975).
22. K. Kendall, *Proc. R. Soc. London, Ser. A* **341**, 409–428 (1975).

Fundamentals of Adhesion and Interfaces, pp. 261–279
D. S. Rimai, L. P. DeMejo and K. L. Mittal (Eds)

Relationship between adhesion and friction forces

JACOB N. ISRAELACHVILI,* YOU-LUNG CHEN and HISAE YOSHIZAWA

Department of Chemical & Nuclear Engineering, and Materials Department, University of California, Santa Barbara, CA 93106, USA

Revised version received 2 March 1994

Abstract—The Surface Forces Apparatus technique was used to measure the normal (perpendicular) and lateral forces between variously prepared surfaces under both dry and lubricated conditions. 'Normal' forces include the force *vs* distance functions, F(D), for surfaces separated by thin liquid films as well as the adhesion forces and energies, γ, for two surfaces in adhesional contact. 'Lateral' forces include the static and kinetic friction forces $\mathcal{F}$ of the surfaces as they slide past each other at a given separation, D. The results show that very thin liquid films confined between two solid surfaces can sustain both normal and shear forces or stresses. The results further indicate that the normal force, F(D) or γ, may be directly related to the *static* friction force, $\mathcal{F}_s$, and simple equations are proposed that relate these forces (by 'static' friction force is meant the lateral force that must be applied to initiate motion, but not necessarily to maintain this motion). In contrast, the *kinetic* friction force, $\mathcal{F}_k$, which is the force that must be continually applied to maintain motion at a given velocity, was found to be related, not to the equilibrium or *reversible* interaction but to the dissipative or *irreversible* part of the adhesion or interaction energy during a loading–unloading cycle. There is a high degree of correlation in the way that normal forces and friction forces are affected by changes in applied load or pressure, sliding velocity, loading–unloading rates and temperature. These systematic correlations can be conveniently represented by non-equilibrium 'adhesion' and 'friction' phase diagrams.

Keywords: Adhesion; friction; surface forces.

1. INTRODUCTION

A high adhesion force may be expected to produce a high friction force, but experience shows that this is not always the case and the question has remained unresolved despite some three centuries of interest. Recent experimental and theoretical developments [1] have suggested that friction may be correlated more with the *irreversible* component of the interaction or adhesion energy during a loading/unloading cycle, and this is the focus of the present paper.

The connection between adhesion hysteresis and friction becomes readily apparent once the role of time (or rate) in adhesion is fully appreciated. Referring to Fig. 1,

*To whom correspondence should be addressed.

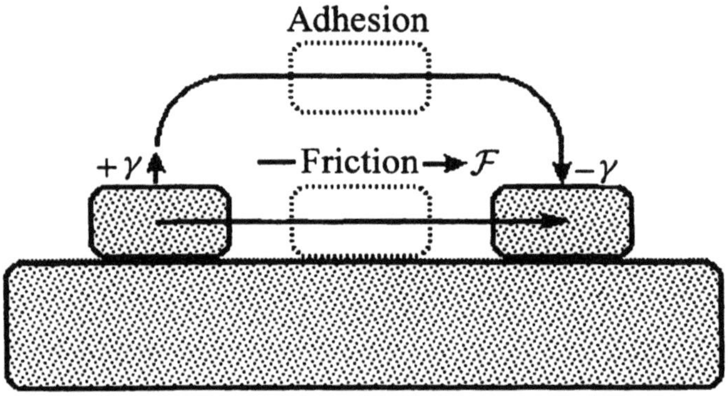

Figure 1. For an adhesion process to be thermodynamically reversible requires that the work energy $+\gamma$ put into the system to separate the surfaces is completely recovered on bringing them into contact again. This, however, does not require that the magnitude of the surface energy (γ) or adhesion force (F_a or L_a) be zero. In contrast, if the two surfaces slide to the same position, the kinetic friction force $\mathcal{F}_k$ must be zero everywhere if the process is to be thermodynamically reversible.

friction involves sliding one surface laterally, as shown. But the same result may be achieved by first separating the two surfaces, which now makes it an adhesion (unloading–loading) cycle. If the adhesion cycle is done at equilibrium, since the initial and final states have the same energy, the *net* energy change would be zero and *independent of the path taken.* This includes the sliding path. Thus, a zero friction force ($\mathcal{F}_k$) should be associated with zero adhesion *hysteresis*, and not with a zero adhesion force or energy. Conversely, a finite adhesion force/energy does not imply anything for the kinetic friction force, which could be large or zero. As discussed below, the situation is different for the static friction force, $\mathcal{F}_s$, which may be related to the equilibrium adhesion interaction.

Recent experimental and theoretical studies on the forces and friction of molecularly thin liquid films (generally films thinner than 5–10 molecular diameters) have revealed that such confined liquids have very different properties from their bulk properties [1–7]: they can support high normal loads or pressures (hundreds of atmospheres), and they can also have a finite shear stress or yield point before they can be sheared.

The cause for this 'solidlike' behavior of thin liquid films may be found by considering how their structure is affected when they are confined between two hard but 'structured' surfaces. The forces between two such surfaces across the liquid is generally oscillatory (Fig. 2), with a periodicity equal to the diameters of the liquid molecules [8]. This indicates that the molecules order into quasi-discrete layers between the surfaces which become harder to squeeze out as the two surfaces are pressed closer together (Fig. 3A). The rheological properties of ultrathin liquid films are also unusual in that they exhibit a critical shear stress and can exhibit stick-slip motion [2], neither of which can be simply described in terms of a shear viscosity, even a modified one, since the term *viscosity* implies that shear motion will occur under the smallest shear stress, which is not the case for these molecularly thin films.

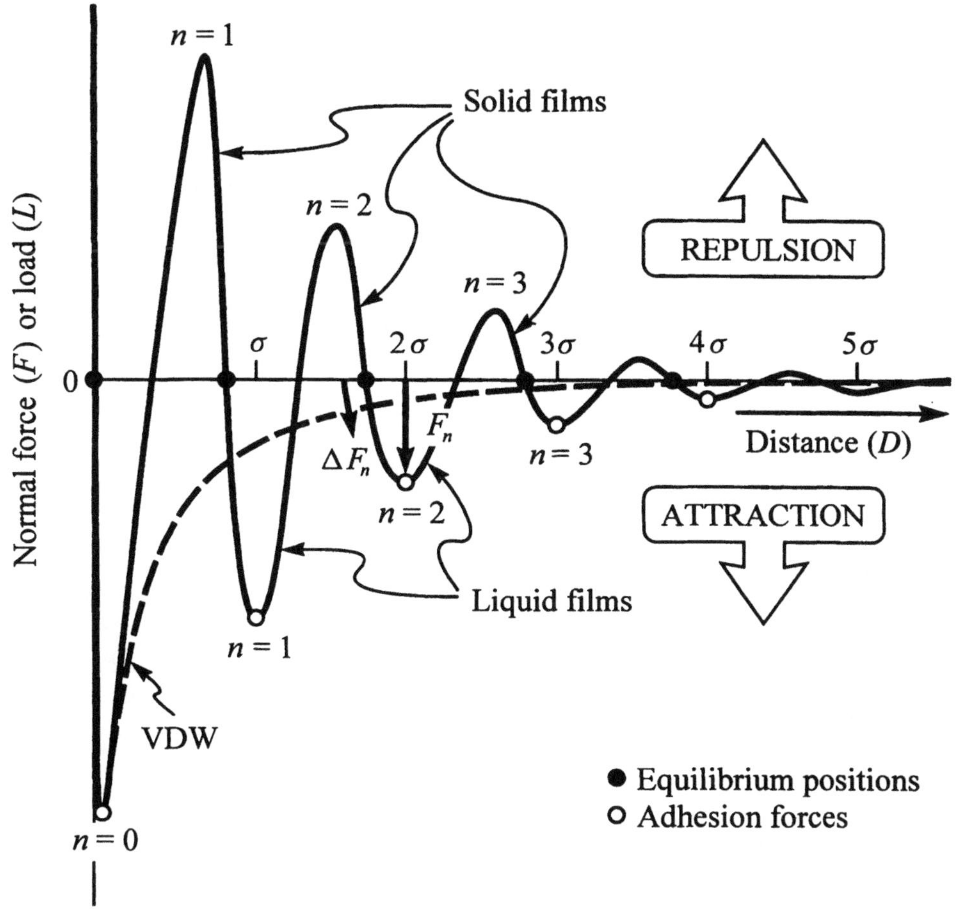

Figure 2. Typical oscillatory force-law measured between two smooth solid surfaces across simple spherical and chain molecules [8, 9]. Under the right conditions of surface-fluid epitaxy, the film be forced to solidity (freeze) and remelt over certain narrow distance regimes within each oscillation, as shown schematically in Fig. 3A. Two surfaces will be in stable mechanical equilibrium at the positions located by the black circles where $F_n = 0$ and where the slope of the force curve is negative. Total separation to an infinite separation will occur if the surfaces are brought to the positions indicated by the white circles, where $-F_n$ is then the 'adhesion' force for that value of n. In contrast, for one surface to slide over the other, it does not have to be separated to infinity, but only by a fraction of a layer thickness, d ($d < D$), requiring the application of only a part of the total adhesion force, ΔF_n ($\Delta F_n < F_n$).

On the theoretical side, recent computer simulations have successfully reproduced some of the results of direct force and friction measurements [1, 5, 6]. They have also provided an intuitive understanding of both the oscillatory force-laws and the yield points of thin films in terms of the induced solidification of the molecules between two surfaces. When confined between two hard, crystalline (and therefore structured) surfaces, the liquid molecules within such films can be forced to epitaxially order and solidify or freeze. Under the right conditions, a first-order phase transition takes place from the liquid state to a solidlike or liquid-crystalline state. Thus, each oscillation of an oscillatory force-function may be considered to represent

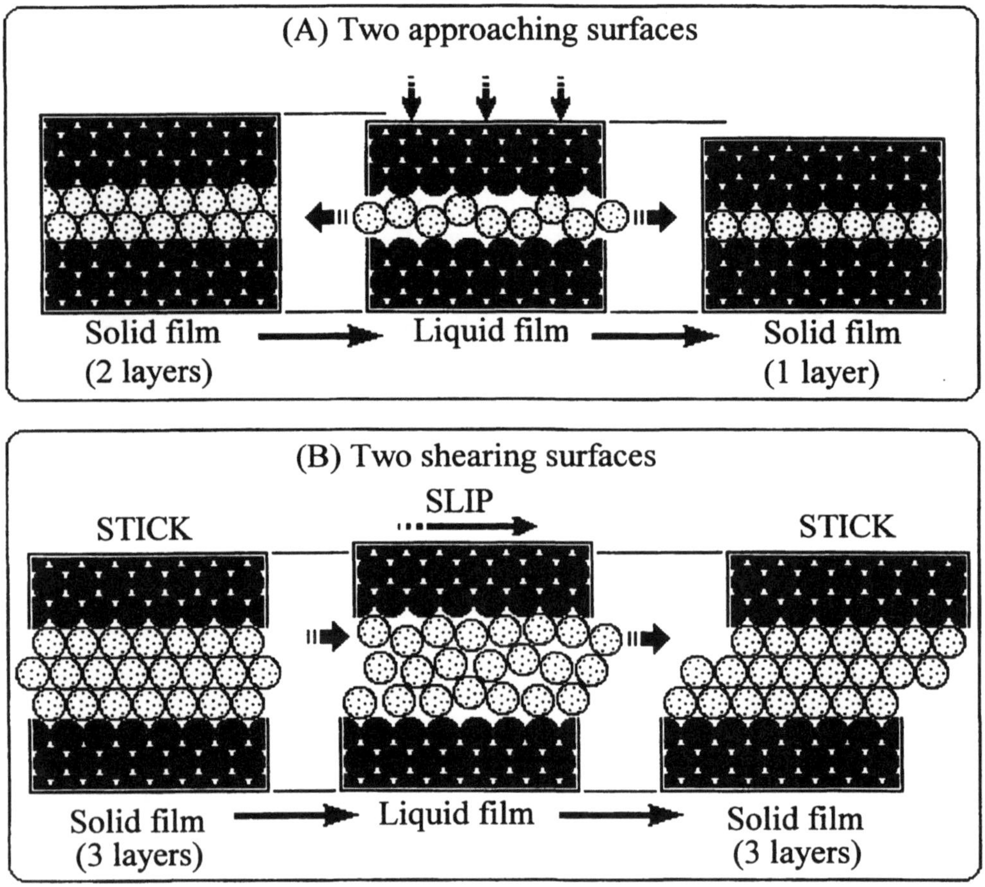

Figure 3. Liquid-to-solid transitions during the approach (A) and stick-slip sliding (B) of two surfaces.

a freezing-melting cycle, as depicted in Fig. 3A. Likewise, stick-slip sliding may also be considered to reflect periodic transitions between the solidlike and liquidlike states of a shearing film, corresponding to the sticking and slipping regimes, respectively (Fig. 3B).

Here, we report the results of force and friction measurements of surfactant-monolayer coated mica surfaces in air (dry, unlubricated surfaces), and of untreated mica surfaces separated by a thin liquid film of tetradecane (lubricated surfaces). A schematic of the contact region is shown in Fig. 4. During each experiment, both the normal and lateral forces were measured on the same pair of surfaces. We also attempt to relate the measured friction forces to the adhesion forces — in the case of the surfactant-coated surfaces, and to the intersurface force-function — in the case of liquid films. Any correlations, if they exist, would be of both fundamental and practical interest. While such data on forces and friction already exist in the literature [1, 2, 9–11], a quantitative comparison could not previously be made because all the earlier measurements were done on different pairs of surfaces, aligned at different crystallographic

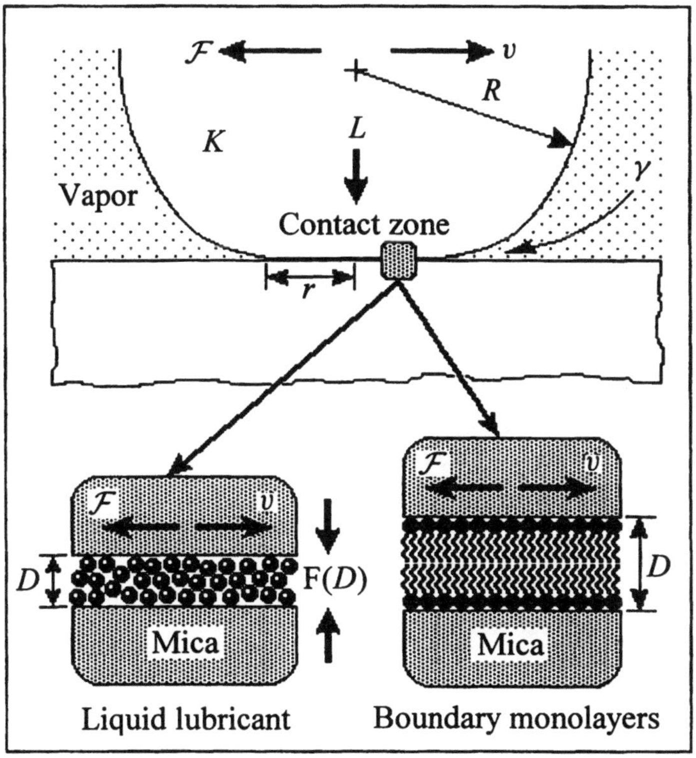

Figure 4. Schematic geometry of contact region of two surfaces in a Surface Forces Apparatus (SFA) during adhesion and friction measurements. R is the initial undeformed radius of the surface (which models an asperity), L is the externally applied load, F(D) is the force-function between the surfaces, $\mathcal{F}$ is the friction force, v is the sliding velocity, r is the contact radius (contact area: $A = \pi r^2$), D is the thickness of the interfacial film which may be a liquid or lubricant film or two surfactant monolayers, γ is the equilibrium surface or adhesion energy per unit area, and K is the elastic modulus of the surfaces or the stiffness of the force-measuring spring. In this paper we discuss the various relationships between r and L, F and D, $\mathcal{F}$ and γ, and $\mathcal{F}$ and L.

angles ('twist angles') relative to each other which is known to affect both of these forces [12].

2. EXPERIMENTAL

2.1. Materials

A variety of surfactant monolayer-coated mica surfaces were prepared either by Langmuir–Blodgett deposition or by self-assembly, as previously described [13]. The surfactants were either ionic or zwitterionic (dipolar), and both single-chained and double-chained surfactants were used [13]. Liquid tetradecane (n-$C_{14}H_{30}$) of molecular length $\sim$ 18 Å and width $\sim$ 4 Å, and n-hexadecane (n-$C_{16}H_{34}$) of molecular length $\sim$ 20 Å and width $\sim$ 4 Å were chosen for detailed study because their interaction forces have been extensively studied [8, 9], and also because they are common constituents of lubricating oils. The liquids, from *Fisher Scientific*, were distilled once before use.

2.2. Force and friction measurements

The Surface Forces Apparatus (SFA) used in these studies was a standard SFA Mark II with a mechanical motor-driven sliding attachment and resistance strain gauge detectors for making friction measurements [2, 14]. Once the two mica surfaces were installed into the apparatus, the chamber was purged with dry nitrogen gas for 12 hours. Some P_2O_5 was also placed inside the sealed chamber to keep the internal atmosphere completely dry at all times. All friction measurements were made with a small but macroscopic droplet (vol. $\sim$ 0.1 ml) of hexadecane or tetradecane injected between the surfaces. When brought together under a large force, the two surfaces would flatten elastically to produce a flat, circular contact zone of radius r (Fig. 4). During the sliding, the following parameters were measured as previously described [13, 14, 18, 23]: normal load L, film thickness D, contact area $A = \pi r^2$, sliding velocity v, and friction force $\mathcal{F}$. Friction traces, giving the friction as a function of time, were recorded on a chart recorder and the optical fringe pattern of the contacting surfaces was recorded on a video recorder for later analysis.

2.3. Adhesion measurements

Adhesion energies were determined by fitting the simultaneously measured contact radius r and applied load L, defined in Fig. 4, to the Johnson–Kendall–Roberts (JKR) equation [8, 13, 15]:

$$r^3 = \frac{R}{K}\left[L + 6\pi R\gamma + \sqrt{12\pi R\gamma L + (6\pi R\gamma)^2}\right], \tag{1}$$

from which one can obtain the loading (advancing) or unloading (receding) energies, γ_A and γ_R, depending on whether the load (or contact area) is being increased or decreased. Receding adhesion energies could also be measured directly from the adhesion force or 'pull-off' force L_a or F_a needed to separate two surfaces from contact from which γ_R may be determined from the equation [13, 15]

$$\gamma_R = -L_a/3\pi R. \tag{2}$$

3. THEORETICAL BACKGROUND

3.1. The Cobblestone Model of interfacial friction

To put the experimental procedure and subsequent data analysis into perspective, we here review the results of recent experiments on, and theoretical modeling of,

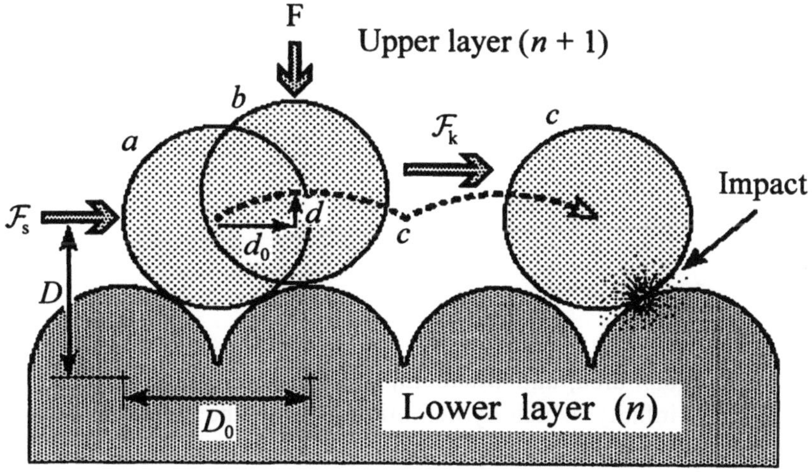

Figure 5. Molecular events occurring when the molecules of the upper surfaces or $(n+1)$-th layer slide over the molecules of the lower surface or n-th layer. This model is based on the Cobblestone Model of boundary friction [16, 17] which is similar to the original Coulomb of Interlocking Asperity models for the friction of rough, worn surfaces.

adhesion forces, **oscillatory structural** forces and **friction** forces. This can be discussed within the framework of the Cobblestone Model of interfacial and boundary friction, which is conceptually similar to the Coulomb and Interlocking Asperity Models except that the former applies to the friction of smooth, undamaged surfaces whereas the latter models apply to the shearing of rough, usually worn, surfaces [16, 17].

Consider two structured surfaces that are initially in adhesional contact. When at rest, the surface molecules will adjust themselves to fit snugly together (Fig. 5, position *a*), in an analogous manner to the self-positioning of cartwheels on a cobblestone road. A small tangential force $\mathcal{F}$ applied to the top surface will initially displace it elastically both horizontally and vertically by distances much smaller than the lattice dimensions, D_0 and D, respectively. To initiate sliding, these displacements must reach some critical values, say d_0 and d (Fig. 5, position *b*). Note that these lateral and vertical displacements must be uniquely related to each other via the geometry of the two surface lattices, their mutual crystallographic orientation (twist angle), and the direction of sliding.

3.2. The static friction force, $\mathcal{F}_s$

Let $\mathcal{F}_s$ be the lateral force that must be applied to initiate motion, that is, to bring the top layer of atoms from position *a* to position *b* (after which they will fall spontaneously to the next lattice position *c* and perhaps continue to move further — discussed later). To bring the upper of molecules to *b* requires that a certain fraction of the van der Waals adhesion energy ΔE_{ad} between the two surfaces be overcome when the applied force $\mathcal{F}_s$ acts over the distance d_0. This energy change goes into lifting the two surfaces from a separation D to a separation $D + d$, and may conveniently be expressed in terms of the Hamaker constant, A, as follows [8]

$$\Delta E_{\mathrm{ad}} = \mathrm{E}(D) - \mathrm{E}(D+d) = \frac{A}{12\pi}\left[\frac{1}{D^2} - \frac{1}{(D+d)^2}\right] = \frac{A}{12\pi D^2}\left[1 - \frac{D^2}{(D+d)^2}\right]$$

$$\approx \frac{2E_0 d}{D} \approx \frac{4\gamma d}{D} \quad \text{per unit area (for } d \ll D\text{)},$$

i.e.

$$\Delta E_{\mathrm{ad}} \approx \frac{2AE_0 d}{D} \approx \frac{4A\gamma d}{D}, \tag{3}$$

where A is the contact area, and where $E_0 = 2\gamma$ is the energy needed to separate the two surfaces to infinity, which by definition is twice the surface energy, γ. Note that since we expect $d \ll D$ the above expression basically states that the adhesion energy that must be overcome is some fraction of the total adhesion energy. If the surfaces interact through some other type of interaction than van der Waals, equation (3) would be modified, but in general, the energy should always be expressible in the form

$$\Delta E_{\mathrm{ad}} = \frac{2\phi E_0 A d}{D} = \frac{4\phi\gamma A d}{D}, \quad \text{where } \phi \approx 1.$$

In addition to overcoming the adhesion energy, the top surface must also move against the externally applied load L, which adds an additional term to the energy required given by $\Delta E_{\mathrm{load}} = L \times d$.

Thus, the frictional energy which goes into the system, $\Delta E_{\mathrm{fr}} = \mathcal{F}_{\mathrm{s}} \times d_0$, is used partly to overcome the adhesion force and partly to overcome the load. The net energy balance therefore gives

$$\Delta E_{\mathrm{fr}} = \mathcal{F}_{\mathrm{s}} \times d_0 = \Delta E_{\mathrm{ad}} + \Delta E_{\mathrm{load}},$$

or

$$\mathcal{F}_{\mathrm{s}} = \frac{2E_0 A d}{D d_0} + \frac{L d}{d_0}. \tag{4}$$

Equation (4) applies only to the static friction force, $\mathcal{F}_{\mathrm{s}}$, the force needed to initiate motion, but not necessarily to sustain it.

3.3. Kinetic friction force, $\mathcal{F}_{\mathrm{k}}$

During continuous sliding, friction energy is continually dissipated as heat into the lattice, increasing the kinetic energy of the molecules surrounding the shearing interface. If the atomic collisions between the individual atoms were perfectly elastic in the sense that all the kinetic energy is 'transmitted' during the impact of the top atom with a bottom atom (cf. Fig. 5, position c) the surfaces would require that the friction force be maintained at $\mathcal{F}_{\mathrm{s}}$ to continue moving. However, if — as is generally the case — only a fraction ε ($\varepsilon < 1$) of the energy gained is transmitted (absorbed or dissipated) by the lattice each time the surface moves by one lattice dimension,

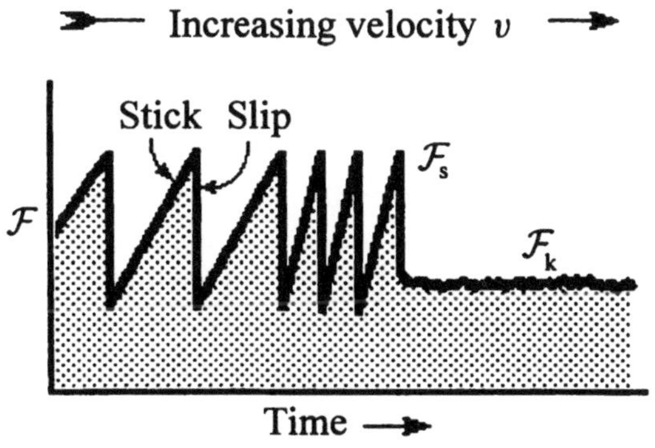

Figure 6. Typical friction trace obtained for simple liquid films shearing between two mica surfaces [2]. On changing the load, both $\mathcal{F}_s$ and $\mathcal{F}_k$ generally change by similar relative amounts, i.e. the ratio $\mathcal{F}_k/\mathcal{F}_s$ remains much the same. With increasing sliding velocity, v, the frequency of the stick-slip increases and the stick-slip amplitude decreases slightly until, above some critical velocity, the stick-slip disappears altogether and sliding proceeds smoothly in the liquidlike, kinetic state [2, 5, 18].

sliding could be maintained at a lower force (Fig. 6). This is the *kinetic* friction force, given by $\mathcal{F}_k = \varepsilon\mathcal{F}_s$. We may therefore write the following general equation for both static and kinetic friction forces

$$\mathcal{F} = \varepsilon\left(\frac{2E_0A}{D} + L\right)\frac{d}{d_0}, \tag{5}$$

from which we also obtain the critical shear stress

$$S_c = \frac{\mathcal{F}}{A} = \varepsilon\left(\frac{2E_0}{D} + P\right)\frac{d}{d_0}, \tag{6}$$

where

$\varepsilon = 1$ for the *static* friction force, $\mathcal{F}_s$,

$\varepsilon < 1$ for the *kinetic* friction force, $\mathcal{F}_k$,

and where A is the area of contact, and P is the externally applied pressure.

The static and kinetic friction forces are therefore expected to bear a roughly constant ratio to each other, independent of load, as often observed [2], and this ratio may be seen as a measure of the fraction of energy that is dissipated as heat during atomic collisions at the shearing interface.

In many cases, the increased friction force $\mathcal{F}$ with load L comes, not from the second term in equation (5), but from the first term due to the increase in the contact area $A = \pi r^2$ with L, as given by the JKR equation, equation (1).

As a test of the above equations, consider a typical hydrocarbon or a van der Waals surface for which $E_0 = 2\gamma \approx 50 \times 10^{-3}$ J/m^2. Other typical values would be: $d \approx 0.5$ Å, $D \approx 2$ Å, $d_0 \approx 1/2D \approx 1$ Å, and $\varepsilon \approx 0.1$–1.0 for kinetic sliding. Using these parameters, equation (6) predicts: $S_c \approx (2$–$20) \times 10^7$ N/m^2. This range of values compares very well with typical experimental values of $\sim 2 \times 10^7$ N/m^2 for hydrocarbon and mica surfaces sliding in air [16, 17]. It appears, therefore, that in principle one should expect to be able to calculate the static friction force from the equilibrium adhesion force or energy, together with a knowledge of the surface lattice structure, the sliding direction, etc. However, this does not appear to be the case with the kinetic friction force, which requires the additional knowledge of the fraction of energy ε dissipated per molecular collision. At present, this cannot be obtained from a knowledge of the equilibrium properties of the materials and surfaces. The calculation of ε is clearly not a simple matter, and sophisticated computer simulations appear to be required to properly model and understand dynamic friction.

Equation (5) shows that the friction of two molecularly smooth surfaces can be described by a general equation of the form

$$\mathcal{F} = C_1 A + C_2 L,$$

where C_1 and C_2 are constants. The first term may be considered as the intrinsic 'adhesion' contribution to the friction force, the second as the load-dependent 'structural' contribution, though both terms depend on the structure of the two surface lattices. In the more common case of rough or worn surfaces, their area of true molecular contact is very small and the adhesion contribution to the total friction force is also usually negligibly small (as manifested by a zero friction at zero load). This leaves the load-dependent structural term, which for rough surfaces takes the form known as Amontons' Law: $\mathcal{F} = \mu L$, where μ is the coefficient of friction. It is noteworthy that this is also the dependence expected from equation (5), which in the absence of the adhesion term simplifies to $\mathcal{F} = (\varepsilon d/d_0)L$, where d/d_0 is defined in Fig. 5. When compared with Amontons' equation, we find that the friction coefficient μ can be equated with the ratio of the vertical and lateral asperity dimensions, d and d_0, respectively. This is consistent with the original Coulomb and Interlocking Asperity Models of friction. However, even though there is always an Amontons' type load-dependent contribution to any friction force, the surface properties that determine the constants C_2 and μ are quite different in the case of molecularly smooth and rough surfaces.

3.4. Friction of lubricated surfaces

The above model strictly applies to dry, unlubricated surfaces, but it may be extended, at least semi-quantitatively, to lubricated sliding, where a thin liquid film is present between the surfaces. If the normal force-profile between the surfaces is oscillatory, as depicted in Fig. 2, it is clearly possible to move the $(n+1)$-th layer laterally relative to the n-th layer, in much the same way as for unlubricated surfaces. Referring to Figs 2 and 6, with an increase in the number of layers, E_0 is replaced by E_n (the adhesion energy minimum of the n-th layer), the film thickness D increases, and d decreases as the film becomes more accommodating to shear displacement.

3.5. Friction force in relation to normal force-law across a liquid film

First we note that the energy of the n-th minimum E_n may be related to the measured force-function F(D) between a curved surface of radius R and a flat surface (the usual geometry of SFA, STM, AFM and pin-on-disc measurements) by the Derjaguin approximation [22, 8]:

$$E_n = F_n/2\pi R. \tag{7}$$

The friction force of a liquid film with n liquid and thickness D may be obtained by substituting the above equation into equation (5):

$$\mathcal{F}_n = \varepsilon\left(\frac{AF_n}{\pi DR} + L\right)\frac{d}{d_0} \qquad (0 < \varepsilon < 1), \tag{8}$$

where all the parameters now refer to the n-th layer, for example, where D is the n-layer film thickness so that $D \approx nD_0$. Equation (8) gives the friction force of a liquid film containing n molecular layers in terms of the n-th adhesion force minimum, F_n or F_n/R, as measured in a typical surface force experiment. F_n is in terms of measurable quantities, except for d which may be estimated to be of the order of 1 Å or less. Further, the contact area, A, can be related to the adhesion energy and applied load via equation (1). We return to test the above equations in the Results and Discussion Section.

4. RESULTS AND DISCUSSION

4.1. Adhesion and friction of unlubricated surfaces

The equilibrium (thermodynamically reversible) adhesion energy for all the surfactant monolayer surfaces studied did not vary much outside the range expected for hydrocarbon surfaces, viz. $\gamma \approx 25\ \mathrm{mJ/m^2}$. However, both the adhesion hysteresis and friction forces varied greatly, depending on the type of monolayer and the experimental conditions. In addition, these variations were found to be closely correlated: larger friction forces occurring between surfaces that exhibit larger adhesion hysteresis during a loading and unloading cycle. An example of this effect is given in Fig. 7 for physisorbed monolayers of calcium alkyl benzene sulphonate which are in the fluidlike state at room temperature because they contain branched chains [21], but which become even more liquidlike when short-chained hydrocarbon molecules penetrate into them from vapor or bulk liquid [13, 21].

4.2. Dynamic phase diagram representation

Both the friction and adhesion hysteresis were found to vary non-linearly with temperature [13], generally peaking at some particular temperature, T_0. Thus, the adhesion hysteresis and friction forces could be conveniently represented by a 'non-equilibrium' or 'dynamic' phase diagram (Fig. 8). It was also found that T_0, and the whole bell-shaped curve, are systematically shifted along the temperature-axis (as well as in the vertical direction) when the load, sliding velocity, etc. are varied, as illustrated in the figure.

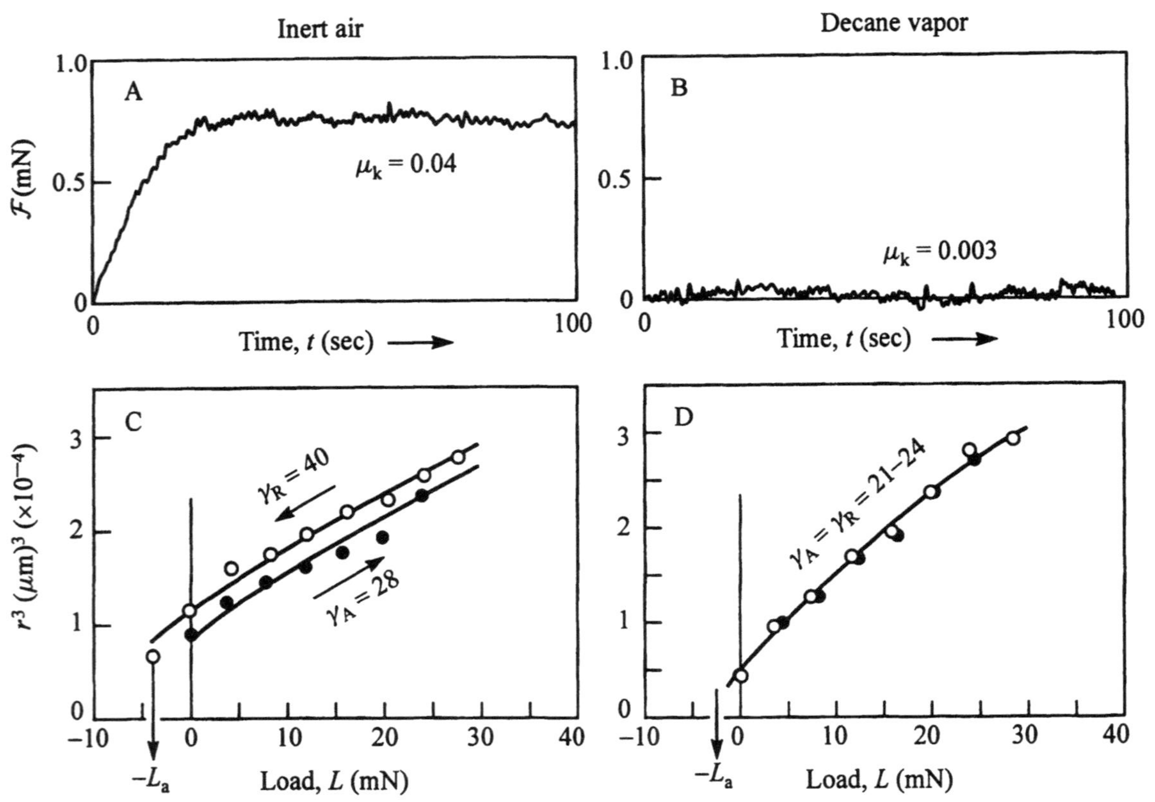

Figure 7. Top: Friction traces for two fluidlike monolayer-coated surfaces at 25 °C showing that the friction force is much higher between dry monolayers (A) than between monolayers whose fluidity has been enhanced by hydrocarbon penetration from vapor (B). Bottom: Contact radius versus load ($r^3 - L$) curves measured for the same two surfaces as above and fitted to the JKR equation (1) — shown by the solid lines. For dry monolayers (C) the adhesion energy on unloading ($\gamma_R = 40$ mJ/m^2) is greater than that on loading ($\gamma_A = 28$ mJ/m^2), indicative of an adhesion energy hysteresis of $\Delta\gamma = \gamma_R - \gamma_A = 12$ mJ/m^2. For monolayers exposed to saturated decane vapor (D) their adhesion hysteresis is near zero ($\gamma_A \approx \gamma_R$), and both the laoding and unloading curves, as well as the pull-off adhesion force, L_a, are well-fitted by the thermodynamic value of the surface energy of fluid hydrocarbon chains, viz. $\gamma \approx 24$ mJ/m^2. (From [23].)

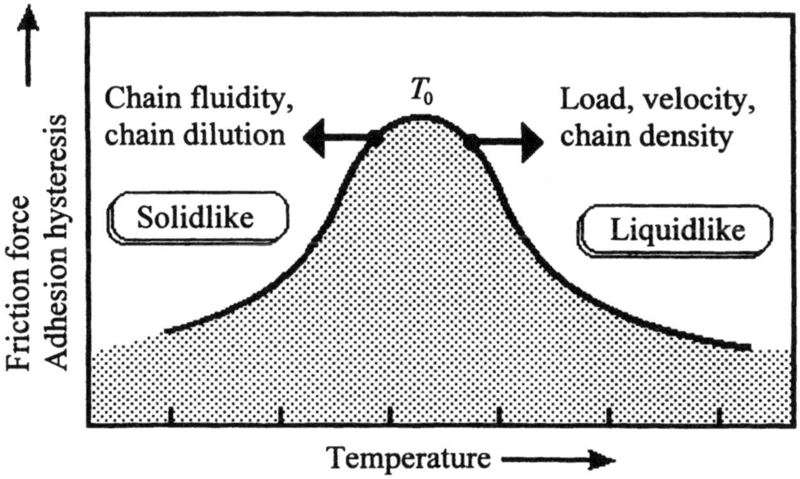

Figure 8. Schematic 'friction phase diagram' representing the trends observed in the friction forces of a variety of different surfactant monolayers [13, 18]. The characteristic bell-shaped curve also correlates with the monolayers' adhesion hysteresis, $\Delta\gamma$, but not with the absolute magnitude of the adhesion energy, γ. The arrows leading from the curve indicate the direction in which the whole curve is pulled when one of the experimental conditions, e.g. the load, is increased.

Other interesting correlations were also found; for example, both adhesion hysteresis and friction were found to change in the same way when the temperature, pressure or humidity are changed, or when the sliding speed and loading/unloading rates are changed. Such effects are also commonly observed in other energy dissipating phenomena such as polymer viscoelasticity [19] and it is likely that a similar physical mechanism is at the heart of all of these phenomena. A possible molecular process underlying the energy dissipation of chain molecules is illustrated in Fig. 9, which shows the three main 'dynamic' phase states of surface monolayers, A, B, and C, whose characteristic properties are suggested to be as follows [13, 18].

(A) *Solidlike chains:* There is little interpenetration of chains across the solidlike, crystalline interface and little energy is transferred (dissipated) between the two surfaces during adhesion or sliding. The adhesion hysteresis and friction forces are medium to low.

(B) *Amorphous chains:* Significant interdigitations occur across the interface with time both at rest (giving rise to adhesion energy hysteresis) and during sliding (giving rise to a large kinetic friction force).

(C) *Liquidlike chains:* The chains of the two monolayers are significantly interdigitated across the interface but, unless the sliding velocity is very high, the system is always close to equilibrium. The frictional force is again low (and similar to the viscous drag of a Newtonian liquid), and the adhesion hysteresis is also low (and similar to the negliglible difference in the surface tension of an expanding and a contracting liquid surface).

The above scenario should apply not only to the organic surfaces studied here, but also to other types of surfaces, including polymer and mineral oxide surfaces [18, 20].

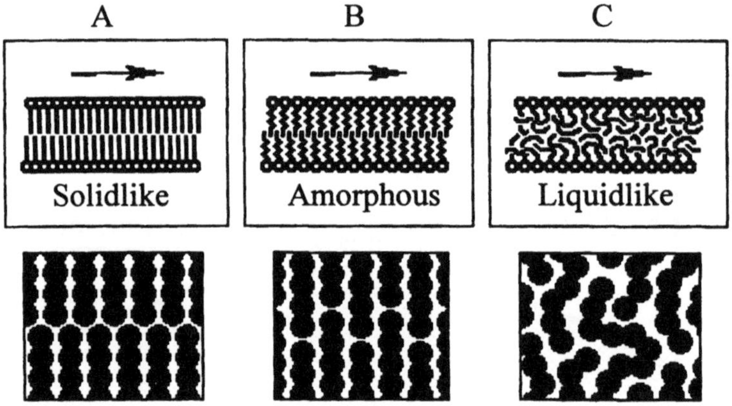

Figure 9. Different dynamic phase states of boundary monolayers during adhesional contact and/or frictional sliding. Low adhesion hysteresis and friction is exhibited by solidlike and liquidlike layers (A and C). High adhesion hysteresis and friction is exhibited by amorphous layers (B). Increasing the temperature generally shifts a system from the left to the right. Changing the load, sliding velocity and other experimental conditions can also change the 'dynamic phase state' of surface layers as shown in Fig. 8.

4.3. Quantitative comparison of adhesion energy hysteresis and friction forces

The following simple model shows how adhesion energy hysteresis and friction of surfaces such as shown in Fig. 7 may be expected to be quantitatively related. Let $\Delta\gamma = (\gamma_R - \gamma_A)$ be the adhesion energy hysteresis per unit area, as measured during a typical loading-unloading cycle (cf. Figs 7C, D). Now consider the same two surfaces sliding past each other and assume that frictional energy dissipation occurs through the same mechanism as adhesion energy dissipation, and that both occur over the same characteristic molecular length scale D_0. Thus, when the two surfaces (of contact area $A = \pi r^2$) move a distance D_0, equating the frictional energy $(\mathcal{F} \times D_0)$ to the dissipated adhesion energy $(\Delta\gamma \times A)$, we obtain for the kinetic friction force

$$\mathcal{F}_k = \frac{\Delta\gamma \times A}{D_0} = \frac{\pi r^2}{D_0}(\gamma_R - \gamma_A) \tag{9}$$

and

$$S_c = \frac{\mathcal{F}}{A} = \frac{\Delta\gamma}{D_0}, \tag{10}$$

which is the desired expression and which has been found to give order of magnitude agreement between measured friction forces and adhesion energy hysteresis [18, 20]. For example, in Fig. 7, we have $A \approx 100\ \mu m^2$, $\Delta\gamma \approx 10\ mJ/m^2$, and $D_0 \approx 1$ nm (assumed), which when put into equation (9) predicts that $\mathcal{F}_k \approx 1$ mN, which is close to the value measured (Fig. 7A).

An interesting result is obtained if we equate equation (9) with equation (5), since $4d/d_0 \approx 1$, and $D \approx D_0$, and we obtain the following relation

$$\varepsilon \approx \frac{\mathcal{F}_k}{\mathcal{F}_s} \approx \frac{\Delta\gamma}{\gamma}. \tag{11}$$

Thus, the fraction of the energy transferred between colliding molecules across an interface during sliding/shear motion is seen to determine not only the static–kinetic amplitude ratio, but also the adhesion energy hysteresis ratio. This is entirely consistent with the overall picture presented here, where the static friction is related to the equilibrium surface energy and where the kinetic friction depends on the dissipative part of that energy. It is important to appreciate, however, that the latter connection is meaningful only if we compare kinetic friction and adhesion hysteresis at the same rates, that is, equation (9) may be expected to hold only when the loading–unloading time during an adhesion cycle is similar to the time it takes two surfaces to slide past the same characteristic molecular length during a friction force measurement. Luckily, since neither effect is very sensitive to velocity or rate, the above equations may be applicable even when friction and adhesion forces are compared at quite different measuring rates.

4.4. Quantitative comparison of oscillatory force-functions and friction forces

Figure 10 shows the measured force–distance curve for *n*-tetradecane between two mica surfaces. The force-law is a decaying oscillatory function of distance arising from the transverse layering of the tetradecane molecules between the two surfaces. The force barriers seen at $D = 4,\ 8,\ 12,\ 16$ and 20 Å (i.e. multiples of 4 Å), indicate that the periodicity in the oscillations remains constant at 4 ± 1 Å. Given that the molecular width is close to 4 Å, it can be said that these chain molecules are induced to lie parallel to the surfaces, as shown in the inset of Fig. 10. The unusually large repulsive force barrier observed between the first layer ($n = 1$ at $D = 4$ Å) and contact ($n = 0$ at $D = 0$) indicates that the first layer of molecules is particularly difficult to squeeze out from between the two surfaces. Similar forces have previously been measured between mica surfaces across tetradecane and other alkane liquids [9].

Under sliding conditions, it is found that the surfaces move while remaining separated by a discrete number of molecular layers, that is, they prefer to slide while remaining in one of the potential energy minima. The small displacements, d, away from the equilibrium thickness that are expected to occur during sliding, if present, were less than 1 Å — the resolution of our measurements.

Figure 11 shows the effects of load on the tetradecane film thickness (Fig. 11A) and friction forces (Fig. 11B). By comparing the results of Figs 11 and 10 it is readily seen that both the normal forces and the friction forces are quantized with the number of layers n, and that both decrease in magnitude as n increases. We may note that this quantization is not as well-defined or sharp as in the case of spherical molecules such as cyclohexane and the silicone liquid OMCTS [2], because each chain molecule probably does not remain confined to a single layer but meanders between layers, as shown in the inset in Fig. 10. Such 'interlayer entanglements' would tend to smear out the discrete layering effects, especially in the case of dynamic interactions such as friction. In contrast, small, isotropically shaped spherical molecules can only occupy a single layer at any time, and the quantization effects are therefore expected to be more sharply defined.

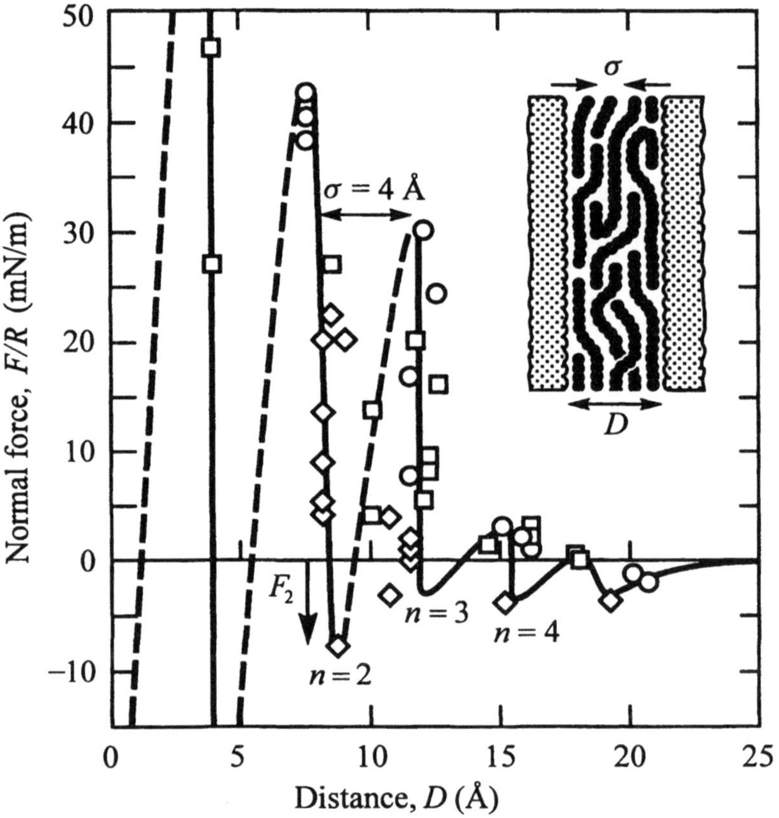

Figure 10. Measured oscillatory force-function between two mica surfaces across n-tetradecane at 19°C. Data points refer to three independent force runs. *Note:* The measured adhesion or 'pull-off' force, F_n or L_n, of the n-th minimum is related to the adhesion energy γ_n of that minimum via equation (2): $\gamma_n = -F_n/3\pi R$ which is accurate to better than 25% regardless of the value of the bulk elastic modulus, K, of the surface materials or the contact area, A, at pull-off [15, 8].

4.5. Quantitative comparison of oscillatory force-functions and friction forces

Concerning any quantitative relation between Figs 11 and 10, we may compare the magnitudes of the adhesion force F_n to the friction force $\mathcal{F}_n$ for $n = 2$ where the force and friction data appear to be most accurately measured. From Fig. 11B, for $n = 2$ the static friction force is given by: $\mathcal{F}_s = 7 + 0.5L$ (in units of mN). This is what equation (8) is expected to describe. Putting the following measured parameters into equation (8): $F_2/R = 8$ mN/m, $A = 3.7 \times 10^{-9}$ m^2, $D = 7.5$ Å, $d_0 = 2.0$ Å, $d \approx 1$ Å (assumed), and $\varepsilon = 1$ (for static friction), we obtain

$$\mathcal{F}_2 = \left(\frac{AF_2}{\pi DR} + L\right)\frac{d}{d_0} \approx (6 + 0.5L)\ \text{mN},$$

which is in excellent agreement with the fitted line of $(7 + 0.5L)$ mN, although this good agreement goes depend on an assumed value for d of 1 Å which is probably an upper bound. We also note from Fig. 11 that the ratio of the kinetic to the

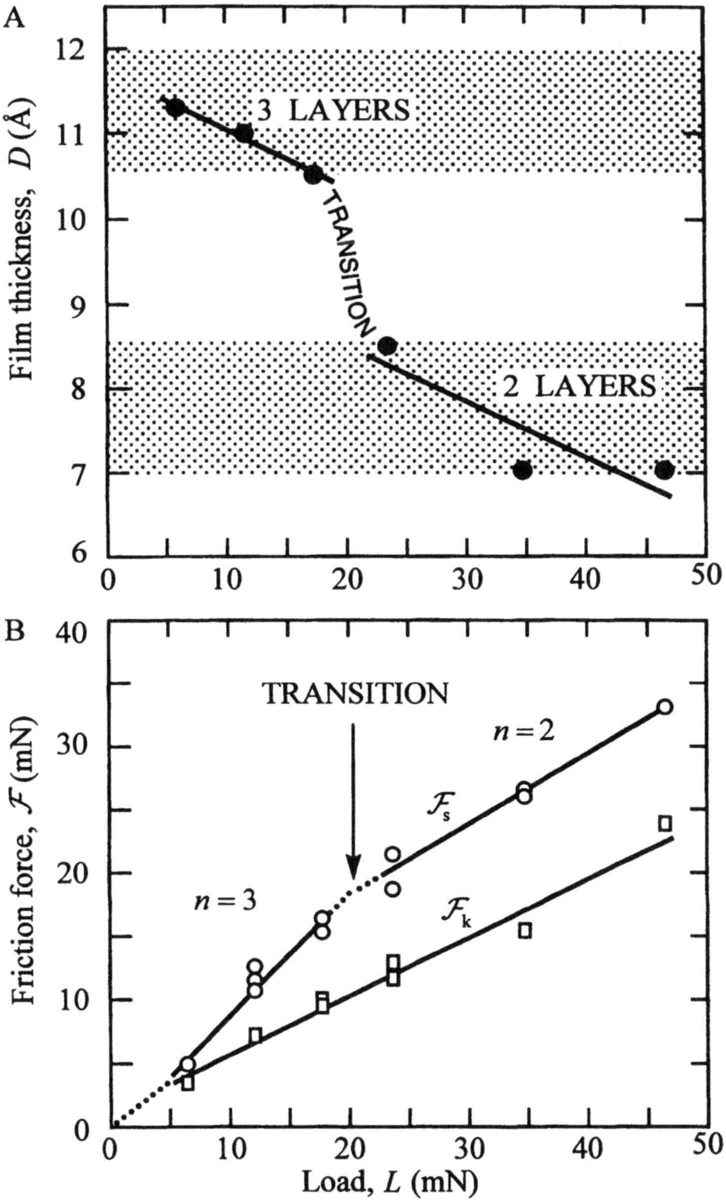

Figure 11. (A) Decreasing film thickness with increasing applied load around the transition from three to two layers occurring between ~ 11 Å and ~ 8 Å (see previous figure). (B) Static and kinetic friction forces for the above tetradecane film as a function of load.

static friction forces $\mathcal{F}_k/\mathcal{F}_s$ remained roughly constant at 0.6 as L changed, suggesting that $\varepsilon \approx 0.6$ in this case. All the above values are reasonable, and when taken together provide strong support for the notion that there is indeed a quantitative correspondence between the normal and lateral forces across simple liquid films.

Returning to the general equation, equation (8), it suggests that if the adhesion forces are strong and of short-range, i.e. if the oscillatory forces are of *large* amplitude and *small* periodicity, then the intrinsic adhesion contribution to the friction force — the first term in equation (8) — will dominate over the second term. In such cases, the friction force will depend more on n, the number of layers, than on the applied load. This is exactly what has been found for films of the quasi-spherical molecules of cyclohexane and OMCTS [2] where the friction forces are highly quantized with the number of layers, and only weakly dependent on the applied load (except for very high loads). In contrast, in the case of alkanes and other chain and irregularly-shaped molecules, their oscillatory forces [8, 9, 21] are much weaker than those of cyclohexane and OMCTS [9, 10], and we therefore expect the first term in equation (8) to be less important, and the friction force to be dominated by the externally applied load. As described here, this is exactly what is observed with tetradecane.

We may also note that a previously reported friction-load curve for n-octane [2] shows that it exhibits both quantized and load-dependent friction, suggesting that shorter-chained alkanes, which are closer to being spherical, fall into the intermediate regime between the two extremes of adhesion-controlled and load-controlled friction.

5. CONCLUSIONS

The results presented here provide further evidence that shows that kinetic friction is correlated with the *irreversible* component of the interaction energy during a loading/unloading cycle. More specifically, our results show that:

- the equilibrium force-law (force *vs.* distance) between two surfaces across a liquid film is quantitatively related to the **static** friction force (or static friction coefficient) required to start shearing the film, and
- the **kinetic** friction force (or friction coefficient), which is the friction force during continuous sliding, is related to both the equilibrium force-law and to the (non-equilibrium) fraction of the energy dissipated to the surrounding lattice every time an atom from the top surface passes, and collides with, an atom of the lower surface.

It is also important to note that both friction and adhesion hysteresis forces can be very sensitive to such factors as temperature, sliding speed, loading rate, relative humidity, the angle between the two surface lattices, and the direction of sliding. But they remain highly correlated so long as comparisons are made under similar experimental conditions, e.g. of humidity, shear rates, etc. These correlations and trends can now be represented quantitatively by equations and graphically by 'dynamic' phase diagrams.

Acknowledgement

This work was supported by the Department of Energy under grant number DE-FG03-91ERR45331, though this support does not constitute an endorsement by DOE of the views expressed in the article.

REFERENCES

1. I. L. Singer and H. M. Pollock (Eds), *Fundamentals of Friction*. Kluwer Academic Publishers, The Netherlands (1992).
2. J. N. Israelachvili, A. M. Homola and P. M. McGuiggan, *Science* **240**, 189 (1988).
   M. L. Gee, P. M. McGuiggan, J. N. Israelachvili and A. Homola, *J. Chem. Phys.* **93**, 1895 (1990).
3. S. Granick, *Science* **253**, 1374 (1991).
4. H. W. Hu and S. Granick, *Science* **258**, 1339 (1992).
5. P. A. Thompson and M. O. Robbins, *Science* **250**, 792 (1990).
6. M. Schoen, C. Rhykerd, D. Diestler and J. Cushman, *Science* **245**, 1223 (1989).
   C. Rhykerd, M. Schoen, D. Diestler and J. Cushman, *Nature* **330**, 461 (1987).
7. J. N. Israelachvili, P. M. McGuiggan, M. L. Gee, A. M. Homola, M. Robbins and P. Thompson, *J. Physics: Condensed Matter* **2**, SA89 (1990).
8. J. N. Israelachvili, *Intermolecular and Surface Forces*, 2nd edn. Academic Press, London & New York (1991).
9. H. K. Christenson, D. W. R. Gruen, R. G. Horn and J. N. Israelachvili, *J. Chem. Phys.* **87**, 1834 (1987); H. K. Christenson and R. G. Horn, *Chemica Scripta* **25**, 37 (1985).
10. R. G. Horn and J. N. Israelachvili, *J. Chem. Phys.* **75**, 1400 (1981).
11. R. G. Horn, *J. Am. Ceram. Soc.* **73**, 1117 (1990).
12. P. M. McGuiggan and J. N. Israelachvili, *J. Mater. Res.* **5**, 2232 (1990).
13. Y.-L. Chen, C. Helm and J. N. Israelachvili, *J. Phys. Chem.* **95**, 10736 (1991).
14. J. N. Israelachvili and G. E. Adams, *J. Chem. Soc., Faraday Trans. I* **74**, 975 (1978).
15. K. L. Johnson, K. Kendall and A. D. Roberts, *Proc. R. Soc. London* **A 324**, 301 (1971).
16. A. M. Homola, J. N. Israelachvili, M. L. Gee and P. M. McGuiggan, *J. Tribology* **111**, 675 (1989).
17. A. M. Homola, J. N. Israelachvili, P. M. McGuiggan and M. L. Gee, *Wear* **136**, 65 (1990).
18. H. Yoshizawa, Y.-L. Chen and J. N. Israelachvili, *J. Phys. Chem.* **97**, 4128 (1993).
19. J. D. Ferry, *Viscoelastic Properties of Polymers*. Wiley, New York (1980).
20. G. Vigil, Z. Xu, S. Steinberg and J. N. Israelachvili, *J. Colloid Interface Sci.* (submitted).
21. M. L. Gee and J. N. Israelachvili, *J. Chem. Soc., Faraday Trans.* **86**, 4049 (1990).
22. B. V. Derjaguin, *Kolloid Zeits.* **69**, 155 (1934).
23. H. Yoshizawa, Y.-L. Chen and J. N. Israelachvili, *J. Phys. Chem.* **97**, 4128, 11300 (1993).

Fundamentals of Adhesion and Interfaces, pp. 281–296
D. S. Rimai, L. P. DeMejo and K. L. Mittal (Eds)

Molecular dynamic modeling of interfacial energy

D. J. QUESNEL,[1] D. S. RIMAI[2,*] and L. P. DEMEJO[2]

[1]*Mechanical Engineering Department University of Rochester, Rochester, NY 14627-0132, USA*
[2]*Office Imaging Research and Technology Development Eastman Kodak Company, Rochester, NY 14653-6402, USA*

Revised version received 21 July 1994

Abstract—Molecular dynamic modeling is used to calculate the changes in the energy between two surfaces when the surfaces are first allowed to approach one another and, subsequently, separated. The equations of motion of the atoms, which were assumed to interact via a Lennard–Jones potential, were integrated using Verlet's algorithm. They were implemented in an environment of periodic boundary conditions, feedback loop temperature and pressure controllers, with direct computation of the stress and strain tensors. This approach allows one to calculate the temperature dependence of the 'leap-to-contact' phenomenon, the thermodynamic work of adhesion, the work needed to separate the surfaces, and the forces of attraction and separation. Effects that occur during approach and separation, such as surface roughening and vacancy formation, were included in the energetics calculations. Sound waves and the resulting thermal transients were also modeled. The adhesion hysteresis and irreversible behaviors during approach and separation that arise from these calculations are discussed in detail.

Keywords: Work of adhesion; interfacial energy; Lennard–Jones potential; molecular dynamics.

1. INTRODUCTION

When a liquid is applied to the surface of a solid, the tendency of that liquid to wet the surface is often described in terms of surface and interfacial energies. If the energy needed to form the interface between the liquid and solid is smaller than the energy gained by eliminating the free surface, the liquid should spread. This thermodynamic argument does not include kinetic effects, however, and only states what is energetically favorable. The rate of wetting/dewetting is controlled by the physical properties of the materials and can include hysteretic effects in the advancement and recession of the liquid on the wetted surface.

When solids are brought into contact, there are the same tendencies to 'wet' which are, again, due to surface energies. Here, however, the finite shear strengths of the solids significantly retard or prevent the achievement of complete equilibrium, even if

*To whom correspondence should be addressed.

the kinetics of the wetting process are ignored. The materials deform either elastically or plastically in response to the large stresses generated by the surface forces. If the surfaces are not flat (as is the case of a sphere approaching a plane) then large scale loads develop and deflections occur which allow a contact area to develop in response to the surface forces and any externally applied loads.

The mechanics of particle adhesion and the deformations resulting from the stresses generated by the forces of adhesion have been studied both theoretically and experimentally for many years. Reviews of the experimental developments have been presented elsewhere [1, 2] and are beyond the scope of this paper.

Adhesion induced deformations were first theoretically postulated, independently, by Bradley [3, 4] and Derjaguin [5]. Derjaguin attempted to calculate the contact radius resulting from the adhesion forces by assuming that the particles acted as Hertzian indentors with the applied load, P° arising from van der Waals interactions. Accordingly, he found that, for a rigid particle in contact with a compliant substrate, the contact radius, a, would be related to the particle radius, R, by

$$a^3 = \frac{3}{4} P^\circ \left[\frac{1 - \nu^2}{E}\right] R, \tag{1}$$

where ν and E are the Poisson ratio and Young's modulus, respectively, of the substrate and P° is given by

$$P^\circ = \frac{\hbar\overline{\omega}}{8\pi z_0^2} R, \tag{2}$$

where z_0 is the separation distance between the particle and substrate (typically about 4 Å) and $\hbar\overline{\omega}$ is the Hamaker coefficient.

Krupp [6] attempted to generalize Derjaguin's model of particle adhesion to allow for cases where the stress exceeded the yield strength of the material. He assumed that the contact area could be divided into two concentric regions: an inner region of radius a_1 which, being subjected to higher stresses, would deform plastically and an outer annulus extending from radius a_1 to the total contact radius, a_0 which, being subjected to lower stresses, would deform elastically. Accordingly,

$$\pi a_0^2 = \frac{P^\circ}{H(t)} + \frac{1}{3}\left[\frac{\pi^{3/2}(1 - \nu^2) R H(t)}{2E}\right]^2 \tag{3}$$

and

$$\pi a_1^2 = \frac{P^\circ}{H(t)} - \frac{1}{3}\left[\frac{\pi^{3/2}(1 - \nu^2) R H(t)}{2E}\right]^2, \tag{4}$$

where $H(t)$ is a phenomenologically determined time dependent hardness. In order to allow the contact radius to reach a limiting size, Krupp assumed that $H(t)$ approached an asymptotic value of the order of $10^{-3}E$ within about half an hour.

The classical theories of particle adhesion, which were just discussed, assumed that all interactions were compressive. Modern understanding, which began with the theory by Johnson *et al.* [7] (hereafter referred to as JKR), recognized that both compressive and tensile interactions can occur. According to the JKR theory, all interactions occur within the contact zone and the resulting deformations are elastic. Approaching the problem from a thermodynamic rather than molecular viewpoint, the JKR model predicts that the contact radius is related to the particle radius by

$$a^3 = \frac{R}{K}\left\{P + 3w_A\pi R + \left[6w_A\pi RP + \left(3w_A\pi R\right)^2\right]^{1/2}\right\}, \tag{5}$$

where P is any externally applied load, w_A is the work of adhesion and is related to the surface energies γ_1 and γ_2 and the interfacial energy γ_{12} by

$$w_A = \gamma_1 + \gamma_2 - \gamma_{12}, \tag{6}$$

and

$$K = 4/\left[3\pi\left(k_1 + k_2\right)\right], \tag{7}$$

where

$$k_i = \frac{1 - \nu_i^2}{\pi E_i} \tag{8}$$

and i assumes the values of 1 or 2 for the two materials involved. According to the JKR theory, upon application of a negative load, separation of the particle from the substrate would occur when an external force, P_s, was applied such that

$$P_s = -\frac{3}{2}w_A\pi R. \tag{9}$$

Moreover, the contact radius at separation, a_s, does not vanish, but, rather, is approximately 63% of the contact radius obtained with no applied load. It is interesting to note that P_s does not depend on the Young's moduli of the materials and does not account for hysteretic effects. Although Johnson *et al.* [7] were aware that hysteretic effects occur, particularly in polymers, they assumed that they were due to non-equilibrium conditions existing under separation conditions. As will be shown later in this paper, such hysteresis is a natural result of adhesion even for the case of elastic deformations.

Derjaguin *et al.* [8] proposed an alternative adhesion model (hereafter referred to as the DMT model), which also allowed for tensile interactions, but approached the problem from a molecular level. According to the DMT theory, the shape of the contact region continues to be Hertzian, as it was in the earlier model by Derjaguin. This assumption results in the DMT model predicting that half of the interactions occur outside the contact region, which is in opposition to a fundamental assumption of the JKR model.

The assumptions and predictions of the JKR and DMT models were compared by Tabor [9], who showed that the DMT model predicts contact radii and separation forces of approximately 1/2 and 4/3, respectively, of those predicted by the

JKR theory. Following some debate in the literature concerning these discrepancies [10, 11], Muller *et al.* proposed a new model [12, 13]. In this model, known as the MYD model, it is assumed that the adhesion forces between a particle and a substrate can be described by a pairwise summation between the molecules of the two materials interacting via a Lennard–Jones potential. Muller *et al.* [12, 13] found that both the JKR and DMT models were special cases of this more general theory, with the JKR model being valid for low modulus, high surface energy materials and larger particles, whereas the DMT model was valid for more rigid materials having lower surface energies and for smaller particles. The transition between the JKR and DMT models has recently been studied by Maugis [14, 15]. Others studying the contact region between particles and surfaces include Tsai *et al.* [16], Dahneke [17], Attard and Parker [18, 19], and Maugis and Pollock [20].

The dynamics of particle adhesion has also been the subject of recent studies. Rogers and Reed [21] have explored the effect of particle–substrate impacts, resulting in plastic and elastic deformations, on adhesion. This area was further investigated by Reed [22] and by Wall *et al.* [23]. In these models it is assumed that the energy loss mechanism occurs only through the plastic, not the elastic, deformations. More recently Johnson and Pollock [24] explored the effect on adhesion of a particle impacting a surface, using the JKR model.

The afore-mentioned theories have been instrumental in advancing current understanding of particle adhesion. However, each treats the mechanical response of the materials as factors which are totally independent of the parameters which affect adhesion. In reality the intermolecular potential of a material determines both the mechanical properties and surface energy. It should, at least in theory, be possible to construct an adhesion model in a more holistic manner from first principles. Second, recent experimental advances have shown that the process of particle adhesion is more complicated than previously assumed. For example, as has been shown by Mizes *et al.* [25] and by Schaefer *et al.* [26, 27] using atomic force techniques, the process of a particle impinging onto a surface becomes complex when the particle is in close proximity to that surface. At that time the attractive potential is sufficiently large so as to cause the particle to jump into contact with the substrate, with the resulting energy being dissipated in the two materials. In addition, current theories do not always predict the correct power law dependence of the contact radius on particle radius. For example, it has been found that the contact radius of micrometer size particles in contact with elastomeric substrates can vary as the particle radius to the 3/4 power [28–30] rather than the 2/3 power, as predicted by the JKR theory.

This paper reports results obtained from a first principles approach to the problem of particle adhesion. As such, it is part of a continuing series of papers in which the mechanical and adhesion properties of materials are determined. In addition, the interactions between surfaces and energy loss mechanisms in the presence of elastic deformations are studied. In two previous papers [31, 32] the Poisson ratio and elastic constants were determined for an FCC lattice interacting via a Lennard–Jones potential. In this paper the interfacial energy between two surfaces is reported. In addition, the snap together and snap apart phenomena, which have been observed in the previously discussed atomic force studies, have been simulated. Energy loss mechanisms in the presence of elastic deformations are modeled. Finally, defects such as surface roughening and defect formation upon separation are proposed. In a subsequent paper the physics of adhesion between two-dimensional particles and surfaces will be discussed [33].

2. THE MODEL

As previously discussed, this work builds on earlier studies where it was assumed that particles interact via a Lennard–Jones potential, i.e.

$$\phi = -4\varepsilon\left[\left(\frac{\sigma}{r}\right)^{6} - \left(\frac{\sigma}{r}\right)^{12}\right], \tag{10}$$

where ε is the binding energy between one atom and its nearest neighbor and σ represents the spacing between atoms when the potential crosses the axis. It is recognized that the choice of this potential is idealistic. However, much of the physics of adhesion can be readily obtained, at least qualitatively, using an idealized potential. Moreover, the use of such a relatively simple potential makes this and similar calculations practical on workstations, rather than requiring supercomputers. Finally, by simply substituting the appropriate potential into the code and running it on a more powerful machine, more quantitative predictions of the actual behavior of materials can be made.

Details of the model and computational techniques have been presented elsewhere [32]. However, these are presently summarized for the benefit of the reader. Moreover, additions to the model which permit the calculation of the adhesion affects are also discussed here.

In order to explore the phenomenon of adhesion within the context of this model, it is first necessary to create atomically flat surfaces in a step-wise manner. This allows the energetics associated with the creation of a surface to be determined. A pair of these surfaces were brought together and allowed to form a bond. Subsequently, these surfaces were separated in a constrained tension test to form two fracture surfaces. During these processes, the potential energy of the system and the axial stresses used to produce the displacements were monitored.

The computational model used in the present investigation consists of a rectangular parallelepiped of atoms, arranged in an FCC array, which is similar to that used in the aforementioned earlier studies [31, 32]. Velocity components of each atom were selected by a seeded random number generator and transformed to provide a Maxwell–Boltzmann distribution of initial velocities. The atoms were released and allowed to interact via the Lennard–Jones potential. When the atoms were initially released, the system was obviously not at equilibrium. Rather, all of the energy was in the form of kinetic energy, which rapidly equipartitions into kinetic and potential energies. Newton's second law of motion was integrated at each iteration using the commonly used Verlet algorithm [33, 35]. Independent temperature and pressure controllers, which could be switched on and off, were also used. This allowed for testing under constant volume or adiabatic conditions, as desired.

The computational model utilized in the tension, compression, and shear testing also permits examination of the stresses produced when free surfaces are made to approach one another. This is accomplished by producing a free surface within the specimen that otherwise might be used for mechanical property evaluation and then computing the stresses that are developed across that surface. Figure 1 shows a schematic of a typical test sample used in this study. The dimensions are independently adjustable in the X, Y, and Z directions. It was assumed that the boundary conditions of the samples were periodic. The sizes of the periodic boundary conditions were adjusted so as to continuously attach the material to itself in the Y and Z directions. In the X direction, however, a gap was introduced that was larger than the cutoff radius beyond which it was assumed that the atoms would not interact. In this way the atoms on the end planes of the sample would not 'feel' the presence of the surface that was placed above them by virtue of the periodic boundary conditions. This produced the free surface.

For the majority of the computations reported in this paper, the system was comprised of 768 atoms in an FCC structure aligned with the [100] directions parallel to the X, Y, and Z directions. This results in eight layers of atoms in the Y and Z direc-

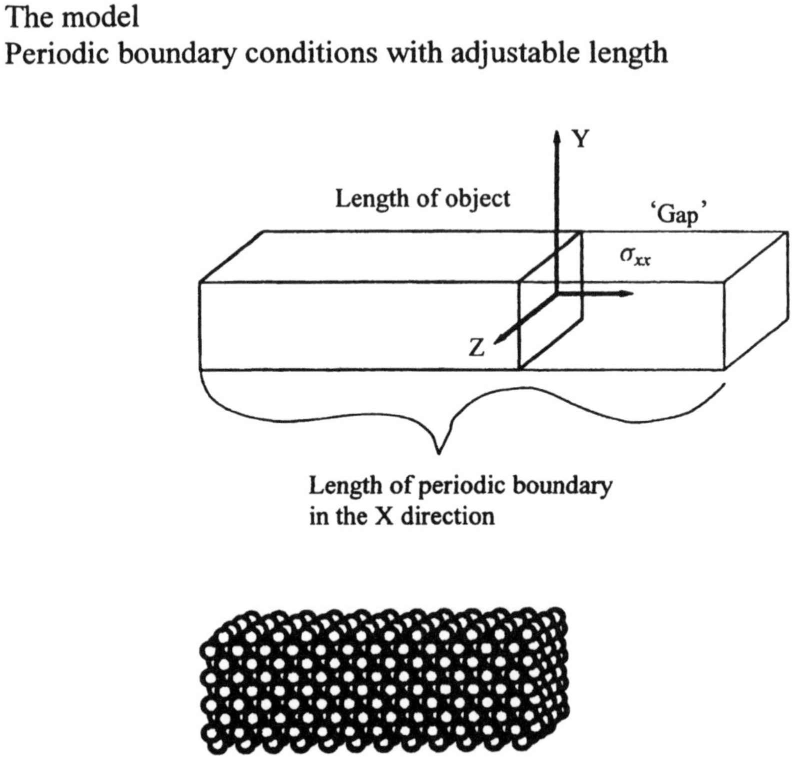

Figure 1. Model system showing the length of the periodic boundary. The atomic positions are shown pictorially.

tions and 24 layers in the X direction, thereby producing a rectangular specimen such as that shown in Fig. 1. Verlet's algorithm was employed with a dimensionless time step [34] of 0.01. All data generated at each time step, excluding the detailed positions of the atoms, were stored for evaluation following a run. This allowed the use of filtering techniques and statistical methods, discussed in detail in [32], to examine the computed results in order to extract the maximum amount of information.

If one wishes to model interfacial interactions, it is first necessary to form the interfaces. This was accomplished in a manner similar to that used in the previous studies [31, 32], by having all 768 atoms in their lattice positions with a Maxwell–Boltzmann velocity distribution corresponding to twice the intended temperature of the run. The periodic boundaries in each direction were set to equal the unstrained dimensions of the system. The computation was run for 100 iterations without any intervention. During those iterations, the apparent temperature and pressure (as measured by one third of the negative sum of the normal stresses) were measured.

By the time the first hundred iterations were complete, the system temperature had typically fallen to half of its initial value. As discussed before, this is due to the equipartitioning of the energy into kinetic and potential contributions. At this time the temperature and pressure controllers were activated.

The temperature controller uses a digitally filtered weighted average as the measured (computed) kinetic energy per atom. This is then compared with the desired or command signal to scale the velocities in a manner which reduces the measured error. This is computational analog of a feedback controller used in many experimental set-ups. The pressure controller works in a similar fashion. The error signal between the measured (computed) pressure and the desired or command pressure was used to 'photographically' scale the size of the periodic boundaries in all three directions. By appropriately selecting the feedback gains in these two controllers, they can be made not to interact. This allows continuous control of temperature and pressure during the run. The system was allowed to evolve in this manner for an additional 900 iterations, thereby producing a stable equilibrium. The pressure was initially set to zero and the temperatures were chosen for each specific run.

In order to examine the stresses on the soon to-be-created free surfaces, it was necessary to disable the pressure controller. This fixed the lateral dimensions of the surface which was about to be created. The size of the periodic boundary was then increased, thereby creating a gap at the end of the sample. To avoid stretching the sample, the atoms were 'put back' into their previous positions after each computation, thereby producing a gradual increase in the size of the gap without actual movement of the atoms. This effectively disabled the temperature controller and suppressed the usual atomic motions, thereby allowing the program to computationally track the apparent potential energy of the system as the surfaces were separated. The potential energy per atom, as a function of iteration number during the creation of the surface is shown in Fig. 2 in the region marked 'stepping'. Once the gap size has been increased to the size of the cut-off radius, there is no further in the potential energy with increasing gap size.

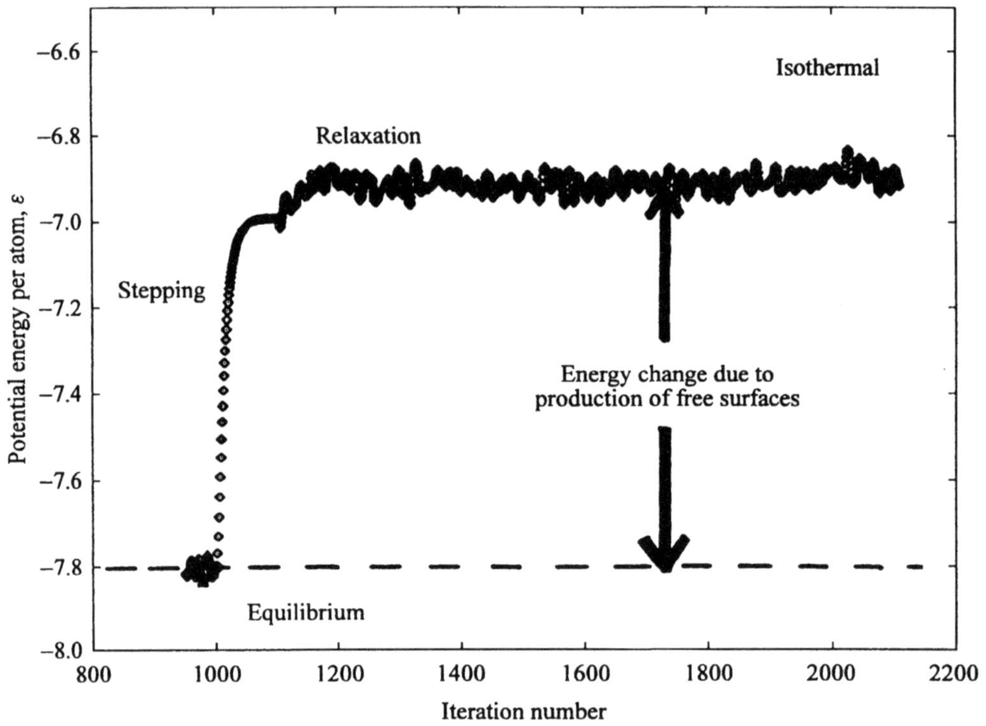

Figure 2. Potential energy per atom as a function of iteration number during the creation of two free surfaces. The separation distance between surfaces is proportional to the iteration number and depends on the rate of separation.

Once the gap has been established, the atomic motions and temperature controller were reactivated. This allowed the system to relax into its new equilibrium. As can be seen in Fig. 2, there is a slight increase in the potential energy per atom over the next 1000 iterations. Intuitively, one might expect the potential energy per atom to decrease. However, the equipartion of kinetic and potential energies requires that the kinetic energy must also increase in an amount equal to the increase in the potential energy from its ground state. The static creation of the free surfaces raised the potential energy while keeping the kinetic energy constant. When the atoms were released, the system re-equilibrated, thereby producing the apparent increase in potential energy. Surface rearrangements may also contribute to the increase in potential energy. Further study is needed to more fully understand this computational observation.

The energy needed to produce two free surfaces at constant temperature and lateral dimension is also shown in Fig. 2. This represents the interfacial energy associated with the production of two [100] surfaces in an otherwise perfect FCC lattice. The constant lateral surface dimension is a realistic constraint if it is assumed that, for given physical surfaces, the underlying material will hold the lateral dimensions fixed.

3. APPROACH OF THE SURFACES

Once free surfaces in equilibrium have been computationally generated, it is relatively simple to examine the behavior as the gap in the periodic boundary is reduced. This, in effect, brings the surfaces closer to one another, thereby allowing changes in the potential energy and the resulting surface tractions that develop at constant temperature and lateral dimension to be examined. This computation is analogous to bringing two large, flat surfaces together in a precise way that allows for the parallel approach of the surfaces. Although this obviously differs from the physical approach of non-smooth surfaces or, for example, the impingement of spheres, rollers, or materials possessing asperities, much can be learned by examining the ideal case of the flat surface.

An appropriate approach rate was selected, such that 10 000 iterations were needed to reduce the gap by one cut-off radius. This slow approach was chosen so as to avoid complications arising from impact energies. Therefore, after these 10 000 iterations the initial periodic boundary size was re-established.

This process is shown in Fig. 3 for a temperature of 0.001 ε and a sample of 256 atoms, which is approximately 1/3 the typical sample length. The component of the axial stress that develops between the surfaces is shown as a function of the iteration number, which is proportional to the spacing change between the surfaces.

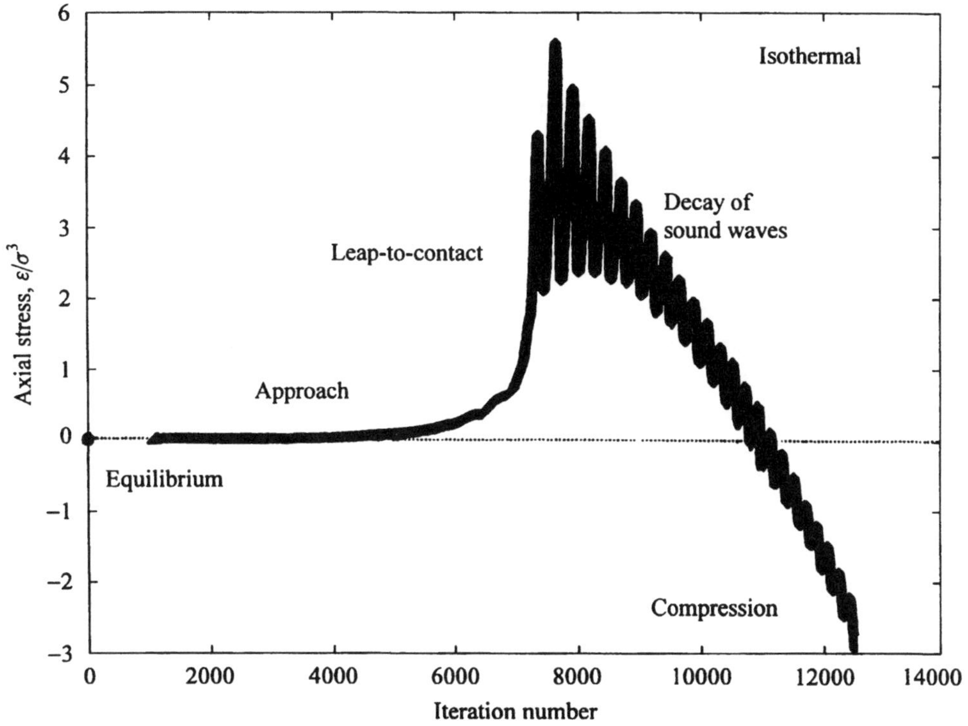

Figure 3. Axial stress as a function of iteration number as the surfaces are brought together. Features evident include the leap-to-contact phenomenon and the generation and decay of sound waves.

During the approach, the mutual attraction between the surfaces was observed to increase monotonically until a critical stress level was reached. At that critical stress, the surfaces leapt into contact with one another as the strain energy increased because the rate of energy storage due to deformation of the elastic system became equal to the rate at which energy was provided by the attraction between the surfaces. This effect is, at least qualitatively, similar to the leap-to-contact between a particle or cantilever tip and a planar surface observed experimentally using atomic force techniques [25–27].

When the surfaces were pulled into contact, the stresses were not uniform. Rather, traveling waves were generated. Additional traveling waves were generated when the surfaces actually struck each other. These waves passed smoothly through the small periodic system, interacting with one another. The decay of the sound waves is a result of the dispersion of these waves into other modes. These waves, which correspond to atomic level kinetic energy, can be interpreted as an increase in temperature. Thus, the acoustic energy is eventually damped to zero by the temperature controller and the dispersion of the system. In a real system these effects would imply that, even in the case of only elastic deformations resulting from the forces of adhesion, there is an energy loss mechanism. Not all the energy is stored elastically and, therefore, recoverable upon separation, as suggested by Krupp [6].

4. SEPARATION OF THE SURFACES

Separation of the surfaces was accomplished the sign of the change in the gap size on each iteration. This corresponds to testing in a tension mode. By separating the surfaces which were just brought together, hysteretic effects could be studied. Figure 4 shows the potential energy as a function of the separation distance for the full size 768 atom sample. The approach and leap-to-contact are also shown in this figure to facilitate comparison.

Several differences between the approach and separation behaviors should be noted. In particular, the separation process occurs over a large region of uniform extension as the entire specimen is uniformly elongated. It was not until the interatomic spacing of some region exceeded the critical value, such that further extension reduced the energy, that fracture of the specimen actually occurred. This is an extremely violent process which occurred over a very small number of iterations, appearing nearly instantaneous of the scale of Fig. 4. It is labeled 'Fracture' in that figure.

Once the sample has broken, the acoustic waves that were generated traversed the sample, reflecting from the free surfaces. As before, they decayed with time, as the acoustic energy changed to thermal energy, and were transported out of the system via the temperature controller. Again, energy loss mechanisms exist even in the case of elastic deformations.

Figure 5 shows a schematic of the stress displacement loop that was obtained by folding the load versus iteration curves to form a loop. This clearly shows the hysteresis associated with the axial stress as a function of displacement (as characterized

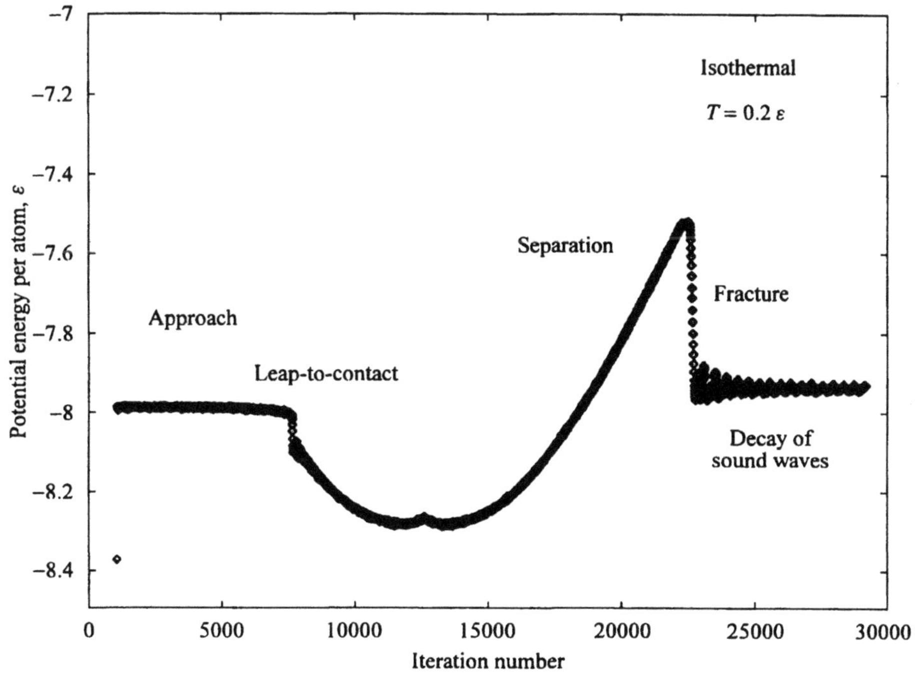

Figure 4. Potential energy per atom as a function of iteration number during approach and separation. The temperature of the system was $0.2\,\varepsilon$.

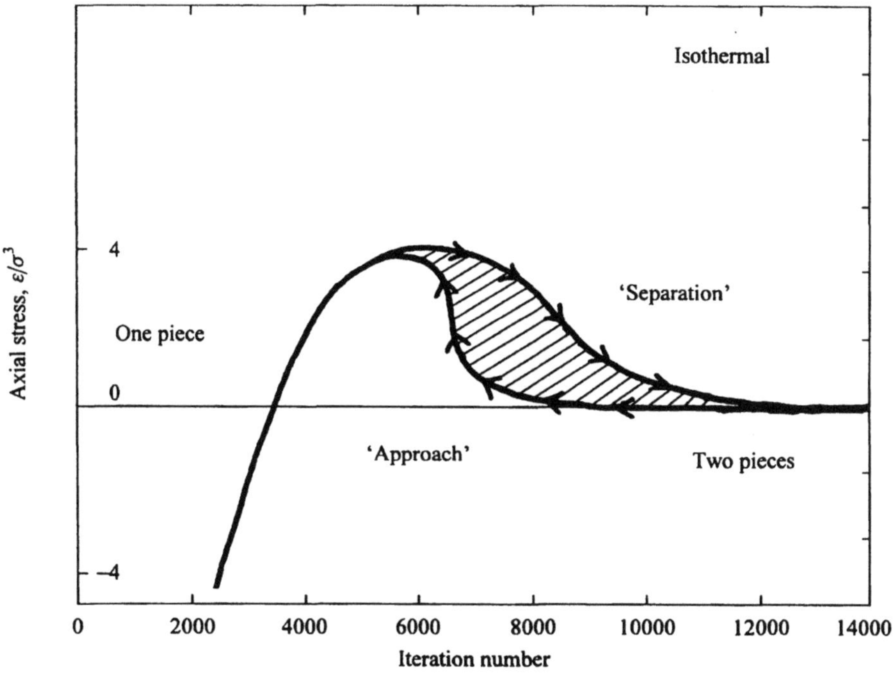

Figure 5. Axial stress as a function of iteration number, showing hysteresis as the surfaces are brought together and subsequently separated.

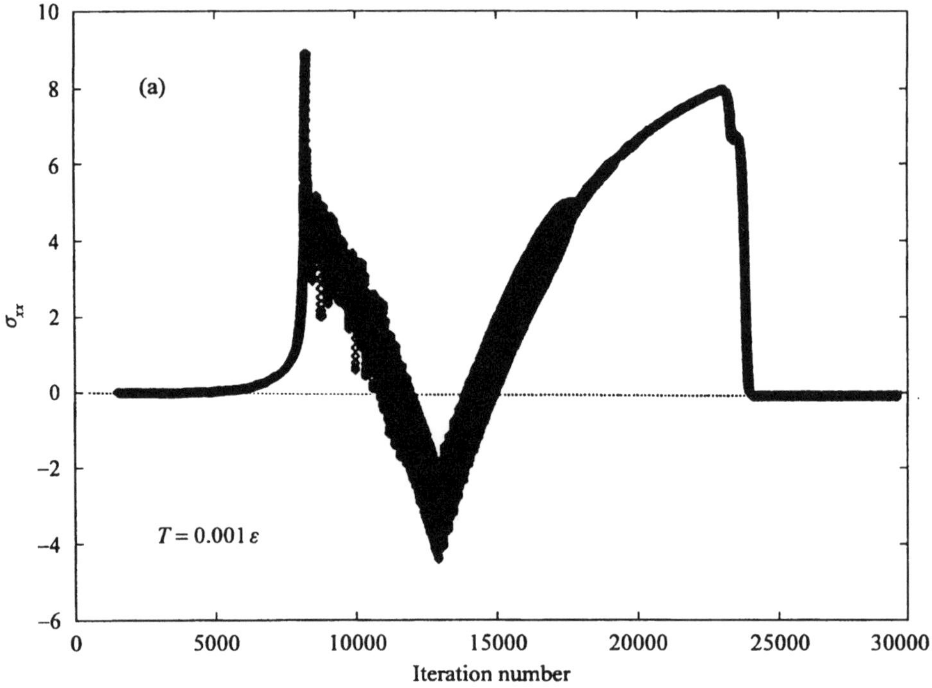

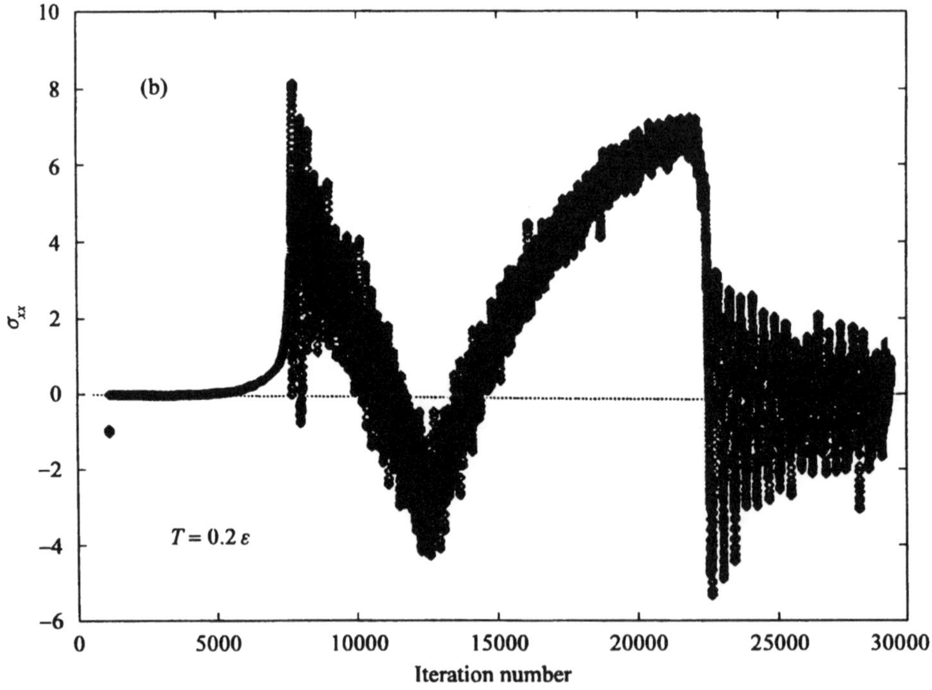

Figure 6(a–b). Stress as a function of iteration number at temperatures corresponding to $0.001\,\varepsilon$ (a), $0.2\,\varepsilon$ (b).

by iteration number) with the leap-to-contact upon approach and the gradual straining during separation. This hysteresis is important in understanding the adhesion of materials, as well as related properties such as friction [36]. Moreover, this hysteresis probably accounts for the effect of Young's modulus on particle adhesion, which is not predicted by the JKR model [7].

5. TEMPERATURE DEPENDENCE

Figure 6 shows the axial stress as a function of separation, again characterized by iteration number, for three different temperatures corresponding to 0.001 ε, 0.2 ε, and 0.3 ε during approach and separation. Figure 7 shows the potential energy per atom as a function of displacement at temperatures corresponding to 0.001 ε and 0.3 ε. This figure complements the data shown in Fig. 4, which corresponds to a temperature of 0.2 ε.

The level of zero stress is indicated in Fig. 6 by a dotted line. As expected, the average stress both before and after the surfaces were brought together was zero. As can be seen by comparing Figs 6a, 6b, and 6c, the maximum stress experienced during the leap-to-contact decreased with increasing temperature. In contrast, the average stress, shortly after the leap-to-contact, was much less sensitive to the temperature.

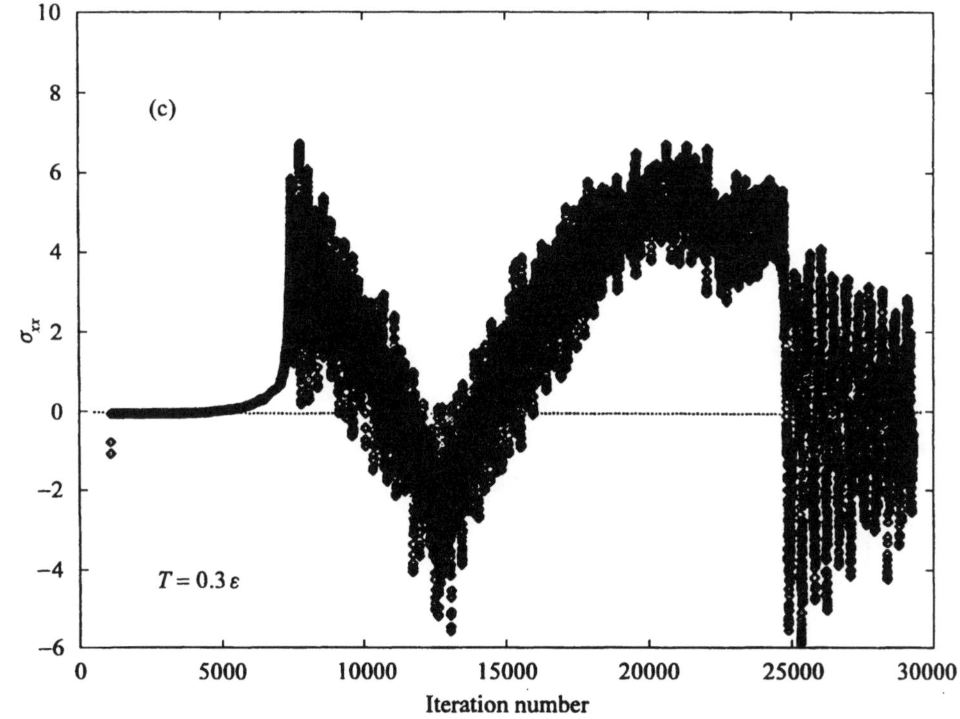

Figure 6(c). Stress as a function of iteration number at temperature corresponding to 0.3 ε.

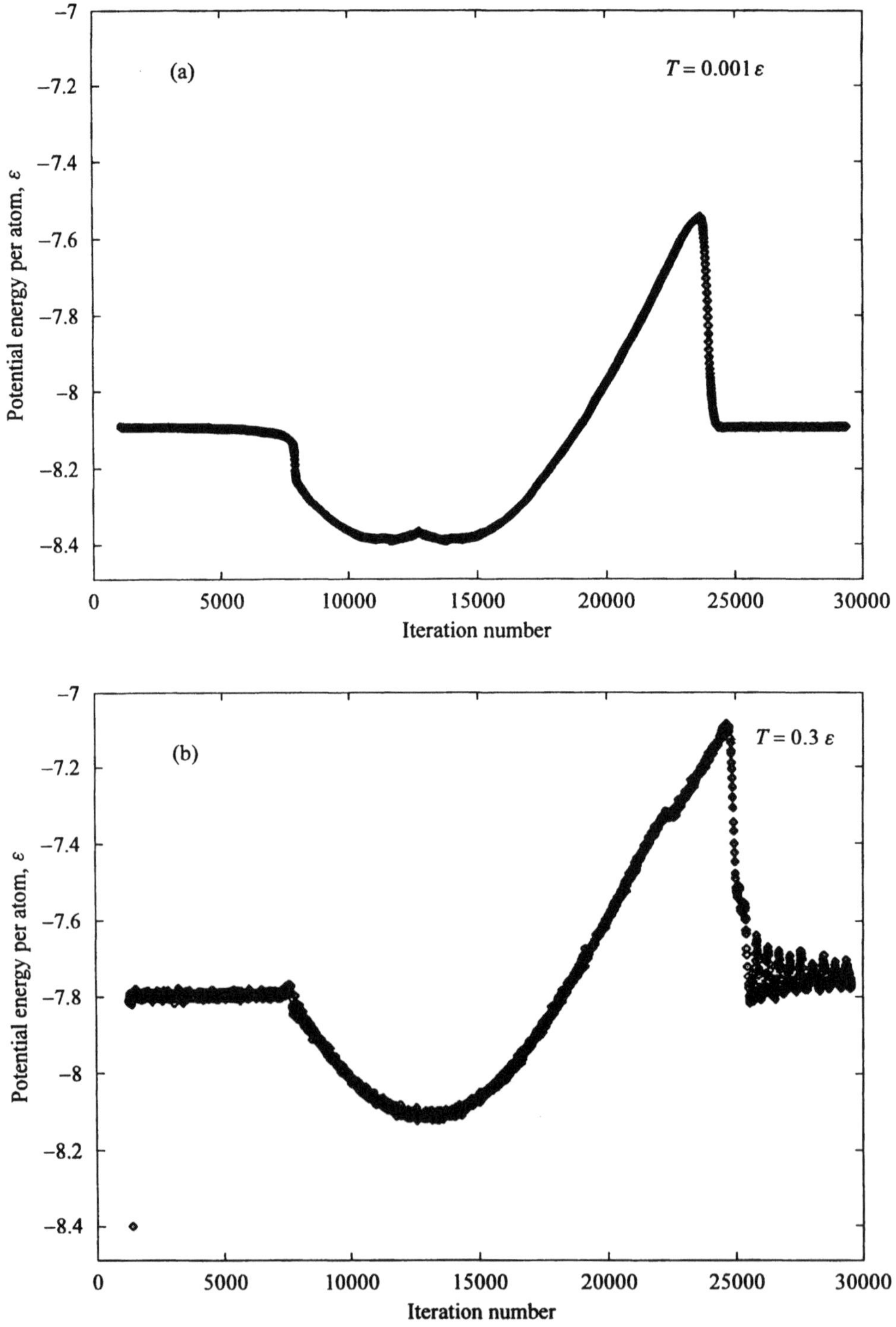

Figure 7. Potential energy per atom as a function of iteration number at temperatures corresponding to 0.001 ε (a) and 0.3 ε (b). An intermediate temperature is shown in Fig. 4.

All curves pass through the axis as the system is compressed following the leap-to-contact. Initially, upon dilation, they follow essentially the reverse path. However, the loading curves are stable until the instantaneous peak stress, including contributions from thermal noise, is about equal to the peak stress associated with the leap-to-contact. At this point separation occurs. This produces a hysteresis between approach and separation in these otherwise deterministic processes.

At maximum applied stress, the stress versus separation curves, at each temperature, show a small drop initially during separation followed by a catastrophic drop into a ringing condition, as the acoustic waves generated by the fracture become attenuated. The small drop may be due either to slip or to vacancy formation during the fracture process.

Figure 7 shows the potential energy curves during approach and separation. At the lowest temperature (Fig. 7a) little, if any, ringing was observed upon separation. Moreover, the potential energy after fracture equaled that of the system before approach. Although fracture occurred in a physically different position in the sample than where the initial surfaces were positioned (as is entirely plausible), the microstructure of the interfacial regions was the same. This was not the case at the higher temperatures, such as those shown in Figs 4 and 7b. Rather, discontinuities were observed in the slope where plastic behavior (which may be slip or vacancy formation), presumably, would be present. Ringing, as characterized by the damp oscillatory behavior, occurs in the potential energy after fracture. Finally, there was a distinct offset between the initial and final potential energies which is due to microstructural changes in the interfacial regions. In particular, the fractured samples at the two higher temperatures contained 25 layers of atoms. However, as determined from the graphical representations of the atomic positions, the two layers near the surface contained numerous vacancies. This resulted in the observed increase in potential energy.

6. CONCLUSIONS

Adhesion effects have been modeled and interfacial energies calculated using small scale molecular dynamics for an FCC lattice where the atoms interact via a Lennard–Jones potential. Hysteretic behavior between approach and separation, similar to that observed using atomic force techniques, has been found. This hysteresis can play a significant role in the adhesion of materials and related problems such as friction. The generation and decay of sound waves, associated with the leap-to-contact and separation, were also observed. These effects can allow for energy loss even in cases of adhesion induced elastic deformations. Although the potential chosen is rather idealistic and was chosen to simplify calculations, it would be a straightforward extension to model real materials by changing the potential and running the code on a more powerful computer.

Acknowledgement

The authors would like to thank the Eastman Kodak Company for the support for the computational activities that led to these findings.

REFERENCES

1. D. S. Rimai, L. P. DeMejo, R. C. Bowen and J. D. Morris, in: *Particles on Surfaces: Detection, Adhesion, and Removal*, K. L. Mittal (Ed.), pp. 1–31. Marcel Dekker, New York (1994).
2. D. S. Rimai, L. P. DeMejo and R. C. Bowen, *J. Adhesion Sci. Technol.* **8**, 1333–1355 (1994).
3. R. S. Bradley, *Philos. Mag.* **13**, 853 (1932).
4. R. S. Bradley, *Trans. Faraday Soc.* **32**, 1088 (1936).
5. B. V. Derjaguin, *Kolloid Z* **69**, 155 (1934).
6. H. Krupp, *Adv. Colloid Interface Sci.* **1**, 111 (1967).
7. K. L. Johnson, K. Kendall and A. D. Roberts, *Proc. R. Soc. Lond. A* **324**, 301 (1971).
8. B. V. Derjaguin, V. M. Muller and Yu. P. Toporov, *J. Colloid Interface Sci.* **53**, 314 (1975).
9. D. Tabor, *J. Colloid Interface Sci.* **58**, 2 (1977).
10. B. V. Derjaguin, V. M. Muller and Yu. P. Toporov, *J. Colloid Interface Sci.* **67**, 378 (1978).
11. D. Tabor, *J. Colloid Interface Sci.* **67**, 380 (1978).
12. V. M. Muller, V. S. Yushchenko and B. V. Derjaguin, *J. Colloid Interface Sci.* **77**, 91 (1980).
13. V. M. Muller, V. S. Yushchenko and B. V. Derjaguin, *Colloids Surfaces* **7**, 251 (1983).
14. D. Maugis, *J. Colloid Interface Sci.* **150**, 243 (1992).
15. D. Maugis and B. Gauthier-Manuel, *J. Adhesion Sci. Technol.* **8**, 1311–1322 (1994).
16. C.-J. Tsai, D. Y. H. Pui and B. Y. H. Liu, *Aerosol Sci. Technol.* **15**, 239 (1991).
17. B. Dahneke, *J. Colloid Interface Sci.* **40**, 1 (1972).
18. P. Attard and J. L. Parker, *Phys. Rev. A* **46**, 7959 (1992).
19. J. L. Parker and P. Attard, *J. Phys. Chem.* **96**, 10398 (1992).
20. D. Maugis and H. M. Pollock, *Acta Metall.* **32**, 1323 (1984).
21. L. N. Rogers and J. Reed, *J. Phys. D: Appl. Phys.* **17**, 677 (1984).
22. J. Reed, in: *Particles on Surfaces 2: Detection, Adhesion, and Removal*, K. L. Mittal (Ed.), pp. 3–18. Plenum, New York (1988).
23. S. Wall, W. John and S. L. Goren, in: *Particles on Surfaces 2: Detection, Adhesion, and Removal*, K. L. Mittal (Ed.), pp. 19–34. Plenum, New York (1988).
24. K. L. Johnson and H. M. Pollock, *J. Adhesion Sci. Technol.* **8**, 1323–1332 (1994).
25. H. A. Mizes, K.-G. Loh, R. J. D. Miller, S. K. Ahuja and E. F. Grabowski, *Appl. Phys. Lett.* **59**, 2901 (1991).
26. D. M. Schaefer, M. Carpenter, R. Reinfenberger, L. P. DeMejo and D. S. Rimai, *J. Adhesion Sci. Technol.* **8**, 197–210 (1994).
27. D. M. Schaefer, M. Carpenter, B. Gady, R. Reifenberger, L. P. DeMejo and D. S. Rimai, *J. Adhesion Sci. Technol.* **9**, 1049–1062 (1995).
28. L. P. DeMejo, D. S. Rimai and R. C. Bowen, *J. Adhesion Sci. Technol.* **5**, 959 (1991).
29. D. S. Rimai, L. P. DeMejo and R. C. Bowen, *J. Appl. Phys.* **66**, 3574 (1989).
30. D. S. Rimai, L. P. DeMejo, W. B. Vreeland and R. C. Bowen, *Langmuir* (in press).
31. D. J. Quesnel, D. S. Rimai and L. P. DeMejo, *Solid State Commun.* **85**, 171 (1993).
32. D. J. Quesnel, D. S. Rimai and L. P. DeMejo, *Phys. Rev. B* **48**, 6795 (1993).
33. D. J. Quesnel, D. S. Rimai and L. P. DeMejo, (to be published).
34. J. M. Haile, *Molecular Dynamics Simulations: Elementary Methods*, pp. 158–159. Wiley, New York (1992).
35. L. Verlet, *Phys. Rev.* **159**, 98 (1967).
36. J. N. Israelachvili, Y.-L. Chen and H. Yoshizawa, *J. Adhesion Sci. Technol.* **8**, 1231–1249 (1994).

Bei Fragen zur Produktsicherheit wenden Sie sich bitte an:
If you have any questions regarding product safety,
please contact:

Walter de Gruyter GmbH
Genthiner Straße 13
10785 Berlin
productsafety@degruyterbrill.com